The Six Sigma Way Team Fieldbook

Also by Peter S. Pande, Robert P. Neuman, and Roland R. Cavanagh

The Six Sigma Way: How GE, Motorola, and Other Top Companies Are Honing Their Performance

The Six Sigma Way Team Fieldbook

An Implementation Guide for Process Improvement Teams

Peter S. Pande
Robert P. Neuman
Roland R. Cavanagh

McGraw-Hill
New York Chicago San Francisco Lisbon London Madrid
Mexico City Milan New Delhi San Juan
Seoul Singapore Sydney Toronto

17 18 19 20 21 22 23 24 25 26 DOC/DOC 0

ISBN 978-0-07-183166-6

The sponsoring editor for this book is Richard Narramore. Editorial, design, and production services were provided by CWL Publishing Enterprises, John Woods, President, Madison, WI, www.cwlpub.com.

Contents

Dedication

With love to Olga, Stephanie, and Brian.
Thanks for making room on the counter!
—PSP

Preface

A Guide to The Six Sigma Way Team Fieldbook

Who Is It For?

This book is designed foremost for the people who are carrying out the "heavy lifting" of their organizations' ambitious Six Sigma efforts: the team leaders (or "Black Belts") and team members (aka "Green Belts") assigned to improve, redesign, and create efficient, customer-focused business processes.

In addition, we hope business leaders who are guiding Six Sigma projects (often called "Champions" or "Sponsors") can use *The Six Sigma Way Team Fieldbook* to gain insights into the tasks and challenges faced by their colleagues in the trenches of process improvement. While you may not have time to attend weeks of Black Belt training, you can find in these pages a summary of the milestones, tools, and issues that arise in driving Six Sigma improvement.

Finally, anyone seeking to better understand the details of the Six Sigma movement—how it integrates tools and best practices from various disciplines into a more powerful system of management—can, we hope, gain insights from these pages.

What's Inside?

As the name suggests, this fieldbook is a companion to our book, *The Six Sigma Way*, which provides an overview of Six Sigma as a 21st century approach to building and sustaining business success. While this book goes much further in detail and "how to," we've taken the liberty to summarize some of our key points from *The Six Sigma Way* in Chapters 1 through 4 of this fieldbook. If you've read or are familiar with the other work, you may want to skim these chapters or just skip to Chapter 5.

Part Two in *The Six Sigma Way Team Fieldbook* takes an in-depth look at the process improvement model in use by the vast majority of Six Sigma organizations: DMAIC—for Define, Measure, Analyze, Improve, and Control. Each of these five phases is covered in three chapters as follows:

- An overview of the key steps and challenges of the phase. With examples and illustrations, we explain the "what, why, and how" of D, M, A, I, and C.
- A review of "Power Tools" for each phase. These chapters give detailed instructions and job aids for many of the most important tools used in DMAIC projects. (While they're organized by phase, be aware that many tools can be applied in various points of a Six Sigma project.)
- Tips on handling the teamwork and collaboration challenges of DMAIC. Even a very technically adroit Six Sigma team leader can fail if he or she is unable to deal with the people and change management issues that are integral to boosting business performance. These chapters offer guidance on getting a team to work cohesively—and on sustaining effective teamwork under the pressure of Six Sigma projects.

In Chapter 21, we discuss adapting the DMAIC approach to meet the less frequent but more challenging goal of designing and redesigning key processes.

How Do You Use This Book?

The answer is easy: however it works best for you. Our goal for the fieldbook is for it to be effective as a narrative that you can read cover to cover and gain a solid understanding of the Six Sigma improvement process and the DMAIC model.

At the same time, you should feel free to use it as a reference tool, dipping into particular sections, reviewing tools or using some of the worksheets provided as you need in your own projects.

In particular, for team leaders and champions we've included details on how to prepare for and conduct "tollgate" reviews: progress reports that leaders use to ensure projects are moving forward, staying on track, and achieving desired results.

Despite all the information and ideas we're tried to pack into these pages, we are aware—and so should you be—that each Six Sigma project poses unique challenges. No single book can advise you on every nuance or issue that may arise, or provide details on every tool. We sincerely hope our effort gives you at least some of the answers needed to meet your Six Sigma goals, and we apologize for any gaps we've left unfilled.

Acknowledgments

This book has been truly a team effort. (Had we been aware of the challenges it would pose, in fact, we might have passed on the whole idea.) But fortunately, we've benefited from the hard work of some dedicated people without whom it would not have come to fruition.

First and foremost, thanks and great credit go to the editor, repair person, organizer, and diligent worker who guided this project from its confused "first draft" stages to an actual publishable work. Sue Reynard has put in countless hours and the right kind of perspective (i.e., detail-oriented but with a view of the big picture) needed to pull this together. Sue will understand when we give her a big "Hip-Hip-Hooray!" and our heartfelt thanks.

Also contributing many hours, careful input, and creative skills are our Pivotal Resources colleagues Cheralynn Abbott and Julia Oseland, who had to take cryptic instructions and vague lists of graphics and notes to pull together many of the illustrations and job aids on the following pages. Without their work, Sue would have a much steeper hill to climb and no one to help climb it.

Behind all this work, the tireless (though sometimes frustrated) encouragement of Richard Narramore at McGraw-Hill has been a key force. To Richard and his colleagues at McGraw-Hill we owe special gratitude; they've had to suffer from the fact that we're about a year late with this book and have never threatened us (physically or legally), but have simply given gentle reminders as the weeks and months slipped by. Working with Richard, John Woods of CWL Publishing Enterprises, along with his colleagues, Robert Magnan and Nancy Woods, did the final editing and production to create the book you now hold. They went out of their way to make sure everything was as good as it can possibly be.

Enormous thanks go to the people we have the opportunity to work with every day to make Six Sigma efforts pay off: our clients, colleagues, and friends in many countries who've adopted this "new" approach to building enduringly great companies. And along with these people, of course, we thank our families, who allow us the time to help and learn and then share our insights on these pages.

Thanks, finally, to you—our reader and customer—for the time and energy you are devoting to Six Sigma and to advancing the cause of smarter business leadership. We hope you can share some of your "voice of the customer" with us by contacting us at: ssw@pivotalresources.com. Comments and suggestions are one of the ways we learn, and we'd appreciate hearing from you.

The Six Sigma Way Team Fieldbook

Part One

What a Six Sigma Project Team Needs to Know Before It Gets Started

Chapter 1

The Six Sigma System

A New Way to an Old Vision

SIX SIGMA. A new name for an old vision: near-perfect products and services for customers.

Why is Six Sigma so attractive to so many businesses right now? Because being successful and *staying* successful in business is more challenging today than ever before. In today's economy, most people provide services rather than making goods and products. And most of those services operate at levels of inefficiency that would close down a factory in a month if it produced as many defects. Six Sigma provides power tools to improve those services to levels of accuracy and quality seen so far only in precision manufacturing.

Companies like General Electric and Sun Microsystems are flexing the Six Sigma system to create new products, improve existing processes, and manage old ones. Leaders of these and other Six Sigma companies know that Six Sigma encompasses a wide variety of simple and advanced tools to solve problems, reduce variation, and delight customers over the long haul. Six Sigma ...

- Generates quick, demonstrable results linked to a no-nonsense, ambitious goal: To reduce defects (and the costs they entail) to near zero by a target date.

- Has built-in mechanisms for holding the gains.
- Sets performance goals for everyone.
- Enhances value to the customer by exposing "defects" caused by functional bureaucracy and by encouraging managers and employees alike to focus their improvement efforts on the needs of external customers.
- Speeds up the rate of improvement by promoting learning across functions.
- Improves our ability to execute strategic changes.

You can find descriptions of successful applications of Six Sigma in *The Six Sigma Way*. This chapter reviews key concepts introduced in that book—as a refresher for those who have read it and background for those who have not.

What Is Six Sigma?

If Six Sigma is so great, where has it been hiding all these years? Like most great inventions, Six Sigma is not all "new." It combines some of the best techniques of the past with recent breakthroughs in management thinking and plain old common sense. For example, Balanced Scorecards are a relatively recent addition to management practices, while many of the statistical measurement tools used in Six Sigma have been around since the 1940s and earlier.

The term "Six Sigma" is a reference to a particular goal of reducing defects to near zero. Sigma is the Greek letter statisticians use to represent the "standard deviation of a population." The sigma, or standard deviation, tells you how much variability there is within a group of items (the "population"). The more variation there is, the bigger the standard deviation. You might buy three shirts with the "same" sleeve length only to discover that none them are exactly the length printed on the label: two are shorter than the stated length, and the other is nearly an inch longer—quite a bit of "standard deviation."

In statistical terms, therefore, the purpose of Six Sigma is to reduce variation to achieve very small standard deviations so that almost all of your products or services meet or exceed customer expectations.

Variation and Customer Requirements

Traditionally, businesses have described their products and services in terms of averages: average cost, average time to deliver a product, and so on. Even hospitals have a measure for the average number of patients who pick up a new infection during their stay.

Trouble is, averages can hide lots of problems. With the way that most processes operate today, if you promise customers to deliver packages within two working days of getting their order, and your average delivery time is two days, many of the packages will be delivered in *more* than two days—having an *average* of two days means some packages take longer and some take less. If you want *all* packages to be delivered in two days or less, you'll have to dramatically eliminate problems and variations in your process.

Here's an example from *The Six Sigma Way:* you want your "drive to work" process to produce defects (early or late arrivals) no more often than 3.4 trips out of every million trips you make. Your target arrival time at work is 8:30 a.m., but you're willing to live with a few minutes either way, say 8:28 to 8:32 a.m. Since your drive normally takes you 18 minutes, this means your target commute time is anywhere between 16 and 20 minutes. You gather data on your actual commute times, and create a chart like that shown in Figure 1-1.

There will always be *some* variation in a process: the core issue is whether that variation means your services and products fall within or beyond customer requirements. If you want to be a Six Sigma commuter, the problem is that your process produces a lot of defects (late or early arrival times).

So you set about improving your process. You find the route that is *most* reliable (has the least traffic and fewest stop lights), you get up when your alarm clock *first* goes off, you recalibrate your cruise control, etc. After all your changes have been implemented, you gather more data. And *voilà,* you have become a Six Sigma commuter. The new standard deviation of just 1/3 of a

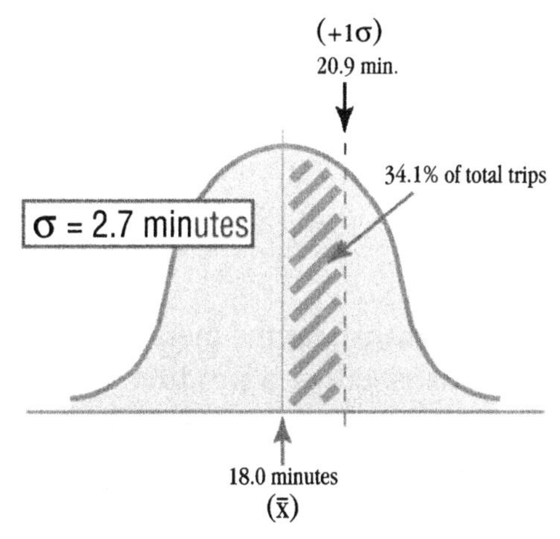

Figure 1-1. Drive times to work

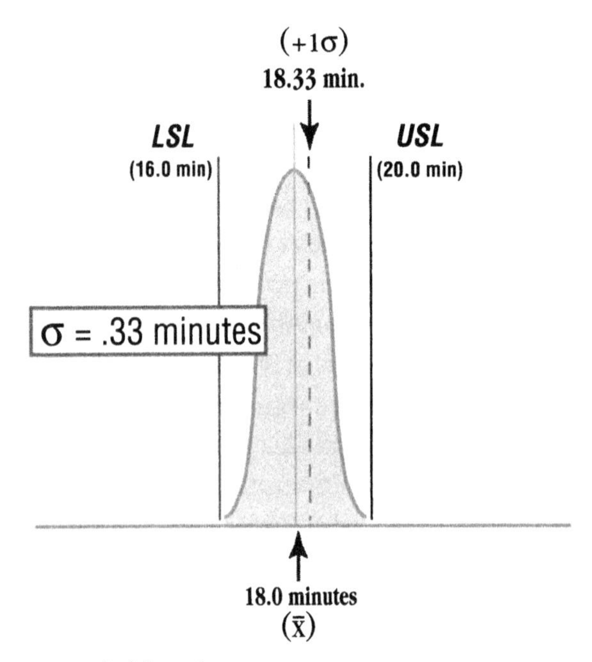

Figure 1-2. Improved drive times

minute means the variation in your process practically guarantees that you will always arrive within 16 to 20 minutes of leaving your house (see Figure 1-2).

This example has direct meaning for the business world. If we promise on-time airline departures, but actual departures vary from 5 to 30 minutes late, customers will understandably be angry and take their business elsewhere (except it might be hard to find an airline that does that well!). And an electric toaster that toasts the bread today but burns it tomorrow—at the same darkness setting—will find its way back to the store, along with an unhappy purchaser.

What happens when we achieve Six Sigma performance? For a Six Sigma commuter, it means predicting commute time very precisely every day. And a "defect"—a commute taking less than 16 or more than 20 minutes—would happen only 3.4 out of every 1 million commutes (may we live that long!).

Defects and Sigma Levels

One virtue of Six Sigma is that it translates the messiness of variation into a clear black-or-white measure of success: either a product or service meets customer requirements or it doesn't. Anything that does not meet customer requirements is called a **defect**. A hotdog at the fair with mustard is a defect if the customer asked for ketchup. A rude reception clerk is providing defective service. A bad paint job on a new car is a defect; a late delivery is a defect; and so on.

If you can define and measure customer requirements, you can calculate both the number of defects in your process and outputs as well as the process **yield**, the percentage of *good* products and services produced (meaning they are without defects). There are simple tables that let you convert yield into sigma levels.

Another approach to determining a sigma level is to calculate how many *defects occur compared to the number of opportunities there are in the product or service for things to go wrong*. The outcome of this calculation is called **Defects per Million Opportunities** (DPMO), which is another way to calculate the Sigma Level or yield of a process.

System Alignment: Tracking the Xs and Ys

Six Sigma companies commonly use shorthand to describe some key ideas when thinking about their own closed-loop business system. For example, "X" is shorthand for a cause of a problem or one of the many variables affecting a business process; "Y" is an outcome of the processes. For a bakery, the quality of the flour used and the temperature of the ovens are some of the key Xs to be measured, with the loaf itself being the key Y, along with the satisfaction of a hungry customer. Identifying and measuring such critical Xs and Ys are basic tasks in Six Sigma organizations.

6σ

Details on calculating sigma levels are provided in Chapter 9. You'll also find a worksheet and conversion table in Chapter 10.

Measuring Xs and Ys isn't an end in itself. Xs or causes have to be connected to critical Ys or effects. For example,

If X is...	*Then Y might be...*
Actions toward goals	Strategic goals achieved
Quality of work done	Level of customer satisfaction
Cycle time	On-time delivery
Staffing level	Time to answer phone
Incorrect information	Defects produced

Many companies and managers have a poor understanding of the relationship between their own critical Xs and Ys as they pedal along. They keep their corporate bike upright through a mixture of luck and past experience, or by making jerky corrections when the road suddenly changes. Six Sigma managers, on the other hand, use measures of process, customers, and suppliers to be more like experienced cyclists who anticipate problems or respond instantly and smoothly to changes around them.

As you start working on a Six Sigma team, try to become more aware of the outputs you want to achieve (your Ys) and what factors will affect how you get there (the Xs). Becoming more attuned to factors and their effects will help you focus your efforts more strategically. Also remember to link your Ys to what your customers really want—not just to what you think they need or what's convenient to you.

Six Ingredients of Six Sigma

The *Six Sigma Way* introduced six critical ingredients needed to achieve Six Sigma capability within an organization:

1. Genuine focus on the customer.
2. Data- and fact-driven management.
3. Process focus, management, and improvement.
4. Proactive management.
5. Boundaryless collaboration.
6. Drive for perfection, tolerate failure.

These ingredients are woven throughout this book and recapped below.

1. Genuine Focus on the Customer

Although companies have long proclaimed that "The Customer is Number One" or "Always Right," few businesses have actually succeeded in improving their understanding of their customers' processes and requirements. Many companies claim to meet customer requirements when they actually spend lots of time trying to convince the customer that what they bought is really what they wanted. (Remember the last time you had a phone or cable TV installed and were told that you'd have to spend a whole morning or afternoon waiting for service—when you'd rather have it done at a specific time? This happens when a service company has not been able to control its processes to the point where it can meet customer requirements!) Even when they have gathered information from customers via surveys and focus groups, the results were often buried in unread reports or acted on long after customers' needs have changed.

Customer focus is the top priority in Six Sigma. Performance measurement begins and ends with the Voice of the Customer (VOC). "Defects" are failures to meet measurable customer requirements. Six Sigma improvements are defined by their impact on customer satisfaction and the value they add to the customer. One of the first tasks of Six Sigma improvement teams is the definition of customer requirements and the processes that are supposed to meet them.

2. Data- and Fact-Driven Management

Although computers and the internet have flooded the business world with data, you won't be shocked to learn that many important business decisions are still based on gut-level hunches and unfounded assumptions. Six Sigma teams clarify which measures are key to gauging actual business performance; then they collect and analyze data to understand key variables and process drivers.

Finally, Six Sigma provides answers to the essential questions facing managers and improvement teams every day:

- How are we *really* doing?
- How does that compare to where we want to be?
- What data do I need to collect to answer the other questions?

3. Process Focus, Management, and Improvement

Whether you're designing a new product or service, measuring today's performance, or improving efficiency or customer satisfaction, Six Sigma focuses on the process as the key means to meeting customer requirements.

One of the most impressive impacts of Six Sigma has been to convince leading managers—particularly in service-based functions and businesses—that mastering and improving processes is not a necessary evil, but an essential step toward building competitive advantage by delivering real value to customers. In one of its first meetings the Six Sigma team must identify the core business processes on which customer satisfaction stands or falls.

4. Proactive Management

To be proactive means to act ahead of events; the opposite of being reactive, which means to be behind the curve. In the world of business, being proactive means making a habit of setting and then tracking ambitious goals; establishing clear priorities; rewarding those who prevent fires at least as much as those who put them out; and challenging the way things are done instead of blindly defending the old ways.

Far from being boring, proactive management is actually a good starting point for true creativity, better than bouncing from one panicky crisis to the next.

Constant firefighting is the sign of an organization losing control. It's also a symptom that lots of money's being wasted on rework and expensive quick fixes.

Six Sigma provides the tools and practices to replace reactive with proactive management. Considering the slim margin for error in today's business world, being proactive is the only way to fly.

5. Boundaryless Collaboration

Coined at General Electric, "boundarylessness" refers to the job of smashing the barriers that block the flow of ideas and action up and down and across the organization. Billions of dollars are wasted everyday through bickering bureaucracies inside a company that fight one another instead of working for one common cause: providing value to key customers.

Six Sigma requires increased collaboration as people learn about their roles in the big process picture and their relationship to external customers. By putting the customer at the center of the business focus, Six Sigma demands an attitude of using processes to benefit everyone, not simply one or two departments. The Six Sigma improvement team foreshadows the boundaryless organization on a small scale, and can teach much about its benefits to the whole company.

6. Drive for Perfection, Tolerate Failure

Six Sigma places great emphasis on driving for perfection and making sustainable results happen within a useful business time frame. As a consequence, Six Sigma teams often find themselves trying to balance different risks: "Is spending two weeks on data collection worth the effort?" or "Can we afford to change the process knowing that we'll likely create more problems in the short term as we work out the bugs?"

The biggest risk teams can take is to be afraid to try new methods: Spending time on data collection may seem risky at first glance, but usually it results in better, more effective decisions. *Not* changing a process means work will go on as it always has, and your results won't get better.

Fortunately, Six Sigma builds in a good dose of risk management, but the truth is that any company shooting for Six Sigma must be ready for (and willing to learn from) occasional setbacks. As a manager in a Six Sigma company once said, "The good kids have got us as far as they can by coming up with the right answers. Now the bad kids have to move us ahead by challenging everything we do."

Moving Forward

We'd be surprised if you didn't say "But we're already doing some of those things!" That's not surprising—remember, much of Six Sigma is not new. What is new is the way that Six Sigma pulls all these things into a coherent program backed by determined management leadership.

As you begin the job of leading a Six Sigma team, be honest about the strengths and weaknesses of your company, and be open to trying new things. You'll get better and faster results if your organization is willing to admit its shortcomings and learn from them.

Review your existing methods to make sure they are helping you improve the delivery of product and service to your customers. If not, you'll have to change past practices. Deciding exactly what changes will mean the most to your organization and its customers is the subject of most of this book!

Eyes on the Prize: Using Six Sigma Teams as a Learning Tool

Six Sigma teams are formed to address specific business issues and improve processes, products, and services. But if that's all they do in your organization, you're missing the bigger picture. It's short-sighted to make "projects" the sole objective of a Six Sigma effort. No matter whether a project is a huge success or fails to reach its goals, you've missed a big opportunity if the participants don't take new skills and habits back to their jobs after the project is complete. As many of our clients have realized, Six Sigma ideas need to become a way of life.

Never forget Six Sigma teams are a learning tool. Champions and Senior Managers should be studying the teams, the DMAIC improvement process, and the data-driven approach—and then applying those tools to their own daily management processes. If they do not, four or five years from now the organization will still be selecting projects, Black Belts, team members, etc.—only the names will be changed to protect those who refused to do things a better, different way. The organization as a whole will not be evolving, nor will it be anywhere near to reaching Six Sigma quality levels in its key processes.

Every leader in the organization should take on the responsibility of exploiting Six Sigma projects to the fullest extent, asking questions such as "What can we learn from these teams? How are they making gains? What can we apply to our everyday work?" Answering those questions will help your organization become dynamic and profitable, with unparalleled efficiency and customer loyalty. That's what it means to achieve Six Sigma levels of quality.

Chapter 2

Three Ways to Six Sigma
Strategies to Improve, Create, and Manage Processes

CUSTOMER KNOWLEDGE AND effective measures fuel a Six Sigma engine with three basic parts (see Figure 2-1), all of which focus on the processes in your organization. The linkage of these three parts together is one of the most important (and least recognized) innovations that Six Sigma offers. The three parts are Process Improvement, Process Design (and Redesign) and Process Management. Because your team's work will be touched by one or more of these, we'll introduce them all briefly here.

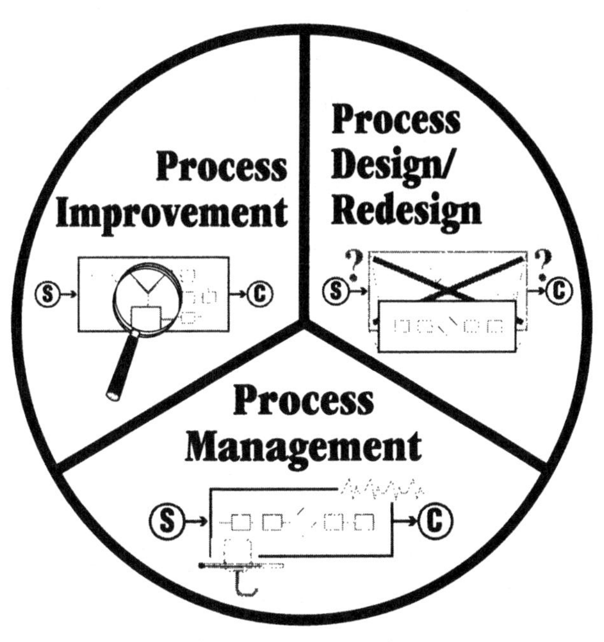

Figure 2-1. Three engines of Six Sigma

1. Process Improvement: Finding Targeted Solutions

"Process Improvement" refers to a strategy of finding solutions to eliminate the root causes of performance problems in processes that already exist in your company. Process Improvement efforts seek to fix problems by eliminating the causes of variation in the process while leaving the basic process intact. In Six Sigma terms, Process Improvement teams find the critical Xs (causes) that create the unwanted Ys (defects) produced by the process.

Process Improvement teams use a five-step process to attack problems:

Define the problem and what the customers require.

Measure the defects and process operation.

Analyze the data and discover causes of the problem.

Improve the process to remove causes of defects.

Control the process to make sure defects don't recur.

This process—often called DMAIC (Da-MAY-ihk)—will be the subject of most of this book. (See also Figure 2-2.)

Case Study: The E-Rock Crisis

The success of the Phernicher Company's electronic rocker-recliner chair over the past three years has been remarkable. However, for the past couple of months, sales have been in a dramatic and alarming slide. Distributors have been flooding the Phernicher home offices with e-mails about returns and customer complaints. As the biggest contributor to the company's profits, any problems with the chair (called the E-Rock) created big worries for Phernicher management.

The top product management group and the senior executive team met to discuss strategies for dealing with the declining sales.

"This product has outlived its welcome," exclaimed the head of marketing. "We need to get going on a whole new generation of E-chair!"

"That's going too far," countered the field sales director. "I can't believe the entire market suddenly decided the E-Rock is outmoded in a period of weeks."

The head of seat engineering weighed in: "We need to give more

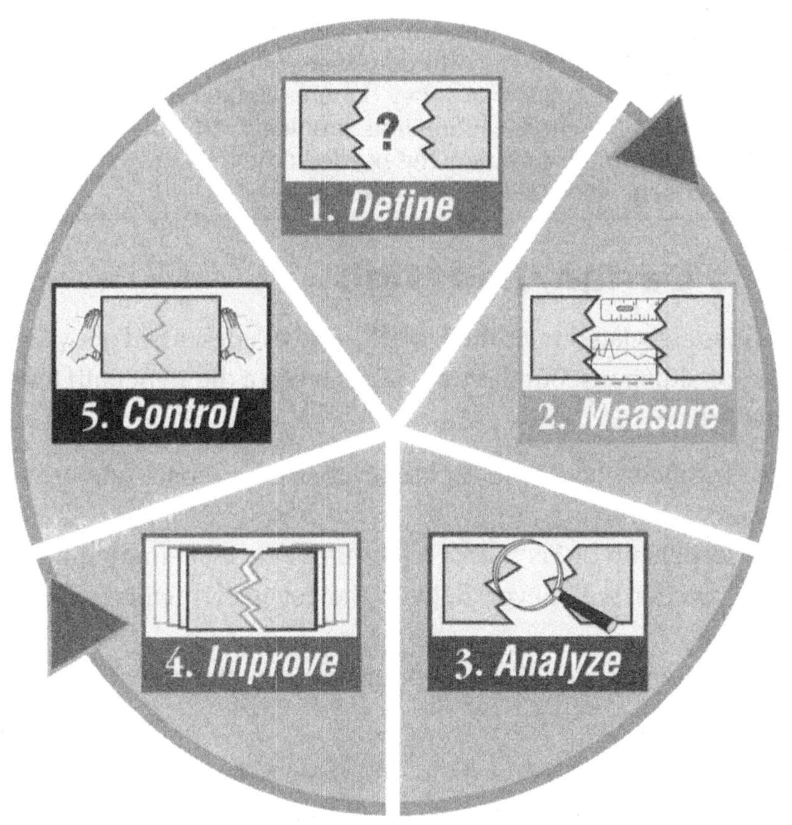

Figure 2-2. Five steps of process improvement

incentives to those distributors or threaten to drop them. It's clear they're getting lazy."

After a few more minutes of discussion (using some better meeting management tools) the group was able to agree to get a Six Sigma DMAIC team started on trying to find out the causes of the sales decline—an Improvement project.

Why was "Improvement" the right call? Here's why:

- The "quick fix" of incentives for distributors or any other short-term solutions would not address the real reason for the sales decline.
- Trying to design, develop, manufacture, and launch a new E-chair would take months. While the idea of a new round of product development was recognized as an issue worthy of more discussion, the leaders recognized that it would not correct the current sales slump.
- The rapid change in sales volume suggests something happened

somewhere in the E-Rock's manufacturing, shipping, or sales process to create problems with customers, who either aren't buying or are returning the chairs they bought. Figuring out the cause will likely help Phernicher address the problem with a permanent solution—giving the company the needed time to start working on the "next generation" electronic rocker-recliner.

2. Process Design/Redesign

The key activities associated with the DMAIC process described above fit a wide range of business situations. But there are times when a different path through the DMAIC is needed:

- When a business chooses to replace, rather than repair, one or more core processes.
- When a leadership or Six Sigma team discovers that simply improving an existing process will never deliver the level of quality customers are demanding.
- When the business identifies an opportunity to offer an entirely new product or service.

In these cases, the business needs to design or redesign core processes. This path goes by a number of names: Process Design or Redesign, "Six Sigma Design" (SSD), or "Design for Six Sigma" (DFSS). In Process Design, teams use Six Sigma principles to create revolutionary new processes, goods, and services built around customer requirements and validated by data and tests.

In Process Design, the specifics of the DMAIC steps are often adapted to focus on identifying innovative, effective ways to get work done:

Define customer requirements and goals for the process/product/service.

Measure and match performance to customer requirements.

Analyze and assess process/product/service design.

Design and implement new processes/products/services.

Verify results and maintain performance.

(Some companies or teams use "DMAIC" for both improvement or design efforts, and simply adjust their activities according to the nature of their project.)

Process Design will usually take longer than Process Improvement, and because it involves the creation and implementation of a brand new product or process, the risk of failure is greater than improving an existing process. (Such failure is often prompted by a lack of vision around the Design goals, or by not having people on the team who have the right skills and aptitude for a design opportunity.)

Though the outcome and goal of Process Design is very different from Process Improvement, much of the project work is very similar (Figure 2-3). Chapter 21 has details on how to adapt the Process Improvement instructions given throughout the book for a Design project.

Six Sigma Improvement Processes

	Process Improvement	Process Design/ Redesign
1. Define	✓ Identify the problem ✓ Define requirements ✓ Set goal	✓ Identify specific or broad problems ✓ Define goal/change vision ✓ Clarify scope & customer requirements
2. Measure	✓ Validate problem/ process ✓ Refine problem/goal ✓ Measure key steps/ inputs	✓ Measure performance to requirements ✓ Gather process efficiency data
3. Analyze	✓ Develop causal hypotheses ✓ Identify "vital few" root causes ✓Validate hypothesis	✓ Identify "best practices" ✓ Assess process design ◆ value/non-value adding ◆ bottlenecks/disconnects ◆ alternate paths ✓ Refine requirements
4. Improve	✓ Develop ideas to remove root causes ✓ Test solutions ✓ Standardize solution/ measure results	✓ Design new process ◆ challenge assumptions ◆ apply creativity ◆ workflow principles ✓ Implement new process, structures, systems
5. Control	✓ Establish standard measures to maintain performance ✓ Correct problems as needed	✓ Establish measures & reviews to maintain performance ✓ Correct problems as needed

Figure 2-3. Comparison of the Improvement and Design Processes

Case Study: Tri-Part's Patchwork Approach

After the merger of three regional equipment leasing companies, business was on the upswing. New leases of construction equipment, printers, copiers, and trucks were pushing Tri-Part Leasing into the ranks of the national leasing companies.

The positives on the "top line" of the business (increasing sales) were being significantly undermined, however, by problems in getting paid. Initial efforts to combine the invoicing and receivables processes of the three merged companies were not successful. It was taking from five to nine weeks to get an invoice out to customers—and another six to ten weeks to get paid.

Initially, Tri-Part Leasing got some partial relief. First, three cross functional "Fast Track" problem solving teams were able to identify some obvious gaps in handling lease documents. These fixes reduced the invoicing time by several days. Next, a process improvement Black Belt, working with a group of trained Green Belts, identified the cause of lost data in the company's accounting system, which cut the cycle time dramatically, but only for office equipment clients.

After reviewing the results of the solutions to date, the VP of Tri-Part's finance group proposed a new course of action: "I think we were dreaming when we tried to patch our different processes together," she commented. "There are so many inconsistencies and nagging problems in how we're doing things that it'll take dozens of small projects to get this thing turned around. Meantime, our cash is running low unless we can get paid faster!"

The conversation that followed was intense. Several of the managers worried that trying to redesign the invoicing and receivables process was too big and risky. In response, the idea of trying to do the redesign in phases helped people feel more comfortable. In the end, the vote to launch the redesign effort was unanimous and the comments enthusiastic.

"Sometimes," noted the billing supervisor, "you just have to throw out the old and bring in the new."

Why was "Design/Redesign" the appropriate strategy in this case?

- The impact of the problem is significant; it could threaten the health of the company. Not only is their cash flow affected, but so too is their image among clients—since it takes so long to get them their bill.
- Concerted attempts to achieve needed change—using both a "Fast Track" approach to weed out obvious flaws in the process as well as a Six Sigma Improvement team effort—have yielded only partial success. A lot more needs to be accomplished.
- The history of the process and merger indicates that a more extensive effort needs to be made to eliminate all the flaws and problems in the invoicing activities. More extensive, aggressive action seems warranted.
- The leadership group is ready to support a design effort. Without that support and the patience it will take to do the redesign, it still might have been better to continue with incremental changes.

(Support is that critical, especially for the Design strategy.) Also, by phasing the design effort, Tri-Part is likely to achieve interim gains while rolling out the full redesign.

3. Process Management for Six Sigma Leadership

One aspect of the DMAIC process improvement strategy that is often overlooked is that the concepts apply to managing processes across the organization, not just working through an improvement project. This third application of DMAIC—Process Management—is the most evolutionary of the three because it involves changes in culture and management throughout the organization that must accompany Six Sigma efforts if their full power is to be realized. As the name suggests, Process Management means that a focus on managing processes across the organization replaces managing individual functions by different (and sometimes competing) internal departments.

Because it requires a fundamental makeover in the way an organization is structured and managed, Process Management is often the most challenging of the three Six Sigma strategies to master. Nevertheless, without Process Management, Six Sigma is often doomed to become just another flavor-of-the-month program. In general, Process Management includes:

- Defining processes, key customer requirements and process "owners."
- Measuring performance to customer requirements and key process indicators.
- Analyzing data to enhance measures and refine the process management mechanisms.
- Controlling performance through ongoing monitoring of inputs/operations/outputs and responding quickly to problems and process variations.

You may have noticed that in Process Management, we can apply most of the major steps outlined above (Define, Measure, Analyze, and Control; we skip Improve)—though the focus is on an entire process, not just a specific problem or design challenge.

Process Management tends to evolve as a business expands its Six Sigma effort and deepens its knowledge of its processes, people, and customers. While it may take five or more years for a company to achieve 5σ or 6σ in some of its core processes, it will probably take twice as long for the company to evolve from a reactionary management-by-the-seat-of-the-pants outfit into an organization that proactively manages its processes with the same precision as it manages its equipment.

Process Management is work that business leaders do to improve their processes for managing the business. As such, it does not usually fall within the authority of a Six Sigma improvement team, and therefore is not covered in this book. But as a Six Sigma team member or leader, be aware your work may be influenced by (or help to influence) broader efforts going on in your organization.

Case Study: Surviving the Shake-Out

Dot Comedy is one of the few survivors of the e-business shake-out. The company, formed by a husband and wife standup team in (where else?) their garage, has grown into a popular provider of gag e-mails and jokes that show up in people's electronic mailboxes and business presentations throughout North America. Global expansion is starting next year, with Dot Comedy Japan gearing up for launch.

As the company grew, however, the founders and new management team became increasingly uncomfortable. "I know we're making money," said harried co-founder Rim Shot, "but that's about it."

His wife, Lota Laeaffs, expressed similar dismay. "I'm afraid the new Global Group and the domestic people are not cooperating with each other. Why did they need two different joke testing services?"

Explaining (well, whining) about how easy and simple the company was in the good old garage days (three years ago), Dot Comedy's owners asked their hand-picked President, Noah Djohk, for ideas.

"I've been preparing for this," Djohk responded. "I think we need to start some Process Management work. What used to be simple tasks that you two could do on your own have ballooned into six departments and 700 people. But no one has ever looked carefully at those departments and how they fit together. Since I've been here I've been worried that we have no good data on how different activities are working; it's like flying a plane with your eyes closed."

"That's how it feels to me," said Shot.

"So, I've tentatively scheduled a meeting to get started on identifying what are our core processes. We need to get some sense of ownership and measures around them," said Djohk.

"Let's do it!" the two comedians exclaimed in unison.

Is Dot Comedy doing the right thing? Yes! And here's why:

- For the moment, the business is enjoying great success. There are (apparent) problems, but financial and customer results are strong.
- On the other hand, the company is growing more complex and the need for information on how things are operating is becoming more critical. Some signs of redundancies in processes are showing.
- By getting a clearer view of the processes in place at Dot Comedy, including their interdependencies, the company will be able to better assess its performance, identify trends and changes, improve its understanding of changing customer requirements, and take action to address problems and opportunities promptly.

Using All Three Strategies

There is a precise parallel between the evolution of process improvement teams and the evolution of management within a company.

Every team will have to define customers and eliminate the defects in product and services that make them unhappy today (Process Improvement). They may have to create entirely new sub-processes to eliminate and prevent defects. When processes reach their ultimate capability and still can't meet increased customer requirements, teams will have to start from scratch to create new processes (Process Design/Redesign) that can deliver quality at the 4σ or 5σ or 6σ levels.

All improvement teams will have to implement some way of managing an improved process and hand it off to other people. The new solution or process will not survive and prosper unless someone is assigned the job of tracking results and maintaining the gains. In essence, the Control phase of DMAIC is a piece of the overall Process Management effort.

This same pattern is true of the overall Six Sigma effort in a company: it's usually easiest and most effective to improve what exists (Process Improvement) before trying to invent a new way of doing business (Process Redesign). Starting with specific projects also helps an organization gradually learn how to organize and manage using new approaches (Process Management).

In a very real sense, then, improvement and design teams are the time machines that foreshadow the future of the organization and the way it will operate at a Six Sigma level. Those doubtful managers and employees who want to know what the organization of the future might look like should be able to visit a well-running Six Sigma team today and see what they're missing.

Choosing a Six Sigma Approach

Many companies have adopted the DMAIC model—Define, Measure, Analyze, Improve, Control—or some variation of it for Six Sigma improvement projects. We'll use these five steps as our preferred model later in this book. However, if your organization already uses or has taught another process improvement or redesign model, it's by no means mandatory that you abandon it in favor of DMAIC.

Many of the various models used in different organizations can serve well as guides to Six Sigma improvement efforts. In fact, all of the models—DMAIC included—are based on the "Plan-Do-Check-Act" (PDCA) cycle, and each has strengths and weaknesses. If your existing model works and is familiar to your staff, changing to DMAIC may be confusing—plus you'll need to teach a whole new model to replace the old. If you want to continue your existing model, you can either adapt the improvement and redesign steps described in later chapters to your own model, or see if your leadership is willing to use the DMAIC model, which is becoming an industry standard.

Bottom line: there is no right or wrong, one-size-fits-all model for Six Sigma. If the Define-Measure-Analyze-Improve-Control steps work for your business, great! If you have another approach now that works, that's OK, too—just remember that this book is organized around DMAIC, and translate those steps into what your organization does.

Either way, Six Sigma will work for your business!

Chapter 3

Organizing for Six Sigma

Meet the Players

SIX SIGMA IS NOT JUST ABOUT DATA TOOLS and defect calculations. Nor is it just about having people work in teams. Teams alone cannot change corporate structures. They must be part of an infrastructure designed to assist in the redesign of the organization, like scaffolding around a building being renovated. One way to understand this renovation structure is to review the roles of people in the evolving Six Sigma organization. There are seven functions and roles that must be developed:

1. Leadership Group or Council.
2. Project Sponsors and Champions.
3. Implementation Leader.
4. Six Sigma Coach (aka Master Black Belt).
5. Team Leader/Project Leader (aka Black Belt).
6. Team Members.
7. Process Owner.

Role 1: The Leadership Group or Council

The leadership council consists of senior managers in the business, gathered in a forum designed to help them learn a new way to manage the business by direct experience with Six Sigma teams. In its natural leadership role, this group plans and executes the Six Sigma implementation plan. In the first stages of the Six Sigma roll-out, the Leadership Group must …

1. **Develop a strong rationale** for doing Six Sigma specific to the company's needs.
2. **Plan and actively participate in the implementation**. In the old story about a breakfast of eggs and bacon, the chicken is *involved* but the pig is *committed*. Likewise, the Leadership group must be committed to direct involvement in the introduction of Six Sigma to their organization.
3. **Create a vision** and an internal "change marketing" plan to sell Six Sigma to the key customers inside the organization.
4. **Become powerful advocates** for Six Sigma as a means to fix problems *and* a new way to do business. Support for Six Sigma from leaders must be strong, constant, and high energy.
5. **Set clear objectives** for Six Sigma that can be translated into action items in the trenches. Goals like "5 Sigma in 5 Years"—if they are realistic—can focus attention throughout the organization.
6. **Hold itself and others accountable** for the success or failure of Six Sigma efforts. At General Electric and other Six Sigma pioneers, managers' bonus money and chances for promotion are tied to the accomplishment of Six Sigma goals. Think about the chicken and the pig again!
7. **Demand solid measures of results**, including measures of defects and yield (Sigma), cycle-time improvements, reduced costs and rework. Other critical measures might involve customer and employee retention, profit margins, and new product sales.
8. **Communicate results—and setbacks.** Constant, honest communication of good and not-so-good results is crucial to the roll-out of Six Sigma. Speeches, awards, newsletters, e-mails, and storyboards all play a part in focusing attention on the changes underway.

While leaders set the tone and direction for the Six Sigma effort, other key players are essential for the success of Six Sigma.

Role 2: Project Sponsors and Champions

In most organizations, a Sponsor or Champion is a senior manager who oversees a Six Sigma project and is accountable to the Leadership Council for the success of that project. The Champion's role is a delicate one. She/he must give the project improvement team clear guidelines on their project, run interference for the team when it meets roadblocks within the organization, but avoid "taking over" the team or dictating a pet solution for the team to implement. The Champion's responsibilities include:

1. **Setting a rationale and goal** for improvement projects that align with business priorities.
2. **Being open to changes** in the project definition and scope as the team gathers data and deepens its analysis of the process.
3. **Coaching on and approving changes** in the team's charter and project scope, when needed.
4. **Finding resources** (time, support, money) for the team.
5. **Advocating for the team's efforts** in the Leadership Council.
6. **Running interference** for the team when it encounters bureaucratic roadblocks along the way.
7. **Working with other managers** to make sure the team's solution is handed off smoothly.
8. **Learning the importance of data-driven management** from the team and applying the lessons to their own management job.

This last point is crucial to the evolution of the company towards Six Sigma. Champions must avoid the trap of "delegating" projects to teams and then taking a hands-off attitude until the team is ready to report its solutions (what we call the "Aloha syndrome"). The Champion should learn about Six Sigma by direct, steady involvement with the team.

Remember: one of the main reasons for having Six Sigma teams is to help Champions to learn a better way to manage the business. An individual project team can take some steps to make sure the lessons learned are transferred to others in the organization, but it is the Champions and other executive leaders—

who are looking across projects and departments—who can have the biggest impact on using Six Sigma tools and methods to strengthen the business.

Role 3: The Implementation Leader

Someone has to manage the day-to-day roll-out of the Six Sigma effort. Depending on the scale of the operation, one Implementation Leader or Six Sigma Director may be enough, or you may need a staff to handle this broad set of tasks:

1. **Support the Leadership Council,** communicate plans, help in project selection and project tracking.
2. **Identify and recruit other key players**, including outside consulting assistance.
3. **Assist in the selection and development of training materials.**
4. **Plan and execute training**.
5. **Support team Sponsors or Champions**.
6. **Document overall progress** of the roll-out, inform Leadership Team of progress and problems.
7. **Execute internal marketing** plans for training and tracking teams.

The energy and talent required for what appears to be an administrative support position is in fact enormous. While this person is more often a "generalist" than a Six Sigma expert, the implementation leader probably has more to do with the success of the Six Sigma undertaking than anyone outside the Leadership Council—of which they should be a full-time member.

Note: Titles used for the Implementation Leader role vary quite a bit. At 3M Corporation, for example, implementation leaders are called "Master Black Belts," though their role is different from the typical "MBB" (see below). Other companies call this a "Deployment Champion." It's the responsibilities that count, not the title.

Role 4: The Six Sigma Coach (Master Black Belt)

The Six Sigma Coach provides expert advice to a number of Process Owners and Six Sigma Improvement Teams in areas ranging from statistical measurement tools to change management and process design strategies. Coaches must walk a fine line between advising and meddling, especially with improvement teams. They are wise to contract with team leaders on the precise nature of their relationship to the teams and how they will provide assistance.

In addition to being firmly grounded in the basics of teamwork and improvement, Coaches should be able to provide guidance on:

1. **Communicating** with Project Champions and the Leadership Council.
2. **Establishing and sticking to a firm schedule** for projects.
3. **Dealing with resistance** to implementing Six Sigma.
4. **Estimating, measuring, and validating dollar and other savings** attributed to improvement projects.
5. **Helping to resolve team and other conflicts.**
6. **Gathering and analyzing data about team activities**. For example, how long does the average team take to complete a full cycle of the DMAIC process? What average savings can be expected from a DMAIC project?
7. **Helping teams promote and celebrate their successes.**

As you can probably tell from this description, MBBs need an unusual combination of people skills and statistical skills. That's why companies often turn to external consultants in the earliest stages of implementing Six Sigma to provide that expertise—then turn these responsibilities over to full-time employees once they developed sufficient expertise. These "internal" consultants often go by titles such as Black Belts and Master Black Belts (see p. 30). It is the Master Black Belts who often serve as Six Sigma Coaches, while Black Belts serve as Team Leaders.

Alternatively, other companies identify separate resources for each skill set: they may have MBBs who are good at people/leadership skills, and have them collaborate with statistical experts (who may be internal or external to the organization).

Role 5: The Team or Project Leader (Black Belt)

The Team Leader or Black Belt is the person who accepts primary responsibility for the routine work and results of a Six Sigma project. The duties are similar to those of the Coach, but specific to one team only. The Team Leader is usually someone familiar with the issues under analysis and will normally be a part of the process they're trying to improve. Their responsibilities include:

1. **Reviewing/revising/clarifying the project rationale** with both the project Champion; helping the Champion understand how Six Sigma techniques apply to everyday operations.
2. **Working with team members** to develop/update the team's charter and implementation plan.
3. **Selecting or helping to select project team members.**
4. **Identifying and finding resources and data for the team.**
5. **Supporting team members** as they learn and implement Six Sigma methods and analytical tools.
6. **Making sure the team uses its time effectively,** such as through the use of meeting management techniques, decision-making strategies, and planning tools.
7. **Maintaining the team's project schedule** and keeping the team moving toward completion of the project on time.
8. **Supporting the transfer of new solutions or processes** into ongoing operations by working with functional managers or process owners.
9. **Documenting final project results** and creating a storyboard to display the work and results of the team, often in the form of a presentation to the Leadership Council.

Role 6: Team Members

Usually chosen because of their work in the process under review, team members bring the brain and muscle for collection and analysis of data needed to improve the process. Since team members seldom work full time on team projects, they will have to contract with their supervisors on how they will coordinate their team work with their regular jobs. Team members must be willing to:

1. **Ask "dumb" questions** and participate actively in the team's work both at and away from team meetings.
2. **Carry out instructions for data collection and analysis.**
3. **Listen actively to others**, and practice good meeting management skills regarding discussions, decisions, and plans.
4. **Carry out assignments** between meetings and be ready to report results to the team.
5. **Review the efforts of the team itself** from time to time to improve the meeting process.

Until their managers routinely apply Six Sigma techniques to normal operations, the project improvement team will be the place where most employees will learn about the new data-driven ways to manage the operation. Once exposed to DMAIC and team meeting skills, team members should expect their own managers to use similar techniques and skills in everyday operations.

Tips on Selecting Team Members

Keep the size of the team manageable: four or five members plus the Team Leader is about right. Others may join you as "team temps" to supply specific data at crucial points in the DMAIC process. Generally speaking, the larger the team, the longer everything takes. So beware of wanting to be too inclusive.

Here are some of the criteria for good team members:

- Good knowledge of process, product, and customer.
- Access to data about the problem or process.
- Willingness to work cooperatively with other members.
- Ability to devote three to four hours per week to data collection and team meetings.
- Ability to challenge the status quo.

Role 7: The Process Owner

When Six Sigma starts up in a functional organization, the Process Owner is normally the manager of a part of a particular function. They are the people who receive the solution created by an improvement team, and become the "owners" responsible for managing the improved process.

Eventually, as Six Sigma evolves, and the organization along with it, the focus on the function will be replaced by focus on core processes. Process owners will then be those people who manage a process (like sales) "end-to-end" across the organization. The emergence of Process Owners is gradual in many Six Sigma organizations. Process ownership only makes complete sense in an organization that has implemented Process Management as its chosen way of doing business, an implementation that may take several years.

Black Belts, Master Black Belts, and Green Belts

Three special roles within a Six Sigma organization have been named using terms inspired by levels of karate skills; two of these have been discussed already.

- **Black Belts:** Originally used at Motorola in the early days of Six Sigma, the term Black Belt has come to mean someone who either coaches or actually leads a Six Sigma improvement team. Black Belts usually get several weeks of training in process analysis and team meeting skills. In technical and manufacturing settings, this training includes a solid dose of statistical tools like sampling, multivariate analysis, and design of experiments. In service businesses, less emphasis is placed on these tools, and more on the mapping and analysis of processes, and the use of such tools as cause-and-effect diagrams, histograms, and Pareto charts.
- **Master Black Belts**: Master Black Belts usually receive in-depth training on statistical tools and process improvement. They perform many of the same functions as Black Belts but for a larger number of teams. Normally, Master Black Belts have successfully managed many process improvement teams and thus have lots of good experience under their belts. They may also serve as change-agent consultants to the Leadership Council and other managers.
- **Green Belts:** Green Belts are usually employees who have received enough Six Sigma training to participate in a team or, in some companies, to work individually on a small-scale project directly related to their own job.

Different companies use these "Belts" in different combinations with Sponsors and Champions to guide teams. Several options are shown in Figure 3-1.

In some companies, the Black Belts and especially Master Black Belts are full-time positions. In such cases, both are expected to support several teams at once. Often the Black Belts are expected to complete a certain number of process improvements in a set time or save a particular amount of money by reducing defects or creating new processes. The precise definition and jobs of Black Belts and Master Black Belts will vary depending on the needs of the organization and its stage of development from a functionally managed to a process-managed company.

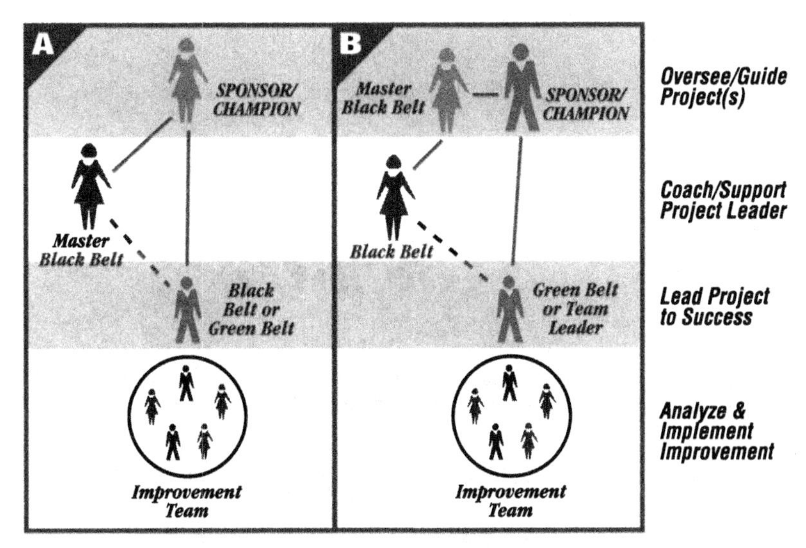

Figure 3-1. Organizing the roles needed to support Six Sigma efforts

Conclusion

Whether senior managers on the Leadership Council or front-line employees going to their first team meeting, everyone involved in launching the Six Sigma ship must have one thing in common: an absolute commitment to using new tools to learn new things in order to create a new kind of organization. It will be a long time before the scaffolding of councils, Champions, Black Belts, Team Leaders, and consultants can be removed to reveal a boundaryless organization that no longer consciously "does" Six Sigma "stuff," but simply has the tools and attitude behind Six Sigma in its bones.

In the meantime, success will depend on selecting the right Six Sigma projects to begin implementation. That's the subject of the next chapter.

Chapter 4

Selecting Winning Six Sigma Projects

Getting It Right the First Time

IN OUR EXPERIENCE—and from the feedback from dozens of senior executives and Six Sigma leaders—project selection is the most critical and most challenging activity in launching Six Sigma. In company after company, leaders have selected world-hunger-size projects for inexperienced teams to solve. The painful and predictable results? Missed deadlines, mountains of frustration, and often partial success at best.

The right way to select projects is a simple equation, really:

A clear, well-selected, well-defined project
+ a well-trained team
+ a committed Champion
= good, fast results for customers

If it's so simple, why do companies get it wrong so often? Among the more common reasons:

1. Senior managers on Leadership Teams think they're already pretty good at scoping out and assigning projects, and refuse to review and learn

more about Six Sigma project selection.
2. Many managers are out of touch with key customers and core processes under their direction, and resort to guessing about what needs improvement.
3. Some managers think that a sprinkling of Six Sigma "magic pixie dust" will be able to solve problems that have stumped them for years.
4. Others believe they already have the solution for the problem and only want to use a team to get the Six Sigma seal of approval on their pet project solution.
5. Teams are assigned projects that focus on issues that do not directly benefit external customers. For example, improving the process of new personnel recruitment is important to the internal operation of a business, but it may be better to focus on more pressing issues that directly affect paying customers.
6. Leadership Councils call for many teams to be formed too soon in the implementation of Six Sigma. Managers compete to see who can send the most teams to training. Projects pop up like mushrooms, and no one can keep track of them all. Learning and results dwindle. Early enthusiasm turns into indifference.

To have a different outcome—a successful project—the Leadership Council and Champions must learn how to carefully carve out doable projects, and then remain open to revisions in the original project as teams collect data and discover what's really going on in the process.

Start Simply When Learning a New Language

Learning DMAIC—how to rigorously Define, Measure, Analyze, Improve, and Control processes with the cooperation of key customers—is like learning a new language for many teams, Team Leaders, and Champions. If you've studied a language other than your own, you may recall that most classes start out with fairly simple tasks such as learning basic structures (the alphabet, numbers, simple everyday conversations, etc.). You don't start off with tough sentences like "If I'd known then what I know now, I'd have made a different decision." Why not? Because the complex tenses and inflections are too difficult for beginners—and besides, you wouldn't need that level of sophistication during your first visit abroad! (And if you tried, you might get yourself into some embarrassing situations.)

The same principle applies to improvement teams learning DMAIC. Until teams are comfortable applying and interpreting basic tools like process mapping, Pareto charts, and histograms, don't ask them to do something like "test a null hypothesis with a two-sample *t*-test." Like college students overwhelmed by a difficult language course, the team may decide to just drop the class—and the project will drift into limbo.

Three Project Selection Essentials

1. Keep the number of teams small until people become fluent in the language of Six Sigma. Think about tossing two basketballs into a group of a dozen people: odds are both balls would be caught. Now think about tossing two handfuls of dried beans into the same group: chances are many beans would hit the floor before being caught.

The same principle applies to selecting initial Six Sigma projects. A limited number of well-designed projects have a greater chance of being completely successfully—on schedule with good results for customers—than do dozens and dozens of teams scrambling around without good direction and guidance. Remember: one of the main reasons for DMAIC projects is for everyone to learn how to manage the business a better way. You can't do that when you're losing your beans!

2. Scope projects properly. Champions tend to underestimate the time it takes to gather and analyze reliable data. If the project is scoped improperly, a Black Belt or team can spend month after miserable month trying to scale down a "world-hunger" issue into something they can actually accomplish in three to six months or less.

Here's a quick example: Suppose a team is asked to reduce the cycle time for a process that currently fails to meet customer time requirements. There is no reliable existing data for the process cycle time, so the team must collect all new data. Currently, the process takes about two days to cycle in a one-shift-a-day operation, but two-and-a-half to three days is not uncommon (there's that evil variation!). How quickly can the team get sufficient data to get a reliable estimate of the average cycle time? As always in data gathering, the answer is "it depends" on the analytical tool the team plans to use. A histogram showing the shape and distribution of the data needs 50 to 100 data points to be reliable, and a run chart requires 25+ data points. Depending on the tool used, the team could spend 50 to

200 days collecting cycle information: way too long if their Champion has asked them to finish the entire project in three months!

Here's another example: Suppose a team is asked to reduce defects in an order fulfillment process. That kind of process could cover anywhere from three or four to perhaps ten or more different functional areas in the company—everything from the customer service rep who takes the order to the truck driver who makes the delivery. Coordinating improvement across such a broad span would require strong support from many organizational leaders, and involvement from representatives of each major area. It might also take half a year to a year or two to implement changes throughout each sub-process. While experienced Six Sigma organizations with a fully developed infrastructure might be able to carry this off, those in the initial stages would be better off focusing on a segment of the process, or a limited type of defect (usually the one that most impacts the customer and/or costs).

There are other ways that a project can be scoped improperly, such as trying to turn a two sigma process into six sigma capability within just a few weeks or months. The lesson is that the Leadership Council must pay attention to the current capabilities of its processes and people, and determine realistic boundaries both in terms of the process steps being studied and in terms of time.

3. Pay attention to external, paying customers. There is a temptation early on to select quick-hit projects that will improve the efficiency of internal processes. Unfortunately, this thinking ignores the most important people in the equation: the external, paying customer. The corporate graveyard is full of tombstones commemorating corporate surgeons who "reengineered" internal processes, only to discover that while they were gutting the organization, their customers had moved to another company who focused on their requirements. There should be a balance between internally and externally focused Six Sigma projects, but early teams should work on processes that directly affect external customers, even if the gains are small at first.

Project Selection Process

One of the challenges of project selection—as in many business decisions—is to choose from among all the possible activities you could undertake, and decide which have the greatest potential and therefore should be done first. Besides needing to make sure that the projects will benefit the business, the other caveat is that those responsible for selecting projects (such as the Leadership Council) should not hurry through this selection process even though they'll be anxious to get things rolling. Making rash decisions is a mistake that will cause much rework later on when the organization Champions and teams struggle to focus on the issues.

To increase the odds that Six Sigma projects will be successful, follow these steps:

1. Review where you are now (based on data).
2. Develop a list of potential projects and describe the "pain," goal, and rationale for each.
3. Screen out those that don't meet basic criteria.
4. Operationalize criteria for the final choices.
5. Apply the criteria and select the project(s).
6. Evaluate the set of projects selected.
7. Draft a charter for each selected project.

At the end of this selection process, you should have a short list of *manageable* projects that will produce *meaningful* results for your business and its customers.

Step 1. Review Up-to-Date Internal and External Sources of Information About Your Business

External Sources. These include voice of the customer, voice of the marketplace, and comparisons with competitors. Customer surveys, marketplace analysis, complaint data, and research on competitors can all stimulate questions that can lead to possible projects:

- Where are we failing to meet customer requirements? What do they complain about the most?
- Where are we behind our competitors?
- Where is our market going?
- What new customer needs might be on the horizon?

Checklist for Sources of Customer Data

Traditional

- Surveys
- Focus groups
- Interviews
- Formalized complaint systems

- Market research
- Shopper programs

New generation

- Targeted and multi-level interviews and surveys
- Customer scorecards
- Data warehousing and data mining
- Customer/supplier audits
- Quality function deployment

Internal/External Sources

Answers to the questions below help to identify challenges your business faces in defining or achieving its customer and market strategies. These questions include:

- What barriers prevent us from reaching our strategic goals?
- What new acquisitions need to be integrated so that we can be profitable and aligned with our image in the marketplace?
- What new products, services, locations do we want to launch to provide better results for customers and shareholders?

Internal Sources

The goal with internal sources is to link them with processes that can be improved in ways that benefit external customers. The frustrations, waste, problems, and opportunities visible inside your organization provide another source of possible Six Sigma projects. We can call these sources "Voice of the Process" and "Voice of the Employee." You might ask:

- What major delays slow our core business processes?
- Where do defects and rework appear most often?
- Where are costs of rework and quick fixes rising?
- What concerns or ideas for change have employees and managers raised?
- What irritates employees most about the processes on which they work?

Step 2. Identify Potential Projects and Describe the "Pain," Goal, and Rationale for Each

Use the data gathered in Step 1 to identify a range of potential projects. Write a brief description of each potential project, capturing:

- The **pain:** who is suffering because of the problem (include both customers and employees)? In what ways do they suffer?
- The **goal:** what would you like to accomplish?
- The **rationale:** why would it make sense to work on this project now (especially when compared to other improvement opportunities)?

Step 3. Screen the Possibilities

Work through your list of potential projects and eliminate any that don't meet the following basic criteria:

- **There is a significant gap between current and desired/needed performance.** Sometimes the gaps between what your processes are capable of doing and what you need to be doing are obvious: If customers want a 24-hour delivery time and you can only do 48 hours or more, you know where you need to improve! Other times, you may need to collect data on your current processes or quality levels and compare them to customer requirements. The key questions are "Where's the pain?" and "What are the symptoms?" To use DMAIC, you need a problem to solve or an opportunity to exploit. In the case of a new process design, you're starting from ground zero to launch a new process or product where none exists now.
- **The cause of the problem is unknown or not clearly understood.** A problem is something whose effects (or defects) we see, but whose cause is unknown to us or at least we don't have positive proof for our causal theories. For example, if your car's front tire suddenly goes flat and you see a large nail sticking out of it, you don't need to do a lot of Defining, Measuring, and Analyzing to discover the cause of "lost air in tire"! You have some obvious choices: put on the spare, call for roadside assistance, or call in sick. However, if the tire gradually goes soft even when you put more air in and there are no obvious holes, you have a problem that has an unknown cause. That's when you need DMAIC to diagnose and cure the problem.
- **The solution isn't predetermined, nor is the optimal solution apparent.** If the cause of a serious problem is unknown, guessing at solutions can be risky and expensive. While throwing money at problems has a long and varied history, Six Sigma offers a way to be sure that the money thrown hits the right target the first time. If quick fixes are satisfactory, don't use the full DMAIC approach. The same holds true if a solution is clearly obvious.

Step 4. Operationalize Remaining Criteria

To make your final project selections, you need to know specifically what costs and outcomes are important for your business. Answer the following questions to identify specific criteria you will apply to the project ideas.

A. **What business benefits are key?** A selected project should further your business strategy or improve one of your company's "core competencies." For example:
   - What financial gains would you like to realize?
   - How urgent is it to address the problem? Is the trend of the problem getting worse, or can we (and our customers) live with it for a while? Are there several issues that have to be resolved in a given sequence? Obviously urgent problems with worsening trends usually find their way to the top of the list.

B. **How will you determine feasibility?** What limits will you set on:
   - **Time:** What timelines can your organization live with?
   - **Cost**: How much cost could each project incur? What overall costs could you afford?
   - **People:** How much flexibility is there in use of staff time?
   - **Other resources:** How many trained coaches and team leaders do you have? What other resources are or are not available to the teams (such as equipment, outside expertise)?

   Two other feasibility considerations are:
   - Which of the criteria you've identified are most important? Would you be willing to live with a project that might take longer but that has a greater potential gain? Or would making some gains more quickly be preferable?
   - Does the project impinge on an area where there is a history of management inertia when it comes to making change? Prizefighters spar a lot with their partners before taking on tough opponents. If feasible, consider selecting projects where teams can do the sparring on their first project in an area where there is clear management support.

C. **What organizational learning or other changes are important?** Are there particular areas of learning that would have a strategic benefit to your organization? For example, it might be worthwhile to invest in a project that could test a promising new technology. Or if your organization is new to using Six Sigma, having a test project that would let you learn how to manage such projects effectively might be the right strategy. Another important internal strategy is often learning how to break down functional walls and barriers—in which case projects that stretch across traditional boundaries might be beneficial.

Step 5. Evaluate Remaining Projects and Select Best Candidates

Apply the criteria you developed in Step 4 to all the project ideas that weren't previously screened out. As you assess each project, factor in the level of experience or expertise of the people likely to conduct the project. For example, a team doing its first DMAIC project will take longer than an experienced team working on its third or fourth. A team that has members experienced in data collection can take on more sophisticated projects than other teams.

During this analysis, you can either perform a simple Yes/No evaluation (the project does or does not meet the criteria), or apply a more sophisticated scoring system (such as rating the ideas on a scale of 1 to 5 as to how well they meet the criteria). Either way, use a matrix such as that shown in Figure 4-1 to help you document both the criteria and outcomes for each alternative project. (Instructions for completing a criteria matrix are included in Chapter 16, pp. 315-317.)

After rating each project idea, determine which ones best meet your criteria.

Step 6. Evaluate the Set of Projects Selected

One last reality check the Leadership Council or others should perform is to look at all of the projects selected as a whole (what we often call a "Project Portfolio"; some companies call it a "Job Jar"). The critical questions are: Does your organization stand a good chance of successfully completing all the selected projects? Have we identified the right mix of projects to address critical needs of the business and customers? For example:

- Are there enough people to staff all the teams that will be needed without impairing the organization's ability to carry on its everyday work?
- If the solution to a problem turns out to be complex, how well could your other processes cope?
- Do you have the skills and expertise needed to do the projects?

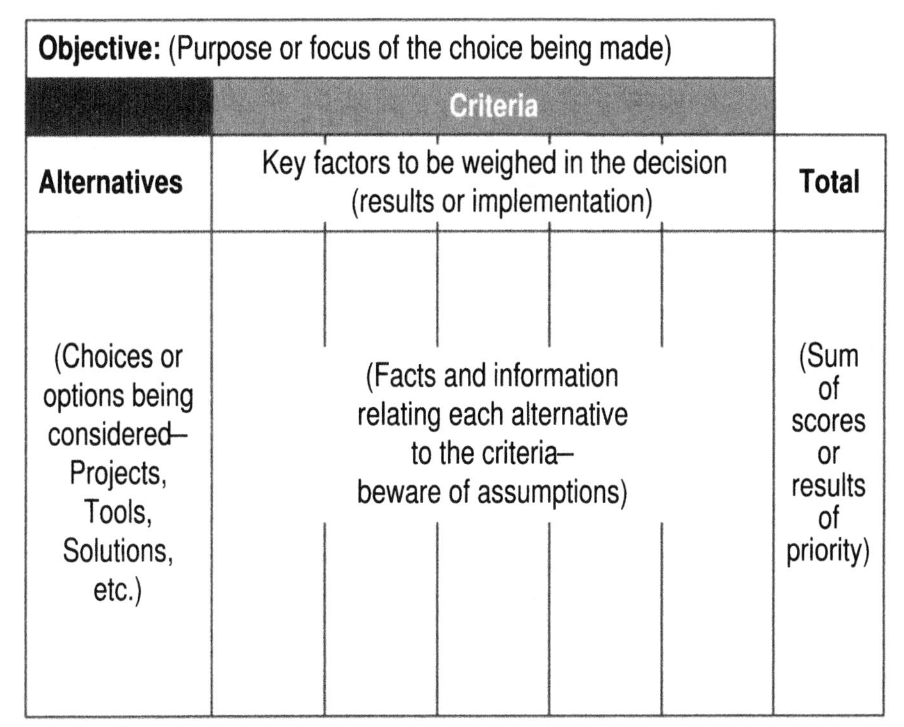

Figure 4-1. Criteria/Decision Matrix

It's the Leadership Council's responsibility to answer questions like these because they make the ultimate decisions about which projects will proceed and which won't. The final mix of projects needs to be ambitious enough to hold the promise of improved business results but safe enough that it doesn't jeopardize the organization's operational effectiveness.

Step 7. Draft a Charter for Each Selected Project

The output of the project selection process (if done correctly) should be a written description of the problem or opportunity, its value to the business (the business case), and a broad improvement goal expected from the team assigned to the project. The Champion will probably draft this rationale in the form of a written (draft) Charter authorizing the team to undertake the project. When writing this Charter, the Champion should remember that the immediate customer for the Charter is the Team Leader, because he or she will use the Charter to select team members and begin the work of the team.

The team members are also customers of the draft Charter, because it will help them understand what it is they are supposed to accomplish. Usually, Charters are revised as the team gathers data, but a clear first draft from the Champion will hasten the process. Over the course of a DMAIC project, the Champion, Team Leader, and team will "play ping-pong" with the Charter as everyone deepens their understanding of the process and the needs of the customer.

See the DMAIC Project Charter Worksheet in Chapter 7, p. 103.

Ideally, the team Champion or Sponsor—often partnering with the Black Belt or Team Leader—will develop the following elements of the DMAIC project Charter:

- **A brief description of the issue or problem and its symptoms.** It's important not to assign a cause, a solution, or blame for the problem here. The Problem Statement merely describes the symptoms we see. For example, "Customer satisfaction ratings for the time it takes to check into our hotels has declined from an average of 85% on last year's survey to 68% on this year's."
- **A broad goal or type of result to be achieved.** Normally this does not include a specific target because precise data about the problem has not been collected yet. When data becomes available, the team and the Champion can negotiate a meaningful target, one that should meet customer requirements. A typical first-draft Goal Statement might read simply "Improve customer satisfaction with our check-in process."
- **The business case: an overview of the value and need for the project.** What are the financial, customer, strategic, and other benefits that come from attacking the problem—and why is it important to do now? Thus, "In today's highly competitive hotel marketplace, we know that customer dissatisfaction with check-in is a direct threat to our financial success. Senior management has made improving customer satisfaction in this key area one of its top five strategic issues for this year."
- **Project scope, constraints, and assumptions.** These will give the team a general idea of the resources available to them, what solutions they may not want to consider, and the like. The Champion might write this: "The team is expected to complete this project within three months, and will be given full Information Technology support in gathering data. The Leadership Council does not expect to hire additional staff in the check-in areas."

The business Rationale/Charter must strike a balance between giving a clear starting point for the team and tying the team's hands before much data is on the table. It's OK not to get it right the first time, though. Champion and Team Leader will discuss the Charter many times over the course of the project.

Do's and Don'ts When Selecting Projects

Do

- **Base your Improvement Project selection on solid criteria.** Balance results, feasibility, and customer impact issues.
- **Balance efficiency/cost-cutting projects with projects that directly benefit external, paying customers.** The customer-focused idea is at the core of the Six Sigma approach. Focusing too much attention on internal savings at the expense of the paying customer sends the wrong message to the teams and, most importantly, to paying customers.
- **Prepare for an effective handoff from Champion to Team Leader.** A clear Project Rationale and Team Charter will get the project off to a good start.

Don't

- **Create "world hunger" projects.** Even more common than "too many projects" are "too big projects." Better to learn from a small project than be frustrated by a monster project that goes on and on and on....
- **Fail to explain the reason a project has been selected.** People like to know they're working on something important to the company and the customer.
- **Start too many projects early in the Six Sigma roll-out.** Improvement teams learning improvement and design methods need lots of care and feeding by their Champions, and Champions need to learn from their teams.

Chapter 5

A Basic Toolkit for Team Leaders

Before You Begin

LEADING A SIX SIGMA TEAM involves skills and methods that haven't traditionally been taught in business schools or in the workplace. Team Leaders wear many hats: conflict mediator one minute, data collection expert the next. They don't have to be experts in every detail of the team's work, but they need to know enough to help the team decide how to best accomplish its mission.

The following areas describe the basic types of tools in the kit Team Leaders will draw from throughout their projects:

A. Understanding team dynamics.
B. Facilitation skills.
C. Data collection.
D. Organizational communication.

A. Understanding Team Dynamics: The Evolution of Six Sigma Teams

It's very important that Six Sigma Team Leaders know something about the workings and evolution of teams. They need to know that the progress of how teams work together is not linear: people don't start at zero at the first meeting and then develop steadily from a group of individuals into a team. Instead, teams evolve in a way that parallels the iterative path of DMAIC itself: they move forward, pause, gather information, back-track, redefine themselves, and then move forward again—if conditions are right! Without the right conditions and coaching, the group may stall out or disintegrate, never reaching the level of teamwork needed for real success.

Along the way, the people on the team will experience highs and lows, good times and bad. Knowing about these experiences in advance, and understanding that they are part of normal development, will help the members to move from a group of individuals looking out for themselves to a team sharing a mutual responsibility to accomplish the project goals. Team development is often described in four stages:

- The initial **Forming stage,** a period of exploration as team members become familiar with each other and the tasks at hand. People generally feel cautiously positive about being involved in an effort that has potential to bring about true improvements, but they are uncertain about how they fit with the team and what role they can play.
- As reality sets in, the team enters a period of **Storming**, when the team's mood swings in the opposite direction: from optimism to uncertainty and/or pessimism. The tasks seem overwhelming, and people may use their energy against one another rather than on the problem under attack by the team.
- With some understanding and, perhaps, coaching by the Team Leader, the Storming phase can be followed by **Norming**, a time where team members accept that DMAIC tools will do the job for them, and that their project scope is limited enough for them to complete it in the time allotted. Respect for and willingness to help other team members usually appears in this phase.
- If all goes well, the team should reach the **Performing** phase as it completes its analysis, and begins to improve and control the process under review. Now the team is really cooking! Real changes are being made, and the team can take satisfaction in its achievements.

These four phases are not a one-way street, and each phase need not last long. Teams can "perform" early on by picking off some low-hanging fruit (for example, solving simple problems in the first few meetings). The team may also have some Storming when it begins to improve things—especially if that improvement involves major changes in how members do their jobs. The Team Leader and team members will have to be alert to both the opportunities and threats that team evolution provides.

The four phases of development roughly match the five stages of DMAIC:

DMAIC Stage	Phase of Team Development
Define	Forming
Measure	Storming
Analyze	Norming
Improve and Control	Performing

Later chapters of this book provide specific guidance on what a Team Leader can do in each of these stages to help the team reach the Performing stage quickly and stay there!

B. Facilitation Skills

No team ever broke down because its members couldn't agree on their interpretation of a histogram, but plenty fall apart because the team members can't stand to be in the same room together after the third meeting.

—The Wise Old Sigma Team Leader

None of the powerful tools that Six Sigma uses to Measure and Analyze flawed processes will be successful if the Six Sigma team itself cannot handle its own human processes (how team members interact and do or do not collaborate—not its biological processes). For many people working on a new team, using new tools will be a challenging experience. It's part of the Team Leader's job to anticipate these challenges, and to channel the stress they create into positive energy that can empower, not cripple, the team.

The first two or three Six Sigma team meetings are likely to be among the toughest the Team Leader will facilitate: the team members may be working together for the first time; the tasks before the team may still be unclear; the team members may not have had much training on Six Sigma—the idea of data-based decision making may be one team members still don't understand. The roles of the team members and their relationship with the Team Leader and Project Champion will be new, as are the expectations of those who have created the team.

With all this uncertainty, the Team Leader is wise to pay as much attention to meeting and facilitation skills as she or he does to the statistics of measurement and analysis.

Meeting Tips for Six Sigma Teams

Ask most people what they think about the meetings they attend where they work and they'll say they're usually a waste of time: the meeting rambles from one point to another, with little resolved at the end other than an agreement to meet again … soon! Some people aren't sure what their role is at the meeting: listener, participant, decision-maker, or what? Others are turned off as one or two people hog the air time sticking to their personal agenda.

It's unfortunate that so many people spend so much time in so many unproductive meetings when following a few simple tips would improve the quality of their meetings—right away.

Tip #1: Create and Follow an Agenda

- **Set the agenda for the next meeting at today's meeting.** This is particularly important to put people on notice who have assignments to complete (and that should be everyone on the team; more on this later).
- **Send the agenda out before the meeting.** Attach any background information that people can read in advance, rather than studying it at the meeting itself.
- **Include the reason for the meeting** ("To identify problems," "To decide on a solution," "To create a plan," and the like).
- **List the items to be covered**, along with a sentence or so about why it's on the agenda. For example: "The reason for holding this meeting is to revise the Problem Statement. The analysis of the sales data from the Western Region shows that we need to focus on Los Angeles customers."
- **Provide time estimates for each item.** A wrong time estimate is better than no estimate at all!
- **State the desired action on each item**. "Select …" or " "Evaluate …."
- **Include a final agenda item where you review the meeting itself.** The team's own process can be improved by taking a few minutes at the end (or near the end) of each meeting to discuss what went well and what improvements to act on at the next meeting.

Keep the agenda visible and use it. At the meeting, post a large version or summary of the agenda on a flipchart or whiteboard at the actual meeting, even if you've sent out the agenda beforehand. One way to keep people on track is to walk over to the large agenda on the wall, tap the item you're on, and remind someone off on a tangent that it's time to come back to earth.

Don't forget to have some fun on the agenda. For example, have a brief (underline brief) ice-breaker activity at the beginning, and/or some "comic relief" in the middle of a long meeting, just to liven things up a little.

Tip #2: Define Meeting Roles

If you took a foreign visitor to their first American baseball game, you might start by explaining the different positions and functions of the players. Because Six Sigma may be a "new ball game" for team members, you'll be wise to explain the positions on this team, starting with your own. (Note that these are team and meeting roles, different from the Six Sigma roles described in Chapter 3.)

The Team Leader ...

- Starts and ends the meeting on time.
- Stays on the agenda.
- Encourages discussions.
- Leads brainstorming sessions.
- Shows team members how to use power tools to improve processes.

The Timekeeper ...

- Monitors how much time the team spends on each agenda item.
- Alerts the team when the allotted time for an agenda item is nearly up and asks the team for closure.
- Gives warnings about the amount of meeting time left and encourages the team to wrap up discussions.

The Record Keeper ...

- Keeps notes on the team's decisions, plans, and actions. The record need not be word-for-word what happened at the meeting, but should capture key decisions, assignments, and updates on the team's progress through the DMAIC.
- Working with the Team Leader, distributes copies of the minutes to all team members and the team's Champion. (Nowadays, many Six Sigma teams keep notes on a laptop computer and copies are often distributed right at the meeting.)

The Scribe ...

- Helps make the team's process visible by capturing ideas on flipcharts (or through some other method) so they can be clearly seen and referred to during a meeting. This really helps keep the team on track. It also reinforces the idea that people's ideas are important enough to be written down in plain view. The scribe should:
  - Write down exactly what people say. Changing what people say leads to confusion. If someone says too much too quickly, the scribe should ask that person to summarize the key points.
  - Check for accuracy whenever he or she is unsure of what someone said.
- If the brainstorm session is fast moving, use two recorders on two flipcharts, alternating one after another so that there's time to record each idea.

Visitors ...

- It's possible that from time to time, the team will have a visit from the project Champion or people who have been invited to join for a day to share specific data. It's a good idea for the Team Leader to explain to such visitors before the meeting exactly what is expected of them. If senior managers visit, there's always a risk that they will dominate the meeting. The Team Leader should explain the team's ground rules and how they guide the meeting ("That reminds us ...").

Tip #3: Set Meeting Ground Rules

Setting ground rules should be one of the first and is probably one of the most important things a team will do together. Ground rules (or guidelines, as they're sometimes called) are the rules that the team agrees to set for itself on how the team will conduct its business and the how the team members will treat each other. (Instructions for developing guidelines appear later as part of the team's first meeting. See pp. 68-69.)

Some teams don't create and post written ground rules because, as they often say, "we're all adults here and know how to act." While it's true you might not need ground rules in the first meeting or so, you probably will as you get deeper into the Define phase of DMAIC because everyone is liable to have different opinions on the "problem"—and often you won't have any data that could sway opposing views. Also, if you don't think about ground rules until the Analyze or Improve phase (when feelings and opinions can also run high) it's too late then to slap rules of conduct on people. So introduce ground rules early on during the feel-good, early stages of team development.

Keep ground rules visibly posted in the meeting space and quickly review them at the beginning of every meeting. Remember that ground rules evolve as the team works through the DMAIC process, so your team will need to discuss rules that no longer seem to apply as they did at first, or create new ones.

You can also use ground rules as the basis for evaluating and improving your meetings. ("How well did we follow our ground rules in this meeting? What can we do to follow them more closely at the next meeting?")

Very important! When you create and clarify ground rules at your first team meeting, ask each team member *directly* whether they agree to follow the ground rules. Getting this agreement is a good way to demonstrate the importance of cooperation among team members. Furthermore, if you don't get agreement early on, later when the team makes some very tough decisions and changes in how people actually do their jobs, you run the risk of people saying, "I never did agree with this whole approach." You don't want to hear that kind of talk three months into a project!

Finally, when new members join the team or visitors come to call, be sure to review the ground rules with them and get their clear buy-in.

Remember: There are rules at every meeting whether they are written out or not. If you don't make them explicit, the unwritten rules may include the old favorite: "Those who have the gold make the rules."

Tip #4: Always Evaluate Team Meetings

Because Six Sigma is about improving processes, don't forget to evaluate and improve the team meeting process itself. At or near the end of each meeting, have team members identify the things that went well, and make suggestions about how the next meeting can be improved. Put the list on a flipchart, discuss the ideas, and act to make improvements. Have team members take turns facilitating this review. Although your meetings may never become defect-free, they will become better and better over time!

Tip #5: Create a Plan for the Meeting

Eventually, your team meetings will become routine, and developing a detailed plan may not be necessary. But early in the project, or when you are holding any special meetings (such as reviews with your project Champion), you will want to have a clear idea of what should happen at the meeting. The Meeting Planning Worksheet (Figure 5-1) can help you cover the essentials.

See the Meeting Planning Worksheet on p. 52.

Basic Discussion Skills

The time a team spends in its meetings is a precious commodity. Much of that time will be spent in discussion: talking about the problem or process being studied, brainstorming ideas and solutions, probing to correctly interpret data, and so on. That's why no business that wants to make fast progress in serving its customers can afford to have its team wasting hours in fruitless discussion.

The discussion facilitation skills described here can help you keep your team focused and ensure that all team members have a chance to contribute. They will be useful throughout your project, so you may want to review this material periodically to strengthen your skills.

Skill #1: Leading Discussions

Team discussions will be more effective if they have some structure. The Team Leader can help by leading the discussions through four steps:

Step 1: Open the Discussion

Begin with a brief background about the item and how it got to be on the agenda. Include what the item is, why it's important, and what the team needs to do about it: Analyze, Define, make a decision or a plan, etc. After this introduction, invite team members to participate in a discussion. Encourage the team to be open to as many viewpoints as possible.

Meeting Planning Worksheet (see next page)

Purpose: To ensure meeting time is used effectively.

Applications: For use by the Team Leader or anyone planning a meeting.

Instructions: The worksheet is self-explanatory. Simply fill in the blanks and you will have covered the main issues needed for conducting a meeting that accomplishes priority tasks.

Meeting Planning Worksheet

What will the meeting cover?

Purpose/Non-Purpose: ______________________

Key Topics/Agenda Items: ______________________

Expected Results/Follow-Up: ______________________

Specific Resources Needed: ______________________

Who should attend?

Who is affected? ______________________

Who will contribute? ______________________

Who will "own" the agenda action items? ______________________

Who will fill key roles? ______________________

When should the meeting be held?

Best time for participants? (time of day, day of week, conflicting meetings) ______________________

How much time? (for agenda items) ______________________

When should reminders/ notifications be sent? ______________________

Where should the meeting be held?

Most convenient site: ______________________

Can we teleconference? ______________________

Is a room available? ______________________

Figure 5-1. Meeting Planning Worksheet

Have a scribe capture the gist of what each person says, perhaps by writing on a flipchart. Making people's words visible to the whole team helps them know their ideas are being taken seriously. Visibility also helps the team stay on track. It's tough to follow a discussion after the fifth or sixth point has been made if there's no list to refer to. A written list will also make the next two discussion steps possible.

Step 2: Clarify Main Points

Ask questions and use the answers to clarify points being made in a discussion, or to refine a brainstormed list of ideas.

Encourage team members to ask questions whenever they are uncertain of another member's intent or meaning. Check for understanding by asking team members to restate or interpret what they've heard each other say. Ask the team what data supports the ideas being generated. Have speakers told the whole story? Are there other relevant points that should be raised?

Be aware that sharp differences of opinion often occur—which is good, because you want people freely sharing ideas based on their own experiences and expertise. Make sure the discussion leader (and the other team members, too) follow the ground rules established for discussions.

Step 3: Summarize Main Points

Having a written list, along with clarifying points, makes it easy to summarize the points made or data analyzed by the group. A quick review of each point will usually do the trick, along with giving people another chance to add to the list itself with other ideas. Part of the reason for this summary is to check that everyone understands what has been said and also that no one's ideas are being distorted. This is especially important if the discussion's purpose is to make a decision and take action on it.

Step 4: Check for Agreement

By this point in the discussion, it will often be the case that some agreement on the analysis of data or a decision will have emerged. If this seems to be the case, it's important to check with the group to make sure. This can be done simply by saying, "It appears to me that the group has decided to...." Then check for agreement by asking each person by name if they agree with this perception. Or you can ask, "Is there anyone who disagrees with the statement?" Either way you're giving people a chance to agree or disagree. Don't rely on body language or guess at people's thoughts. Ask them directly! If they disagree, you may need further clarification.

Skill #2: Generating Ideas

A team doesn't come together simply to apply the same old ideas to the same old problems. Creativity and new thinking are key assets to a team, much of which will come out in team discussion. You'll need to generate ideas—brainstorm—any time you need to create new solutions or find possible causes of problems, ways to measure defects, and so on. Adding just a little structure to a brainstorming session is also a good way to make sure everyone on the team participates.

Brainstorming Process

1. Define the topic you need to think about.
2. Ask each person for at least five ideas.
3. Give people a minute or two to jot down their own ideas.
4. Ask for ideas and list them on a flipchart, without comment.
5. Review/clarify/combine ideas when brainstorming ends.
6. If time permits, revisit and revise the list later in meeting or at the next meeting. Sleeping on ideas often improves quality and brings new insights.

Brainstorming Methods

- Have people write ideas on sticky notes and stick them on a flipchart.
- Have people state their ideas out loud, one at a time. Take turns around the team to make sure everyone gets a chance to give ideas or say "Pass."
- Have each team member write one idea on a sheet and pass it to next person, who can add to or build off other statements. Continue this "write-and-pass" method until each sheet gets to the person who started it. Then review all the ideas.
- Ask the group what ideas a famous person, politician, sports figure, movie star might have on the assigned topic. Review ideas.

Whatever brainstorming method you use, be aware that the first round or two of ideas tend to be predictable. To really get "outside the box," you'll have to push the group and do a number of rounds, asking for even crazier ideas.

Skill #3: Organizing Ideas and Identifying Priorities

The result of a discussion or brainstorming session is often a list of alternative ideas or options. It's the team's job to sort through these options and decide which best meets their goals or purpose. The first step is to clean up the list:

- Hold an open discussion where any team member can make suggestions about how to combine ideas, eliminate duplicate ideas, etc. Make changes only if the people who originated the ideas agree.
- If there are a lot of ideas, you may want to use an *affinity process* to identify the main themes among those ideas.

Voting

Voting on the ideas is one way to quickly shorten a list—as long as you remember that narrowing the list is *not* the same as making a decision. Making a final decision will take more time and old-fashioned discussion until some agreement emerges.

There are two types of voting procedures: "one person, one vote" and multivoting. The former method is straightforward: each team member gets to cast a vote for one of the options. This works well for fairly short lists where the items are clearly differentiated from each other, but you may find it too limiting when there are many ideas from which to choose. In that case, try multivoting (p. 56).

Skill #4: Decision Making

Oftentimes, a team seems to fall into a decision without anyone being aware of exactly how it happened or what it means. Afterwards, support for the decision falters because people don't remember making it in the first place!

Effective decision making is one of the most critical skills a team can learn, and it helps if a team leader understands four basic rules of decision making:

Rule 1: Know What Authority You Have

Your team exists because someone with the authority to allow it to exist thinks it's OK for people to use their time in that way. That individual or group determines how much authority a team has to make decisions. We've all been in situations where we put a lot of time and effort into making a decision only to hear the boss say, "Thanks for the input. I'll take it under advisement." So before your team spends a lot of time deciding how you'll make decisions within the team, you need to be clear about what decisions the team as a whole is authorized to make.

Multivoting

Purpose: To narrow a long list of possibilities to just one that your team can discuss in detail, do the following:

Instructions:

1. Create the list and number the items.
2. Give each team member a number of "multivotes" equal to about one-third of the total number of items on the list. For example, if there are 30 items listed, each team member has 10 votes.
3. Go through the list, asking team members to cast votes for each item. Encourage them to scatter their votes among the items rather than casting all of them on just one item.
4. List the number of votes each item gets.
5. Briefly discuss, then eliminate items with the fewest votes, then repeat the vote. For example, suppose you narrow a list of 30 items down to 10 items. Give each team member three votes, and see if the final list can be reduced to three or four.
6. When the team has a top choice list of three or four items, it's time to stop voting and start some serious discussion to decide on the top possibilities.

The Affinity Process and Diagram

The affinity process is a powerful method for organizing any kind of "verbal data"—lists of ideas, customer statements gleaned from interviews, solution ideas, etc.

Purpose: To gather and organize a large number of ideas/issues from many people, and then summarize them into related groupings or categories.

Applications:

- Encouraging non-traditional connections among ideas/issues.
- Allowing breakthroughs to emerge naturally.
- Understanding the nature of a problem and breakthrough solutions.
- Allowing for easy process mapping when mapping is new to the group. (The "named categories" become the major steps in a process.)

Instructions:

1. State the issue under discussion in a full sentence.

2. Brainstorm 20 or more ideas; follow brainstorming guidelines (pp. 54-55).
3. Record each idea on a self-stick note. (Each person can record his/her own ideas on the notes, then post them on the diagram themselves. This saves time and eliminates discussions.)
4. Place the notes randomly on a flipchart, whiteboard, or other flat surface.
5. *Silently* sort ideas simultaneously into five to 10 related groupings.
   - Silence is critical in this step. You don't want people influencing each other by talking.
   - Keep the clusters relatively small, say two to five or six notes.
   - If any team member disagrees with how an idea is clustered, he or she can move the note back to its original position or place it with another cluster.
   - If a note keeps getting moved from cluster to cluster, it's OK to create a duplicate. Some ideas really do capture several themes, and therefore might belong in several clusters.
   - Focus on the meaning behind and the connection among all ideas.
6. Continue until a natural grouping of ideas emerges. (Remember, keep it silent!)
7. When the process is complete, ask team members to now take a more logical look at each cluster and identify what idea or theme it represents. Capture this theme in a brief title and write it on another sticky note; place this label as a header for its cluster.
   - If you have a hard time writing *one* theme for a cluster, it may be that the group represents two or more themes. Re-examine the group and see if new clusters emerge.
8. Draw the Affinity Diagram connecting all finalized header cards with their groupings.
9. You can stop at this stage, or continue to "cluster the clusters" to create larger groups of ideas. Figure 5-2 shows an Affinity Diagram where the first round of clustering has been completed.

That means working with the "authority" to be clear about what he or she or they want out of the team. For example, are you supposed to…

1. Study an issue and come up with options?
2. Make a recommendation?

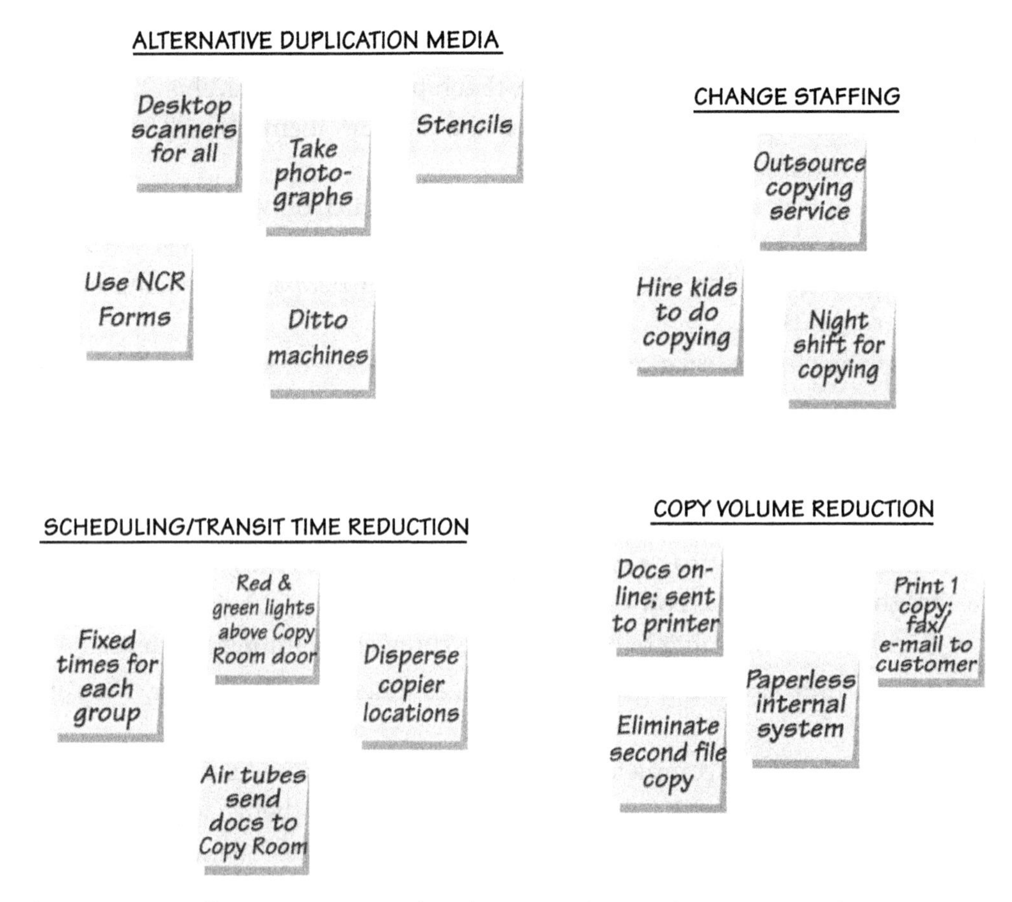

Figure 5-2. Affinity Diagram: This diagram shows the first round of grouping after a brainstorming session to identify problems with getting copies to clients on time

3. Make the decision?
4. Implement it?

You'll also need to know what criteria or limits the authority has placed on the team's outcomes. For example, "implement a new customer survey process with no expansion of staff."

Note: Ideally a DMAIC team will be responsible for developing *and implementing* a solution. But you will always have to get someone's approval (and many people's buy-in) to "green light" your proposed improvement. Understanding *their* decision process is a key to getting that critical support.

Rule 2: Decide How to Decide

How many times have you been part of a team that just kind of let a decision happen? Something seems to emerge, or one person states a strong opinion, and somehow a decision is made. A team will be more effective—and have more confidence in its work and outcomes—if it makes decisions a visible, deliberate part of its work.

Rule 3: Fit the Decision-Making Mode to the Decision

Not every decision is life-or-death, but many have more important implications than what shoes you're going to wear in the morning. (Though if you were planning to hike up a glacier, even what shoes you wear can become important!) Part of applying Rule #2 is adapting the rigor of your decision process and your decision mode (see below) to the criticality of the choice being made. For example, think about what level of involvement by what types of people is needed for a given decision. Deciding on team ground rules or selecting a final solution for implementation, for example, should probably be made by consensus. Deciding what software to purchase may be a decision the team decides to leave up to an expert or delegated to a "subteam."

Rule 4: Do the Groundwork

If the decision is complex or final—such as determining capital equipment purchases or recommending one particular customer satisfaction strategy for implementation—you will need to do a lot of groundwork before making the decision. In some cases, you may want to use a structured decision matrix which includes written criteria and risk assessment. (See Chapter 16, pp. 315-317 for more details.)

Modes of Decision Making

What is a "mode" of decision making? Here are four typical ways to make decisions:

1. **Consensus**: All members of the team agree to support the group's decision—both publicly and privately—even if the selected option is not their first preference. (Note that consensus is *not* a majority vote!)
2. **Voting:** Group members vote on various options.
   Majority vote: For an option to win, it must receive at least 51% of the votes, or perhaps a two-thirds majority.

Most votes: The option with the most votes wins, even if it doesn't have a majority.

3. **Authority decisions:** The decision is made by someone with positional authority (supervisor, manager). Often this option is used when the group can't reach agreement—or, of course, when the leader is convinced the team's choice is wrong. The latter situation should be addressed by the leader to see if the team misunderstood its charter, were unclear about business priorities, or, vice versa, if the team uncovered information the leader was not aware of. Teams usually make "wrong" decisions only when they are lacking critical business information.
4. **Minority decisions:** These are choices the team may defer to someone with a particular expertise or assign to a subset of the team. "Gladys and Joe, you pick the site for the client meeting and let us all know."

As illustrated in Figure 5-3, the objective—no matter which decision mode you use—will always be to drive *toward* consensus. Without consensus (defined as an agreement to support and live by the decision) any majority, minority, or authority decision can be derailed by lack of commitment or compliance.

As a Team Leader, you can facilitate decision making in your team by ...

- Raising the question, "How do we want to make a decision about this issue?"

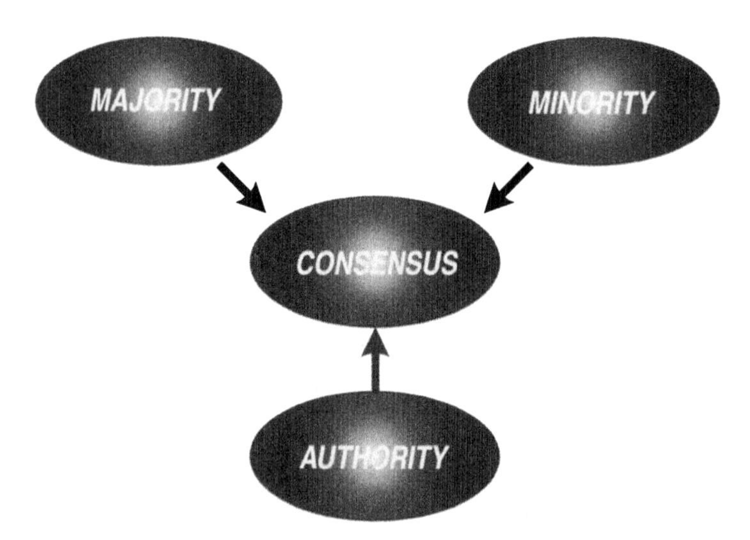

Figure 5-3. Decision Making—all decision-making modes should seek to achieve consensus

- Reminding people of the options available.
- Using your discussion and brainstorming skills to make sure that all ideas are openly discussed and debated.
- Creating an atmosphere where people feel comfortable sharing their honest opinions.
- Helping the team look for a "middle way" if ideas are in direct conflict.
- Managing participation to prevent people who have positional authority (or who are highly opinionated) from dominating the conversations; ask them for their ideas only *after* other team members have given theirs.

As noted above, consensus is the preferred option for most decisions with a team because it means the team is acting as a cohesive unit. The downside is that consensus requires time and participation: the team must thoroughly discuss and debate a short list of choices until a decision emerges that everyone understands completely and can live with. While the team may be impatient and eager to reach a decision, in cutting off discussions too early you may save time up front only to lose it later on when people's unspoken thoughts and doubts at last emerge and slow things down—just when the team is about to take action on the decision. It's a matter of "pay me now, or pay me later." Now is better!

To reach consensus ...

1. Work through a list of items or options one at a time.
2. Ask each person directly to share their ideas on the item.
3. Listen carefully and patiently.
4. Summarize each person's ideas on a flipchart.
5. Allow open discussion only after each person has had an opportunity to speak. Continue to capture ideas on the flipchart.
6. When you think the team is close to consensus, ask a team member to say or write a statement that captures the consensus opinion.
7. Check for consensus around the statement by asking team members, "Will you support this position both inside and outside the team?" If anyone answers "no," the team has more work to do. Ask those who cannot support the statement to suggest alternatives, then check for consensus around those alternatives.

Remember—don't try to hurry! Consensus takes time. During a discussion, include periods of silence so people have time to think. You may even need to table a discussion until another meeting so people have time to consider the options, gather additional data, etc. But, if failure to reach consensus is threatening to severely delay the project, consider shifting to another "mode" (e.g., put it to a vote!).

C. Data Collection: The Right Tool for the Right Time

Using data effectively is the hallmark of a Six Sigma team. Later chapters of this book provide specific guidance on how to decide what data to collect, and on various tools that can be used to display and analyze data. Eventually, those decisions become second nature, but initially it's not always clear. Here are some basics of data tool selection you'll need to keep in mind throughout your project:

1. **Have a clear objective when planning to use a tool.** Never use a tool just because "it's in the book" or "we haven't done one yet." Only pull out a hammer if a nail needs pounding (or if the TV is on the fritz and needs a whack).
2. **Consider your options and select the technique that seems most likely to meet your needs.** With the variety of techniques in the Six Sigma toolkit, there's often more than one method that might be of help. Be careful of which one you try.
3. **Keep it simple—match the detail and complexity of the tool with the situation.** The most basic tools should be used most often. If you're using detailed statistics for every problem or project, it's likely you're over-complicating things.
4. **Look for creative ways to analyze and display your data.** While you mustn't do things like use the wrong kind of data with the wrong tool, you needn't be locked into using the tools in only one way nor using only the tools described in this book. It's OK to create your own variations on a method provided (a) you don't make a change that no one else can understand and (b) you don't end up drawing faulty conclusions from it. You may want to check with a Master Black Belt or other expert in your organization if you have an idea on how to creatively display some data, to make sure your thinking is sound.
5. **If a tool isn't working, stop.** If the train ain't moving, get off and find one that is. Consider every tool you use a "trial"—if you don't get the answer you need or if it isn't working, try something else.

6. **Remember GIGO: garbage in, garbage out!** Be sure to collect meaningful data for the power tool you decide is applicable to your situations. For more on data collection, see Chapters 9 and 10.

Planning for Data Collection

There's nothing more frustrating than spending weeks gathering data only to find out afterwards it can't answer the question you need to answer. You may want to review the data collection process described in Chapter 9 (pp. 145-148) and the accompanying worksheets in Chapter 10 so you can learn to anticipate the types of questions that arise in data collection. As your team's leader, you need to become a champion of creating and using data collection plans.

Using Computer Software in Data Collection

Many of the tools described in this book can be done quickly with a pencil and paper or with simple spreadsheet tools such as Microsoft Excel. However, with the help of your organization's Master Black Belt or other expert, you may want to become familiar with more advanced statistical packages such as Minitab (or perhaps your organization has its own customized software program). Minitab and other programs like it can perform a variety of sophisticated analyses—such as regression analysis, hypothesis tests, and design of experiments—that can increase the power of your data analysis efforts. Again, as a Team Leader, you need not be an expert in these software packages (or in the advanced tools, for that matter), but you should know enough to know when to ask your Coach, Master Black Belt, or statistical advisor for help!

D. Organizational Communication

When Six Sigma is introduced to an organization, the first tangible evidence that it will make a difference is likely to come from the Black Belts and DMAIC teams. However, it's often hard for people outside the team to get a clear idea of "What the heck are they working on and what have they accomplished?" This "information gap" can elevate suspicious and outright paranoia about Six Sigma.

While it's understandable that you, during the course of your project, will want to keep things quiet until you have real results to share, that secretive approach can work against you later as you seek support for your solutions. (It may also cause people to leave the table when you join a group at the cafeteria.) At the same time, even if you want to share information about your project, you need a process for communicating and an idea of who needs to know what (and when).

A good solution to the information gap that you can do on your own—and help leverage your team's activities—is to use portable storyboards to display and share your team's results as your project completes each phase of DMAIC. A storyboard is a pictorial representation of key actions and outcomes in a projects. If possible, get your team's story online on an internal system that's easy to access. Some uniformity in reporting by the teams will make these reports easier to understand for those not on the teams. Here's some more detail on how to create and use the storyboard tool.

Making Progress Visible: Using Storyboards

A storyboard, as you may have guessed, is a series of graphical panels (usually one-page sheets and often done in a presentation software package) that tell the "story" of your project. An example is shown in Figure 5-4. (The tool is borrowed from movies and advertising, where drawings are used to map out a commercial or animated feature before it goes into production.)

The purpose of the storyboard is twofold:

- First, it gives the team a fun way to summarize and display its work. One of our Six Sigma Black Belts said the storyboard reminded him of his daughter's science project display in school. That's exactly what these Six Sigma storyboards are: a public way of showing how we can use data and a scientific approach to solve problems rather than simply throwing money at them or working around them. Teams can spend some meeting time working together on the storyboard, cutting and pasting things to display. This is an excellent team-building activity, far surpassing routine "approval" of meeting minutes!
- Second, storyboards let people outside the team know what's being done on the project. (After all, some of them are helping the team by gathering data and other support. They're curious about what's being done by the team.) Rather than waiting for the completion of the team's work, which may take six or more months, storyboards give a team a way to keep others up to date from the very beginning.

Storyboards come in many shapes and sizes. Some teams compile simple computer printouts; others construct large, colorful flipchart-sized displays. Six Sigma teams use them to display key documents and graphs developed on the

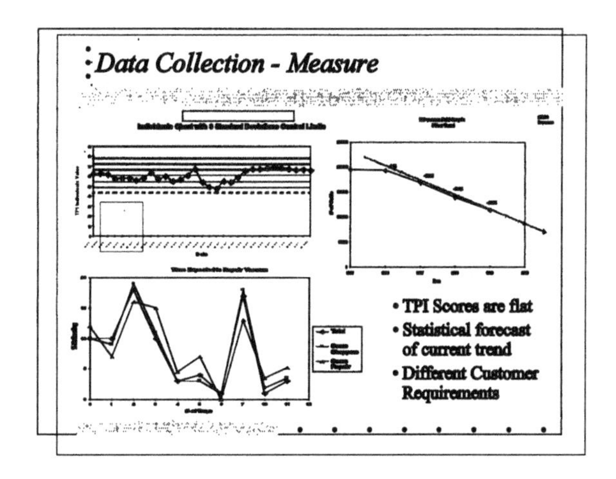

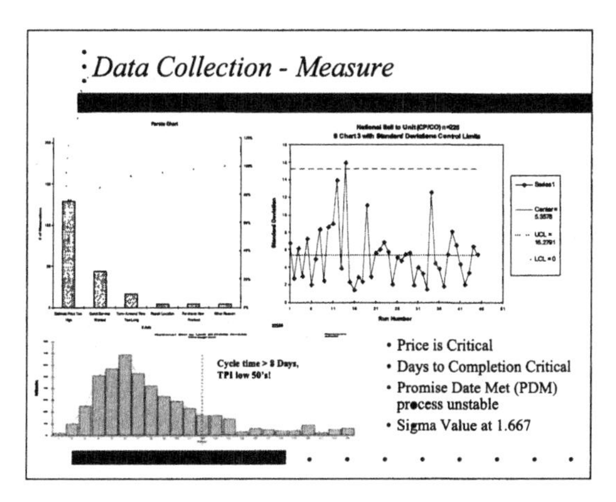

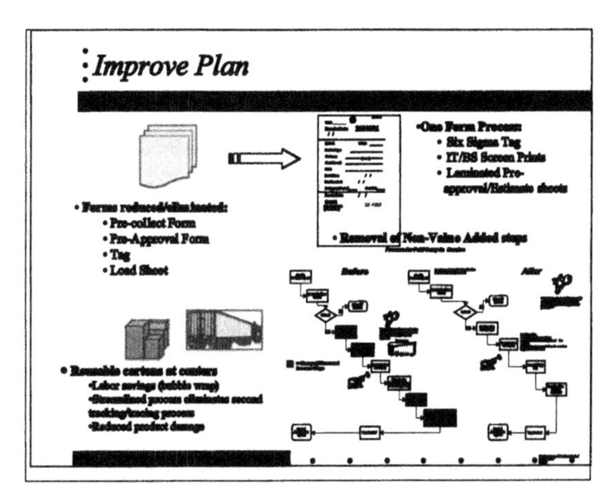

Figure 5-4. Storyboard example

way through the Define-Measure-Analyze-Improve-Control cycle. No matter which option your team chooses, look for ways to share your storyboard with co-workers and management and even involve them in the project. For example:

- Display the storyboard in a breakroom, cafeteria, entrance, or other high-visibility spot.
- Move the displays around so they don't simply become part of the landscape as so many things on bulletin boards do.
- Attach sticky notes to encourage viewers to chip in their own ideas, too. (Six Sigma should not be a spectator sport!)

When your organization gets five or six projects and storyboards completed, you may want to create a "gallery walk" along a well-traveled hallway or cafeteria, complete with "guides" to answer questions about the projects.

All in all, in an age of high-tech e-mail and other "messaging" systems, portable, colorful, cut-and-paste storyboards remain one of the best and most robust ways to keep the changes Six Sigma brings in front of the organization every day. Black Belts need to use their imagination to find ways to make sure that the gains are remembered and incorporated into the organization's culture—and aren't just quickly forgotten like those trees that fall when no one's around! Throughout this book, you'll find reminders to update your team's storyboard at the end of each DMAIC stage, along with tips on what to include to best capture your team's work.

Getting Started

New Team Leaders usually have a lot of questions about the basic logistics of working with a team. For example:

How often should teams meet? The first three or four meetings should be every week in order to keep momentum building. Also, there's a lot to do in these earliest meetings. Later, when assignments will include gathering data which requires some time, meetings can be spaced out to ten days or more. We recommend mini-meetings of some team members in these longer intervals to maintain focus.

How long should meetings last? Three hours for the first couple of meetings, settling down to two hours for later ones—or however long (or short) it takes.

What if someone can't attend a meeting because of sickness or some other reason? Meet anyway! Don't lose momentum. Ask the absent person to communicate their ideas and concerns to another team member who can act as their agent.

What if our team is spread around and can't all meet in person? This is a common challenge you have to work around. Teleconference, supplemented by e-mail, is a good (not perfect) substitute. Videoconferencing can be a little better, provided the logistics aren't too complicated. In today's virtual and global business environment, many DMAIC teams have successfully completed projects with a fairly limited number of "face-to-face" meetings.

What if we've picked the wrong people for the first set of team members? The core team members who must be at every meeting are people who know the process being studied very well, and who will probably inherit the solution the team produces. If you have people on the team who don't meet these qualifications, excuse them as soon as possible and get the right people. Some process experts may be called in later to share special knowledge as required.

What if people say they're too busy or their boss says they can't be on the team? The Team Leader should try to work it out, or have the Champion flex a little muscle. But only *if* the person really is vital to the team's work.

Conducting the Six Sigma Team's First Meeting

Since your Six Sigma team doesn't really exist, per se, until after it begins meeting, most of the preparation for the first meeting will be left up to you. That first meeting is critical because it sets the tone for the whole project and introduces team members to each other, to the DMAIC process, and to the new expectations they face. Here are some tips to get you started.

The goals of the first team meeting are to:

- Review the Project Charter.
- Understand the Project Plan.
- Understand the DMAIC approach.
- Get clear on team members' roles.
- Develop team guidelines.
- Get to know one another better.

Before the meeting, secure a meeting room equipped with flipcharts, tape, pens, and the other things the team will need. Develop and send out an agenda ahead of time. As described on pp. 47-48, the agenda should include items to be discussed—location, start and stop times, etc. You may also want to invite the Project Champion to visit the first meeting to go over the Project Charter and state their commitment to the project before leaving the Team Leader in charge.

First Team Meeting Agenda

1. **Review agenda** (Team Leader).
2. **Brief introductions** of the Team Leader, team members, Champion.
3. **Discuss roles** of Team Leader, members, others.
4. **Set ground rules.** The Ground Rules Worksheet (Figure 5-5) lists typical guidelines that team set for themselves. Use these as a guide for your own team. Decide whether the examples shown are appropriate for your team, and modify them as needed. There is also space for adding guidelines on other topics unique to your project.
5. **Review the draft Project Charter.** Reviewing the first draft of the Charter is the heart of the first meeting; doing so allows the team to start taking ownership of the project. Have a complete version of the Charter on the wall, and review it step by step. If the Champion and Team Leader have done a good job drafting the Charter beforehand, it should contain a good Business Case, a Problem Statement describing the symptoms of the problem, a Goal Statement, some guidelines on the process to be studied, and the Champion's assumptions about the project. Allow at least two hours for the meeting, so each item can be discussed and wordsmithed.
6. **Review DMAIC process and project time lines.** Team members will want to know how long the project will last and how much time they will have to spend at meetings and in between meetings working on assignments.
7. **Make assignments for the next meeting.** No one should leave any meeting without an assignment! For the second meeting, team members will probably have to start collecting information about the process to be studied, its customers, and their critical requirements of the process outputs.
8. **Evaluate the meeting.** Near the end of each meeting, be sure to include time to discuss what went well and what needs to be improved in future meetings.

Ground Rules Worksheet			
Use for our team? Yes No	Topic	Example	Our Team's Version
	Attendance	All members agree to attend every meeting. When someone cannot attend, that person agrees to contact the Team Leader 24 hours in advance of the meeting, if possible.	
	Participation	No substitutes for team members. Participate fully both inside and outside meetings.	
	Interruptions	100 mile rule–attend meetings as we would if we were 100 miles away. Allow interruptions to address emergencies only.	
	Preparation	Come to all meetings with assignments completed, prepared to productively contribute to discussions and decisions.	
	Timeliness	We will start on time if at least 80% of the team is here.	
	Decisions	We will discuss the best decision-making model for each situation. We will support decisions made by the group.	
	Data	We will use data whenever possible as the "ultimate authority."	
	Conflict	Honest disagreements welcome as long as people treat each other with respect. If a conflict cannot be resolved, we will ask a facilitator for help.	

Figure 5-5. Ground Rules Worksheet

The Path Forward

There is no such thing as a "typical" improvement team. But there is a typical process for attacking problems: DMAIC (see Chapter 2). Given a serious business problem, most teams would normally Define the problem, Measure the extent of the problem, Analyze the data to discover the causes, Improve the existing process by getting rid of the causes, and then Control the improved process to make sure the old problem didn't reappear in the future.

This DMAIC process provides the backbone for most of the remainder of this book. The blending of tools and methods needed to improve sigma levels are presented as they relate to each of the DMAIC steps. You'll find extensive background information on what's important in each step, instructions on carrying out key tasks, how to avoid or overcome the challenges associated with each step, and the power tools you'll use to do the job.

If your team is being asked to totally redefine a process, then the DMADV version of DMAIC—Define, Measure and redesign, Analyze process design, Design and implement new process, Verify results and maintain performance—is more appropriate. You'll find instructions for customizing DMAIC to the DMADV model in Chapter 21.

The team will start by defining the problem, and then redefine it after you've collected some data and then again after analyzing the data. So if you start to feel like you're backtracking, don't worry. It's just the normal process of honing your project and focusing on the deep causes of problems.

Part Two

Leading a Six Sigma Project Team Through DMAIC

Chapter 6

Define the Opportunity

Scoping Six Sigma Projects

IN THE DEFINE STEP, a team refines its Problem Statement and goal, identifies the customers served by the process being studied, defines customer requirements, and writes the plan of how to complete the project. Throughout this work, the team should also keep in contact with its Champion, to ensure that it stays aligned with business goals, priorities, and expectations.

This chapter describes the steps needed to complete the Define work and produce three outputs:

1. Team Project Charter and Work Plan
2. Measurable Customer Requirements
3. High-Level Process Map

Two other chapters will help you guide your team through its Define work:

- Chapter 7 provides more detail on the tools referred to below.
- Chapter 8 provides guidance on working through typical team dynamics seen in a team's early meetings, along with troubleshooting tips for avoiding common pitfalls often seen in Define.

Define Step 1: Update and Expand Your DMAIC Project Charter

Portions of your DMAIC Project Charter will likely have been drafted by your Champion or Sponsor, so part of the work in this step may be reviewing what has already been provided to your team. But you and your team also need to fill in a few missing blanks. The instructions here cover all the elements of the Charter, in case one has not been provided to the team, but do not start from scratch if you have a draft from the Champion.

Team Charters have the following elements:

a. Business case
b. Problem/opportunity statement and goal statement
c. Project scope, constraints, and assumptions
d. Team guidelines
e. Team membership
f. Preliminary project plan
g. Identify important stakeholders

See the DMAIC Project Charter Worksheet on p. 103.

A. Business Case

The business case for doing your project should come from your Champion or Leadership Council (see Chapter 4, p. 42). It provides the broad definition of the issue assigned to the Black Belt or team, as well as a rationale for why this particular project should be a key business priority.

B. Problem/Opportunity Statement and Goal Statement

The Problem Statement is a one- or two-sentence description of the symptoms arising from the problem to be addressed. It will often parallel the Business Case quite closely (sometimes they are almost if not exactly the same), but just as often the Problem Statement will be more specific and focused than the Business Case.

Example

- A Business Case might read: "Sales have fallen from last year's levels for three quarters in a row, reducing cash flow to record lows."
- A Problem Statement would then focus on a key element of that larger issue: "Sales of high-end products have fallen X% since last year, contributing to a significant reduction in cash flow."

Problem Statements usually answer these questions:

- What's wrong?
- Where is the problem appearing?
- How big is the problem?
- What's the impact of the problem on the business?

A measure of how big the problem is may not exist when the team starts its work, but as soon as it can, the team should collect enough data to create a baseline. (That's why the team said "X%" in the declining sales example above—they'll have to fill in the exact percentage later.)

Here are some things the Problem Statement should *not* do:

- **State an opinion about what's wrong.** A Problem Statement must focus on a pain or symptom that can be objectively *observed* and *measured*. "The new database is too hard to use" is a bad Problem Statement because it is based on a value judgment about "hard to use." It may be hard to use (maybe not), but the question is: What pain or trouble do you see or feel? A better statement would be: "Usage of the new database is only 50% of forecast, measured by the number of people issued passwords."
- **Describe the cause of the problem.** In Six Sigma, a Problem Statement describes the effects or symptoms of a problem whose causes are unknown. So, building on the example above, the following Problem Statement would be incorrect: "Usage of the new database is only 50% of forecast because, as indicated in initial user interviews, the system is difficult to understand and use."

  Here is another poor example: "Hotel occupancy is down because service is poor." Reduced hotel occupancy is probably a symptom of several underlying problems. Hotel occupancy may be down because of poor service, but there may be other causes. To make sure other important causes aren't excluded, simply say: "Hotel occupancy is down." This description of a problem symptom would include such other possible causes as rates being too high, lack of parking, time of the year, poor advertising, and many others.
- **Assign blame or responsibility for the problem.** Blaming the very people who will probably have to solve the problem is not a good idea. If you name a group or department in a Problem Statement, it's possible you're making a "blame" error.

- **Prescribe a solution.** If existing, reliable data confirm the cause of the problem, go ahead and implement the solution—you do not need to have a team work through DMAIC just to do what a project manager could do. So beware of Problem Statements like this one: *"We should set up a web site to increase sales of our insurance products."* The web site is a solution, but what's the problem, and what's causing it? Which insurance products aren't selling, and why? Until these questions are answered, it may be risky to jump to solutions. One way to test Problem Statements as being descriptions of symptoms is to check the word order in the statement: good Problem Statements follow a simple noun-verb sequence, naming the thing or service followed by a verb and some description of the symptoms as in these examples:

  "Customers experience poor service."
  "Sales fell 20% this quarter compared to last year."
  "Flight cancellations increased 31% since May 1."
  "Three out of four rocket chambers collapsed at 10,000 psi."

See the DMAIC Project Worksheet on p. 106.

- **Combine several problems into one Problem Statement.** Remember the KISP principle: "Keep it simple, please." Finding the buried causes of one problem at a time is hard enough.

Goal Statement

Problem Statements and Goal Statements are a matched pair: while the Problem Statement describes the symptoms of a hidden cause, the Goal Statement defines the "relief" expected from the team's work. Goal Statements usually have three elements:

See the Problem/Opportunity Worksheet on p. 104.

- **A description of what's to be accomplished.** The Goal Statement generally starts with a verb: "Reduce," "Increase," or "Eliminate" followed by the name of something. For example, "Reduce defects on the customer application" would be a good basic Goal Statement at the very beginning of a Six Sigma project.
- **A measurable target for desired results**. The target should put a number on the expected cost savings, defect elimination, or reduction in cycle time, etc. This measured target will eventually become one of the measures of the

team's success. To continue the example used above: "Reduce defects on the customer application by 50% [or X%]." *Be aware that the team may not be able to put a number to the goal until it has completed defining and measuring the problem.*

- **A projected completion date to reach the Goal.** This date may change, but setting some date at the outset will help the team get down to business and may shorten improvement team cycle times which often stretch out if no limit is set. So our example now reads: "Reduce defects on the customer application by 50% by October 31, 2002."

The Goal Statement should *not* say how the goal will be achieved. That would amount to a solution to the problem. For example: "Reduce defects on the customer application by 50% by October 31, 2002 *by installing a Web site.*" The italicized words are a solution. We need to know what's causing defects before we install a solution, or we may fail to solve the original problem (and we'd have spent a lot of time and money on a Web site).

Many teams say that getting agreement on the Problem and Goal Statements is one of the hardest parts of their Six Sigma project. The team Champion's views may differ from those of the Team Leader, and team members from different parts of the organization will have their own viewpoints. This is OK, for one of the lessons that Six Sigma teaches is that without data our early statements of problems and goals are often little more than guesswork.

Typically, the team members and Champion will work together to revise the Charter several times over the course of the project. Making these revisions is an excellent chance for Champions to learn more about DMAIC and the process under review.

C. Project Scope, Constraints, and Assumptions

The Champion normally outlines the scope of the project—usually defined in process terms (where does the project start and stop?)—along with constraints and assumptions.

- **Constraints** usually refer to limits placed on resources to be devoted to the project. A common constraint is the time that team members will devote to the project.
- **Assumptions** might include how often the Champion expects to meet with the Team Leader and team, how he or she will support the team when it runs into roadblocks, and the freedom the team has to implement solutions

without the approval of the Champion. These assumptions amount to a list of the Champion's expectations as one of the customers of the team.

Not all the elements in this category need be limiting, either. A Champion's assumption might be that "The team will make all key decisions about solutions to be implemented." Or "the Finance Department will provide one full-time person to help the team collect Cost of Poor Quality data." Other assumptions might include the time needed to complete the project, the role of the Champion, etc.

If your team has not been given any constraints or assumptions, generate a list of questions that you can ask of your Champion, such as:

- How long can the project take? What is the deadline?
- How much of our work time can be devoted to the project each week?
- Do we have any authority to spend money?
- Can we draw on other people inside the organization? Outside the organization?

You can either invite your Champion to a team meeting to clarify these issues, or send a delegation (such as the Team Leader and/or other team members) to meet with the Champion off-line. Either way, provide a list of questions to the Champion before the meeting.

D. Team Guidelines

Expectations of how the team will do its work can be included, too. Common guidelines include team ground rules for attendance and absence; how much time team members should devote to the team's project; where to meet; and how decisions are made.

Ground rules were already discussed in Chapter 5. See p. 69.

E. Team Membership

The Project Charter should list the name of the Champion responsible for the outcome of the Project, the Team Leader, team members, and the Master Black Belt, Green Belt, or other people assigned to help the team. Team membership may evolve over time, but remember to keep the size of the core team small: it's difficult to manage meetings and assignments if the size of the team gets above six or seven, including the Team Leader. The team members should represent all of the parts of the process under analysis.

F. Preliminary Project Plan

A final deadline for the project that's six months out won't help much in keeping teams on track as they work their way through DMAIC. Listing monthly or weekly milestones will help the team maintain a sense of urgency, and prevent completion dates from slipping behind. You'll need to prepare a summary of your plan for communicating with your Champion and others (see the DMAIC Project Plan Worksheet), but you might also want to use other planning tools, such as Gantt charts, to help the team manage its workflow.

G. Identify Important Stakeholders

A team project is just like a process in that it has customers who expect some particular output, and its work affects others in the organization; it also has suppliers who provide everything from resources to implementation support. The people who can influence or are affected by your project, both inside and outside your organization, are typically called Stakeholders. Understanding who these people or groups are and what they want or expect from your team can help you prepare to deal with their needs and concerns. The earlier you consider and begin to plan how you will "market" your solution, the less likely you'll be surprised by huge waves of opposition to your proposed improvement.

6σ
See the Project Stakeholder Worksheet on p. 110.

Note: Later in your process—as you near implementation—you will perform another analysis, this time to confirm what people or groups will be affected by your project results, and how to sway decision makers to support your recommendations.

Charter Do's and Don'ts

Do

- Make Problem Statements as specific and measurable as you can. Focus on symptoms you can actually observe, not on suspicions or assumptions about what may be going on.

Before moving on...

At the end of *Define Step 1: Update and Expand Your DMAIC Project Charter*, you should have:

✔ Reviewed, drafted, and/or revised each element of your project charter (business case, problem/opportunity and goal statements, scope, constraints, assumptions, etc.).
✔ Reached consensus on team guidelines (ground rules).
✔ Finalized team membership and roles.
✔ Developed a plan.
✔ Identified and contacted stakeholders.

- Use the Charter to set direction and gain agreement on the project, goal, and Project Plan. Take the time to address questions about the project *early*. You'll save lots of time later on if you do.
- Use placeholders for data that will be added in later stages (such as "The CSI is currently _______.").

Don't

- Describe suspected causes or assign blame in the Problem Statement. A key assumption in Six Sigma improvement is that you don't know the cause of the problem at the beginning of the project—even if you do have some guesses.
- Inflate preliminary goals for the project. It's OK to set ambitious goals, as long as they don't lead to impossible expectations.
- Over-"wordsmith" the Charter. The Charter is always going to change as your understanding of the process grows, so don't worry about getting the words "just right." Get the data right!

Case Study: MidwestAuto Tries a Different Approach

Throughout this book, a number of fictionalized case studies will help you understand how teams actually do the things described in DMAIC. These cases are based on actual teams, but the names, locations, and practically everything else have been changed to protect the innocent and not-so-innocent participants. The first case study in this chapter comes from the world of manufacturing.

Death to Dings!

MidwestAuto has been a part of a global car manufacturing giant since the 1950s. Its stamping operation produces the bodies for one of the parent company's best-selling SUVs. The plant manager of MidwestAuto, Helen Tookey, describes her plant as a "Mom and Pop operation with 2,000 people working at it." Everybody knows one another at the factory, with many employees coming from the same family.

One of the long-timers whose family has had many members employed at MidwestAuto's stamping plant is Jerry Traviano, line supervisor of the department that stamps out the doors for the best-selling RoboCar model. A self-described "car guy," Jerry can be found most days running between a computer monitor and the shop floor trying to get a handle on a problem that's always around in stamping operations, but seems to be getting worse in Jerry's corner of the world: "dings."

Depending on which side of the sheet metal you're looking at, "dings" are either depressions or raised pimples that appear on the surface of metal sheets after they've gone through the mighty presses that turn flat metal into curved doors for the RoboCar. Practically any kind of dirt or minute debris that finds its way inside the presses will leave its mark on the surface of the doors, such is the enormous pressure exerted by the press on the dies inside it.

"You're always fighting dirt and dings around a place like this," Jerry said one Monday morning in Helen Tookey's cramped office overlooking the shop floor. Jerry had removed the unlit cigar he normally had clamped between his teeth, and was staring at it thoughtfully.

"No, Jerry, I'm not always fighting dings around here. You are. I'm fighting the finance people who say these dings on the RoboCar doors are costing a fortune in rework and the marketing people who say the dings getting through to the showroom are really ticking off customers. In fact, the marketing people say these dings might just slow down the launch of the new models later this year." Jerry heard a vague warning in Helen's frustrated comments. He knew that the number of dings on the doors coming from his presses was up, and it was causing delays in the painting and assembly areas—and that customers in showrooms were starting to point out the "pimples" on the exterior surfaces.

Jerry shifted in his chair before replying.

"OK, so we've gone from two to three to 18," he said referring to the number of "dings" counted as defects on a truck door. "I'm working on it right now. I think the problem's the metal we're getting from Tuttanhammer Steel. They were barely meeting spec last year, and since we switched over to the new high-pressure presses and quick die change four months ago the dings have been popping up like mad. It's bad steel, I think. I'm talking to their contract people already. If they can't give us better steel, we'll have to go to another supplier, that's all."

Helen scratched her chin, looked at Jerry over her glasses, and spoke.

"Before I mess with the contract with Tuttanhammer I want to be darned sure that it's the steel and not something right in your own area that's causing this problem to get worse."

Jerry leaned forward to protest, but Helen held her hand out like a traffic cop, and continued.

"We've got to get those dings back down to spec or better before the end of the quarter, or you and I are going to be harvesting aluminum cans along the interstate. The people at Corporate are talking about this Sigma thing as you heard last month at the end-of-quarter meeting, and they're serious this time. So I want you and your crew to find out why your part of the shop holds the record for dings. Fast."

Jerry had lots more to say, but Helen had already turned to her computer screen to check her e-mails. Jerry knew the discussion was over. Returning his cigar to his mouth, and inserting the ear plugs worn by everyone on the noisy shop floor, Jerry walked back towards his own work area, already making a list in his mind of the team he would put to work on this ding thing.

A few days later, with the help of Jerry and a Six Sigma consultant from the corporate training office, the "Death to Dings" improvement team had its Project Charter drafted and on Helen Tookey's desk (see Figure 6-1). Helen liked what she saw.

Define Step 2: Identify Customer Requirements

Having reviewed and refined the Project Charter in one of its first meetings, your team is ready to undertake the important job of identifying the customer(s) affected by the problem being studied.

If your organization already has an effective system for translating the Voice of the Customer (VOC) into measurable customer requirements, it may be easy

DMAIC Project Charter Worksheet

Project Title: **Death to Dings Improvement Team**

Project Leader: Iris Washington, Team Leader Business Case: RoboCars represent a key market niche for the company. Charge-backs to our stamping operations for these rejects cost nearly $150,000 a month; costs for lost sales are not known.	Team Members: Susan Terragon, Black Belt Nat Collins, Day Shift Lead Trace Blumenthal, Swing Shift Lead "Roscoe" Smith Alice Smith
Problem/Opportunity Statement: Dings on RoboCar doors have increased in the last six months from 2 or 3 to 18 per door on average. This leads to a growing number of rejected doors once the doors are painted and dings become clearly visible. In addition, dings not caught and found on vehicles in showrooms are starting to hurt RoboCar sales.	Goal Statement: Reduce the number of dings to spec level of no more than two per door by October 15, 2003
Project Scope, Constraints, Assumptions: No money is available for buying new equipment or assigning new people to the operation. Team can implement any decisions based on good data after discussion with Champion.	Stakeholders: Jerry Traviano, Team Champion Moline operations RoboCar purchasers Stamping Operations employees

PRELIMINARY PLAN	Target Date	Actual Date
Start Date:	July 7	
DEFINE	July 15	
MEASURE	July 29	
ANALYZE	Aug 26	
IMPROVE	Sept 23	
CONTROL	Oct 15	
Completion Date:	Oct 15	

Figure 6-1. Death to Dings Project Charter

for the team to validate customer specifications and begin gathering data immediately. Without such data, however, getting the relevant customer input will take some time and effort.

At the very least, team members should be prepared to call the customers affected by the problem—or the people who work with them—and ask them how they view the problem, which of the customer requirements are not being met, and what they regard as defects.

These inquiries can be frustrating for team members. Customers can be vague about what they need: "I'll know it when I see it." Sometimes they offer solutions instead of specific requirements of their own: "You need to hire more people. Then things would be on time." It's not unusual for customers themselves to be unclear about their own processes, and their own requirements. Be prepared for this uncertainty and vagueness, but don't be upset by it. You're talking to the customer now, and that's usually a worthwhile thing to do!

The team will need to be on the lookout for two critical categories of customer requirements.

- **Output Requirements:** these are the features of the final product and service delivered to the customer at the end of the process.
- **Service Requirements**: these are the more subjective ways in which the customer expects to be treated and served during the process itself. These requirements answer the question, "How should we interact with and treat customers during our transactions?" Some service requirements surface at "Moments of Truth," which are any time that your paying, external customers come in direct contact with your organization's products, services, and people—and reach a judgment about the quality they receive.

Figure 6-2 shows examples of output and service requirement statements.

Getting to Specifics: Customer Requirement Statements

A Requirement Statement is a brief, thorough description of the performance standard established for an Output or Service transaction. It's the way we give some reality to the idea that "the customer defines quality."

Requirement Statement Guidelines

First let's establish some goals for a well-written Requirement Statement or performance standard, then we'll look at how to actually compose good statements. An effective Requirement Statement will:

Service Requirements		Output Requirements	
Process	Typical Req'ts.	Output	Typical Req'ts.
Auto Sale/ Purchase Process	• Prompt Attention (<2 min) • Lack of pressure (check with cust every 10 min) • Ability to test drive (all cars available to exit lot)	**Automobile**	• Engine starts in .5 seconds • Gas mileage equal to or better than rated • Door locks operate properly
Mortgage Loan Application/ Approval Process	• Complete loan application per customer's schedule • Include checklist of necessary documents with application • Notify applicant of decision within 15 days	**Mortgage Loan**	• Funded upon close of escrow • Accurate data on loan papers • Favorable interest rate
Wholesale Packaged Foods Ordering Process	• Customer-friendly order process (faxable form) • Notify customer when shipment leaves dock (call or fax) • Follow up with customer to ensure satisfaction with order (on-time arrival, product undamaged)	**Shipment of Packaged Foods**	• Delivered by date requested • Full pallet load • Intact (undamaged) product

Figure 6-2. Example of Service and Output Requirements

1. **Link to a specific Output or "Moment of Truth."** A requirement won't be meaningful unless it describes issues related to a specific product, service, or event. General customer issues or interests are fine in market research, but not when defining process performance standards.
2. **Describe a single performance criterion or factor.** It should be clear what the customer is looking for or will be evaluating—speed, cost, weight, taste, etc. This is usually not difficult. However, there's a temptation to lump factors together. For example: "Industry standard compliant and cross-platform compatible" ties together two criteria that should be defined separately.
3. **Be expressed using observable and/or measurable factors.** Here's more of a challenge. A requirement can't be assessed if it can't be observed. For less tangible requirements, it can take some effort to translate it into something observable. If you can't imagine a way to observe whether or not a requirement has been met, it's still too vague.

4. **Establish a level of "acceptable" or "not acceptable" performance.** The requirement should establish the standard for a "defect." Some will be "binary"—it's either met or it's not. Others will need clear definition of the customer's specifications, in a range often called "specification *limits*" (e.g., more must weigh more than 2 and less than 3 pounds).
5. **Be detailed but concise.** One of the big shortcomings of Requirement Statements comes from being too brief. It can be hard to assess a process or service based on "shorthand" requirements. At the same time, if they're too wordy, no one reads them. The key, of course, is a balance.
6. **Match—or be validated by—the Voice of the Customer**. Most importantly, the requirement or specification needs to fit the need/expectation of the customer. Each requirement inside the process should likewise be able to be *linked* to an external customer requirement (or why is it a requirement?).

Some questions to ask to test your Requirement Statements are:

- Does this requirement really reflect what's important to customers?
- Can we check whether and/or how well the requirement has been met?
- Is this stated so it's easily understood?

Figure 6-3 shows some examples of poorly written and well-written Requirement Statements. The poorly written ones are too general and not measurable.

Six Steps to Defining Customer Requirements

In order to define the Customer Requirements for your project you should:

See the Service or Output Requirement Instructions on pp. 110-111.

1. **Identify the Output or Service situation.** In this step, you answer the question "Requirement for what?" If necessary, use the Service or Output Instructions to help you specify what it is you'll be writing a requirement for.

2. **Identify the customer or customer segment.** Who will receive the product or service the process provides? The more narrowly you can focus, the easier it will usually be. When thinking of external customers, be sure to differentiate between distributors or supply chain partners and "end users" or consumers. (Many companies insist on

Poorly Written	Well Written
Rapid delivery.	Orders delivered within three working days of Purchase Order receipt (POs must be received by 3 P.M.)
Treat all patients like family. [This is fine as a guiding principle but not as a requirement statement.]	• Greet patients within 20 seconds of entry into waiting area. • Address all patients by "Mr." or "Ms." and last name. • Address patients by first name if permission is given by patient.
Make products easy to assemble and not requiring too much technical expertise.	All model 1200 bicycles able to be assembled by any adult in 15 minutes or less, using only a wrench and screwdriver.
Liberal returns policy.	Any returned item retailing for less than $200 accepted with no questions and for full cash refund.
Simple application.	Application form length maximum of two pages.

Figure 6-3. Sample Requirement Statements

using the term "customer" only to the *end user* in a supply chain. Other players—such as retailers in a consumer products business, resellers or independent reps in a computer or financial marketing supply chain—are best viewed as "partners." There are instances where this will not work, but the key is to remember that *all* participants in a supply chain need to focus on serving that *final* customer, or they will all stand to lose.)

3. **Review available data on customer needs, complaints, comments, surveys, etc.** Use objective, quantifiable data, where available, to define these factors. You may need to conduct a mini-survey of customers if no reliable data is available. *Don't* guess at customer requirements, or base what you think is important to them on hearsay.
4. **Draft a Requirement Statement.** This is where you must translate what customers say they want into something observable and measurable. When you aren't sure how to define a clear, objective requirement, try these questions:

6σ

See the worksheet on p. 88 for translating customer comments into requirements.

- What clues or observable factors will indicate we're achieving this requirement?
- Will we be able to objectively observe and/or measure this factor?

Another way to translate the customer needs into requirements is by associating the need with a powerful image. Some examples are shown in Figure 6-4. (Instructions for using a form like this are in Chapter 7.)

After drafting the statement, show it to others within or outside the team to make sure it's clear, specific, observable/measurable, easy to understand, etc.

5. **Validate the requirement.** Check back with customers, survey data, sales people, call center reps, etc. (anyone or any source that can help verify the requirement as defined). The question: Does this statement accurately describe what customers really want?
6. **Refine and revise the Requirement Statement.** When there's a gap between what customers require and what you deliver, it may be most appropriate to

CUSTOMER COMMENT	IMAGE OR ISSUE	REQUIREMENT
"How come you don't have any information on our vacation home under 'Asset'?"	Seeming lack of interest in customer's life. ...... Fear of loan being turned down without all assets known.	Care and appreciation demonstrated for each customer. ...... Clear explanation of factors used in underwriting decisions.

Figure 6-4. A worksheet for translating customer comments into Requirement Statements

negotiate a requirement that *is* feasible (better to ensure the customer has expectations you are able to meet than to surprise them when you don't). When realignment of the requirements is not possible or leaves you with less-than-thrilled customers, that's when DMAIC process improvement becomes really critical! After the requirement has been finalized, distribute and/or communicate it to ensure everyone knows the performance expectations and measurement.

Building a good set of customer requirements *is* a daunting task! It will take time to get clarity and agreement on these requirements, but without them you're just guessing and hoping that what you do will make customers happy. Certainly not a "Six Sigma" way to run your process or business!

Analyzing and Prioritizing Customer Requirements

Not all customer requirements are created equal, nor do customers regard every defect as equally serious. You may be unhappy when your plane leaves late, but you're even more unhappy if it arrives late at the next stop, and you miss the last plane home.

So you and your team will need to think a little about how to categorize and prioritize performance standards and how they affect customer satisfaction. What satisfied your customers last year will probably not satisfy them next year, one more reason to keep listening to the Voice of the Customer.

One approach to prioritizing customer requirements is called a Kano Analysis (Figure 6-5), based on the groundbreaking work of Noriaki Kano, a key figure in the Japanese quality movement. Dr. Kano realized the importance of dividing customer requirements into three categories:

1. **Dissatisfiers or Basic Requirements.** Dr. Kano often called this type of requirement a "Must Be": these features or performance requirements *must be* present to meet the minimal expectations of customers. The customer probably won't notice if these features or performance standards are met, but they *will* notice—and be extremely unhappy—if they are missing. If you turn on your TV and see a picture, you don't say "Wow! Great television!" Seeing a picture is a minimal expectation; you will likely judge the quality of the TV on other features (clarity of that picture, size of the screen, special programming capabilities, and so on).
2. **Satisfiers or Variable Requirements.** The better or worse you perform on

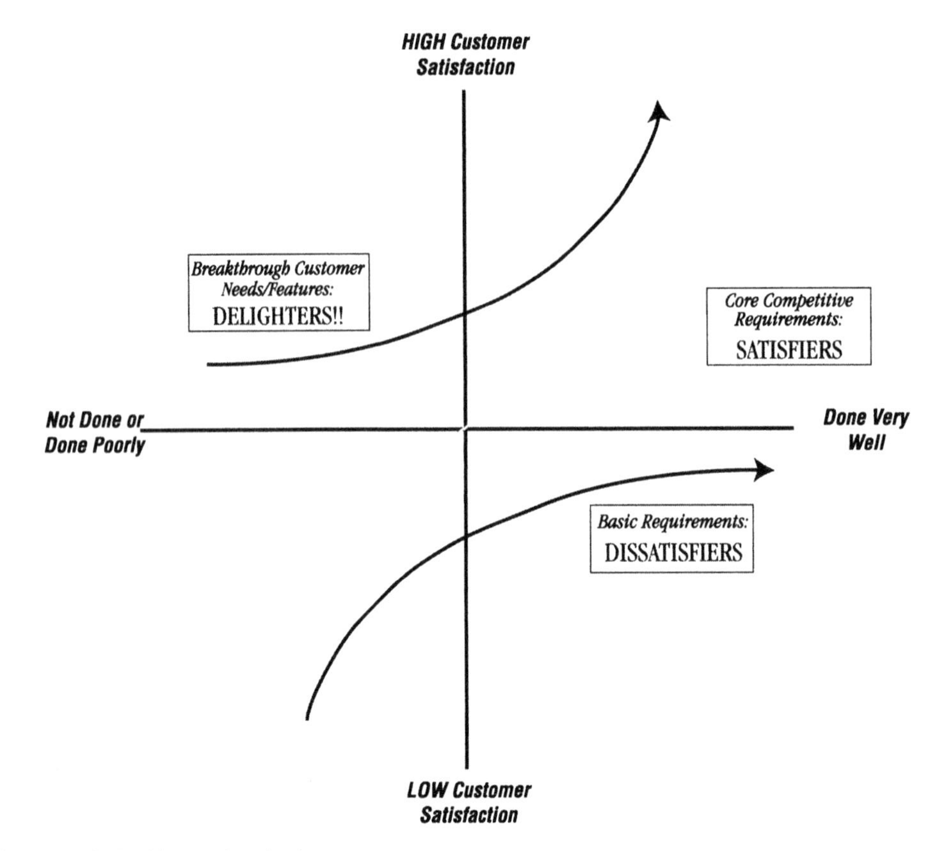

Figure 6-5. Kano Analysis

these requirements, the higher or lower will be your "rating" from a customer. Price certainly is the most prevalent of the Satisfiers: the less a customer has to pay for a given set of features or capabilities, the happier the customer (usually, at least). In Kano's terms, these are the "more is better" category—the more a customer gets of these features, the more satisfied they are. Most day-to-day competition takes place over these factors, features, or capabilities. Assuming that your organization is meeting the customer's Basic requirements, many of your process improvement priorities will likely fall within the Satisfier category.

3. **Delighters or Latent Requirements.** These are features, factors, or capabilities that go beyond what customers expect or that target needs the customers can't express for themselves. No customer ever said, "Give me a Palm Pilot." But having access to a portable electronic organizer has certainly delighted many people. In service industries, the delighters are often

unexpected services that go the extra mile: Finding a mint on the hotel bed at nighty-night time, or a free bottle of spring water or even a basket of fruit has delighted many a hotel guest. Adding automatic preset seat adjustment buttons is a delighter to a couple used to complaining to each other that "You drove my car and now I can't get comfortable!"

6σ See Kano Analysis instructions on pp. 112-113.

A Kano Analysis is not something you can do once a decade! Customer requirements and expectations change quickly, so even if your organization has gone through this analysis recently, your team may want to repeat it. For one thing, features or capabilities that were originally Delighters have a tendency to slip into Basic Requirements: It wasn't too many years ago that having air conditioning in a car was a true luxury; now it is expected by most customers.

Alternatives to the Kano Analysis

Another way to identify priority customer requirements is to let your customers tell you what they think! Show them a list of requirements you've developed and ask them to rate them individually on a scale of, say, 1 to 5, or rank them sequentially from most to least important. In fact, a priority weighting of requirements is a common step in a tool called the "House of Quality" used in process and product design projects.

Getting Measures for Priority Requirements

Knowing your customers' priorities will be of little use if you can't tell whether or not you're meeting their requirements. For practical purposes, therefore, you will probably want to focus your attention on the top two or three requirements, and work with your customers to assign specific measures. For example, a customer might require that "Shipments are received on time." Probing deeper will let you determine what the customer means by "on time," and also where the delivery should be made. Then the requirement can become a measure: "Shipments are received between 0630 and 0700 Monday through Friday at our warehouse on Pomona Avenue."

Gathering, translating, and putting measures to customer requirements is probably as much an art as a science, but it is an important art—for any failure to meet customer requirements is a defect, and the measurement and elimination of defects is what Six Sigma is all about.

Death to Dings Team Nails Down Customer Specs

Identifying customer specs wasn't difficult for the "Death to Dings" team: the immediate customer for the RoboCar doors was Midwest-Auto's painting division in Moline where doors (along with the rest of the RoboCar bodies) were shipped after stamping. The Quality Control/Engineering Specifications Department in Moline demanded no more than two dings visible per door after painting.

When the Team Leader, Iris Washington, checked these specs with the Moline people, she discovered that Moline's definition of a "ding" differed from that used at the stamping plant.

The stamping operators counted as dings anything visible to the eye under normal light at a distance of four feet. Because dings are almost twice as visible after being painted, the Moline people counted anything visible at a distance of *eight* feet as a ding. That meant that a component that was judged acceptable in the stamping plant (viewed unpainted at four feet) was judged unacceptable in the painting facility. To lessen defects at Moline, the stamping plant would have to look even harder for dings than in the past.

On the other hand, the Moline people told Iris that they were happy with the just-in-time deliveries of doors.

"So they're getting dinged doors right on time. Big whoopee!" said Roscoe Smith, the team's sour comic. "That's not even part of the problem."

"That's right," said Iris. "That's why we're not going to say any more about it, right, Roscoe?"

Roscoe nodded agreement, and the team went back to the next agenda item, documenting the stamping process.

Defining Requirements Do's and Don'ts

Do

- Pay equal attention to Service and Output requirements. A company with Six Sigma products but lousy service and customer relations may survive—but only until customers find a better supplier.
- Make the effort to create clear, observable Requirement Statements. Even if the requirements are fuzzy at first, the learning—and discipline—achieved by building measurable requirements is essential to understanding your customers and measuring your own process performance.

Don't

- Close your mind to new information about what customers *really* want. Customer data may contradict what you've always believed. Don't go into denial or reject their assumptions. It's OK to challenge the data, but it isn't smart to ignore what customers really want. Remember, the customer is always ... powerful!
- Turn new requirements into rigid "Standard Operating Procedures." Be prepared to see customer requirements change soon...and fast. Build in reviews for your process to continue translating the Voice of the Customer into your work.
- Fail to measure and track process performance against customer requirements. The team has taken the trouble to define requirements so that it can measure how well we're meeting them today.

Before moving on...

At the end of *Define Step 2: Identify Customer Requirements*, you should have:

✔ Defined customer needs (both service and output).
✔ Validated customer requirements.
✔ Identified which requirements are *priorities.*

Define Step 3: Identifying and Documenting the Process

The final step in Define, which will probably be on the agenda of your third or fourth team meeting, will be to develop a "picture" of the process involved in the project by creating a "high level" flowchart. Some teams are tempted to skip this step, but without it, the team will take even longer to focus on the problem.

Choosing a Process Diagramming Method

We can define a process as "a series of steps by which one thing becomes another thing." For example, a potential customer becomes a real customer as they pass through our sales process *if* we meet their critical customer requirements.

Sounds simple enough, but sometimes processes are varied and complex. During one of your early team meetings you will need to map your process with just enough detail to help you get started with measurement and analysis. In general, Six Sigma teams start with a high-level process map with only a few details, as if the team were taking a snapshot of the process from 30,000 feet up, with only the major steps showing. Later they can swoop down for a closer look at suspicious parts of the process.

We call this high-level general map a **SIPOC** diagram (see Figure 6-7), where the letters stand for

Figure 6-6. SIPOC model

> **S**uppliers—the people or organization that provides information, material, and other resources to be worked on in the process
>
> **I**nputs—the information/materials provided by suppliers that are consumed or transformed by the process
>
> **P**rocess—the series of steps that transform (and, we hope, add value to) the inputs
>
> **O**utputs—the product or service used by the customer
>
> **C**ustomer—the people, company, or another process that receives the output from the process

Often, key requirements for Outputs are added, in which case the diagram is called "SIPOC+R." You can include requirements for the Inputs as well.

SIPOC can be a big help in getting people to see the business from a process perspective. Some of its advantages include:

1. Displays a cross-functional set of activities in a single, simple diagram.
2. Uses a framework applicable to processes of all sizes—even an entire organization.
3. Helps maintain a "big picture" perspective, to which additional detail can be added.

Death to Dings Team Is Flying High to SIPOC Process

"One more meeting like that," said Iris Washington, the Team Leader, to Susan Terragon, the team Black Belt, "and I quit!"

The two women were in the car park after the team's third and, so far, noisiest meeting. They had put together a SIPOC map of the door-stamping operation and there had been sharp differences of opinion about where the process started and ended and what should be included.

"I told them we only needed a few high-level steps," said Iris, shaking her head, "and they get all detailed!"

"Well, remember, Iris," said Susan, "it's the first time they've done one of these maps. We'll need more detail at the next meeting. So a lot of what got written down we can use then. So it worked out OK."

"I guess so," said Iris, unlocking the door of her own RoboCar. "Anyway, thanks for your help in getting it together. I've got to get going, and pick up my kids. See you on Thursday."

See SIPOC Map instructions and worksheet on pp. 114-115, 117.

(The SIPOC map Iris's team came up with looked like Figure 6-7.)

Suppliers	Inputs	Process	Outputs	Customers
Tuttenhammer Steel Co. Steel cutters	Sheet steel Cut steel	Receive cut steel Unbundle Cut the steel Check dies Check press Feed cut steel to press Press steel Remove doors from press Inspect doors Load doors on shipping racks	Doors	Press Area Moline Paint Operation

Figure 6-7. Example SIPOC Diagram

SIPOC Do's and Don'ts

Do

- Be clear about where the process starts and ends.
- Involve the full team in developing a SIPOC map. Everyone must agree on the project boundaries.
- Ask your Champion, Sponsor, or Leadership Council for advice if you're not clear about where your responsibilities begin and end.

Don't

- Be too ambitious in setting the boundaries of the process you're studying. Trying to tackle too much at once is a good way to spread resources too thin and lose track of important customer requirements.
- Get into too much detail at this stage. Stick to a high-level map of the process.

Targeting a Suspicious Process Step

With the SIPOC map complete, the team now can focus on the step(s) where the suspected causes of the symptoms in the Problem Statement are thought to be. Based on their experience, the team members will make an educated guess about which step is worth investigating in detail. Some caution must be used here, because educated guesses and pet theories look a lot like one another.

There'll be lively discussion in the team around these points, but don't let it go on too long—only data and analysis can support or kill these early theories. (On the other hand, experience should count for something, and we do have to start somewhere!)

OPTIONAL: Using a Detailed Process Map

The high-level SIPOC map is usually sufficient detail for this stage in the process. However, more detailed process maps may also come in handy. These tools are described in more detail under the Analyze stage (see Chapters 12 and 13). If you choose to go this route, here are some questions you can ask to help you narrow the focus of your project:

- Are there any steps that team members suspect may be the source of the problem?
- Where can we take measures to learn more about the problem? Pay special attention to steps that loop back. What's causing them to loop back?

- Do we already have data on file for how this sub-process operates? What kind of data is it?
- Where are the points where the process touches the customer directly? These "Moments of Truth" are ones where customers judge our service or product immediately. Are we getting them right?
- Roughly how long does it take—on average—for this sub-process to cycle from start to finish?
- Are there any steps that obviously don't add any value to customer requirements? What would happen if they were eliminated?

If you find obvious errors that can be easily fixed, make the needed change. But don't make too many changes yet! In future steps, you'll be learning a lot more about what is really happening in the process. For example, the team will have to validate the map by walking through the process to make sure that what you've mapped is in fact what actually happens in the process now.

6σ

See the Define Checklist on p. 116.

Your main job here is to identify areas that may be contributing to the problem. You also need to think of points in the process where you can effectively measures its operation. This can be done with some colored dots or some other means of showing what and where you want to "take the temperature" of this or that suspect process step. We'll say more about this in Chapter 9, on measuring.

Iris's Team Drills Deeper into Its Process

Although Iris thought the previous meeting was noisy, it was nothing compared to the latest one. She and the team drilled farther down into the process, and put up a lot of sticky notes. There was loud debate about the sub-steps in the process, and even louder debate about the suspected trouble spots in the process.

Iris had to remind the team several times that these were just guesses at the moment, and that they'd have to collect data to see if their guesses were right.

Despite the debates, the team was able to agree that they needed to know more about how the steel was being cut before it came to their area. (Here the theory was that the dirt and debris was getting on the steel sheets while they were being trimmed to size.) They also decided to get data about the quality of the steel sheets arriving from

Tuttanhammer Steel Company, a major supplier to the whole stamping plant. (Jerry Traviano thought the cause of the problem was bad steel from Tuttanhammer. There wasn't much support for this theory up front, because, as Trace Blumenthal, who had been collecting the data, pointed out, if the cause was bad steel from "King Tut," why did they seem to have more dings on doors every 30 minutes or so in a kind of cycle? Still Jerry was the Project Champion, and their boss, and a nice guy and all....)

There were other steps the team thought might be contributing to the problem, but they agreed to highlight (with a Hi-Liter fluorescent pen) four steps where they could take measures and collect data. Iris said she'd bring Jerry up to date the next morning. Then, after a quick review of the meeting itself during which Roscoe said they needed more doughnuts, the team called it a wrap at five o'clock.

Before moving on...

At the end of *Define Step 3: Identifying and Documenting the Process*, you should have:

✔ Defined the major elements of your process.
✔ Identified the process boundaries.
✔ Created a process diagram.

Finishing Your Define Work

By now your team has completed the four essentials for defining its project:

- Project Charter with problem and goal statements
- List of measurable customer requirements
- SIPOC map of the process being analyzed
- A more detailed map highlighting some suspect areas for measurement

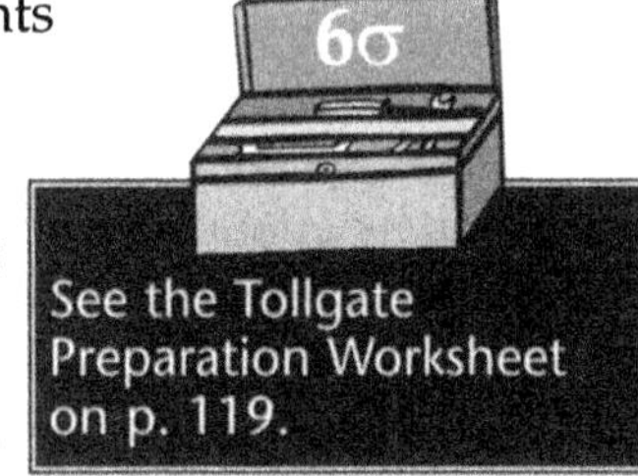

See the Tollgate Preparation Worksheet on p. 119.

There are three last steps to completing the Define stage:

1. Update your project storyboard.
2. Prepare for your tollgate review by your Sponsor or Leadership Council.
3. Review progress and adjust your ground rules, project plan, etc.

1. Update Your Storyboard

As described in Chapter 5 (pp. 64-66), the storyboard will be an important communication tool for your team. To decide what goes on your storyboard, have your team discuss what information will capture the most important themes or outcomes from your Define work. Many teams decide to include information such as:

- Team member names and positions
- The original Project Charter with Problem Statement and goal
- A list of measurable customer requirements
- A high-level SIPOC map
- A list of existing measures in the process itself

2. Prepare for the Tollgate Review

Because the work done by Six Sigma teams is part of strategic decisions made by the organizations, most companies will institute a formal review procedure at the end of each phase of DMAIC. The purpose of that review is for the Champion or Sponsor, Leadership Council, or other managerial group to review the team's progress and give a formal go-ahead for the next phase (or to ask that the team do more work in Define before proceeding). These types of reviews are often called *tollgate reviews* (though many organizations have different names for them) because the team must pass through the "gate" successfully in order to proceed on their path.

Tollgate reviews are ideally conducted in joint meetings between the full team (*not* just the team leader) and the appropriate oversight group. The team makes a presentation summarizing their work; the reviewers ask questions and probe for more detail, then decide whether to allow the team to proceed to the next step.

The tollgate review for Define is often the first time the team will have any formal feedback on its work, so naturally people get nervous ahead of time. The only way to lessen that anxiety is to make sure you are prepared. The Tollgate Preparation Worksheet on p. 119 can help your team complete its preparations; here are a few additional tips.

- **Involve all team members in the preparation and the presentation.** Assign specific responsibilities to individuals or subgroups.
- **Focus on the essentials.** The managers who will be involved in the tollgate review will only become frustrated if you try to describe every decision the

team made. Show them respect by using their time wisely: have your team identify the most essential messages to share with the reviewers, then develop clear visuals or handouts to summarize those messages.

- **Make the decision clear to the reviewers.** The question facing the reviewers is whether they think the team is ready to proceed to the next stage of DMAIC. You need to make it clear just what they are approving: "If we get the green light today, our next steps are to _____."
- **Do a dry run**. Many people on Six Sigma teams are not called upon regularly to do presentations, so their public speaking skills may be rusty! Doing a dry run for a key manager or perhaps a group of colleagues is a safe way to test the presentation and to help people get more comfortable with their roles.

3. Review Your Progress

The heart of continuous improvement is reviewing what we've done, comparing it to what we thought would happen, and making adjustments accordingly. You can be as simple or elaborate as you like. For example, allot 15 minutes in a team meeting to ask, "What have we learned in the Define stage?" Or assign subteams to do a detailed review of each aspects of the team's work to date (ground rules, project goal/charter, process mapping, etc.) and come back with suggested improvements.

Once these steps are complete, your team is ready for its *Measure* work (see Chapter 9).

Chapter 7

Power Tools for "Define"

Getting It Right at the Beginning

THE TOOLS, WORKSHEETS, AND OTHER JOB AIDS in this chapter will help you implement the actions described in Chapter 6:

- Part A will help you organize and manage the team.
- Part B gives you the tools for identifying customer requirements.
- Part C focuses on SIPOC and other tools that help you develop a high-level process map as well as more detailed flowcharts of targeted process steps.
- Part D includes two checklists to help your team finalize its work in Define.

There is also an addendum describing a set of advanced tools that may be appropriate for some teams.

A. Team Management

How well a team operates in the Define stage determines how well it will operate throughout the project. The team needs to be very clear about what it is being asked to do and how its work will impact the organization and its customers. The tools in this section will help the team clarify what the project is supposed to accomplish, and why it is critical that the organization make improvements in this area now, given other competing priorities.

DMAIC Project Charter Worksheet

Purpose: Help the team understand what it is supposed to accomplish, and identify areas that require discussion or clarification with the Sponsor(s).

Application: A job aid for teams that are just starting up.

Related tools: Use the Charter Worksheet (Figure 7-1) in conjunction with the Problem/Opportunity Statement (Figure 7-2).

Instructions: Some of the information needed for a Team Charter should be provided by your team Champion or Sponsor, but if not, most of Chapter 6 covers instructions for generating the information needed to complete this worksheet. See pp. 74-80.

Problem/Opportunity Statement Worksheet

Purpose: To clarify the situation and provide a clear description of a process planning or improvement opportunity (Figure 7-2).

Applications:

- Setting priorities.
- Determining a goal.
- Guiding efforts for determining solution(s).
- Following up and verifying results.

Related tools:

- Used to help complete the Project Charter (p. 103).

<table>
<tr><th colspan="4">DMAIC Project Charter Worksheet</th></tr>
<tr><td colspan="4">Project Title:</td></tr>
<tr><td colspan="2">Project Leader:</td><td colspan="2" rowspan="2">Team Members</td></tr>
<tr><td colspan="2">Business Case:</td></tr>
<tr><td colspan="2">Problem/Opportunity Statement:</td><td colspan="2">Goal Statement:</td></tr>
<tr><td colspan="2">Project Scope:</td><td colspan="2">Stakeholders:</td></tr>
<tr><td>PRELIMINARY PLAN</td><td>Target Date</td><td>Actual Date</td><td rowspan="8"></td></tr>
<tr><td>Start Date:</td><td></td><td></td></tr>
<tr><td>DEFINE</td><td></td><td></td></tr>
<tr><td>MEASURE</td><td></td><td></td></tr>
<tr><td>ANALYZE</td><td></td><td></td></tr>
<tr><td>IMPROVE</td><td></td><td></td></tr>
<tr><td>CONTROL</td><td></td><td></td></tr>
<tr><td>Completion Date:</td><td></td><td></td></tr>
</table>

Figure 7-1. Charter Worksheet

Problem/Opportunity Statement Worksheet
Project Title:
What is the area of concern? What first brought this problem to the attention of your business?
What impact has this problem already had? What evidence do you have that it is really a problem worthy of attention?
What will happen if the business doesn't address this problem?
Summarize the above information in a concise statement.

Figure 7-2. Problem/Opportunity Statement Worksheet

Instructions:

1. **Define the area of concern or opportunity.** State the current situation objectively and clearly.
2. **Describe the impact** (consequences, potential benefit) of the concern or opportunity. Include measurable information and data—or add when available.
3. **What opportunities will be lost or what problems will increase if the issue isn't addressed now?**
4. **Identify market forces** that may be important to consider (such as what your competitors are up to).
5. **Draft a statement,** then review it for clarity. Avoid the temptation to assign blame, offer your own analysis of the situation, or prescribe solutions.

DMAIC Project Plan Worksheet

Purpose: To help a team organize its work and ensure that necessary tasks are completed.

Applications:

- Update throughout a project to make sure critical actions are completed on time.
- Use documentation on actual results to learn how to do better the next time around.

Instructions:

1. **Identify the milestones** associated with the DMAIC phase your team is in.
   - Define, for example, might include "approve and endorse Charter" or "complete SIPOC analysis."
2. **List the milestones** on a form like Figure 7-3.
3. **Assign responsibility** for each milestone to a specific individual on the team.

   Note: "Being responsible" for a milestone does *not* mean that the individual does all the work by him- or herself. It means making sure that the work needed to reach that milestone is completed—helping to identify necessary actions, involving others in the work, checking on progress, and so on.

DMAIC Project Plan Worksheet			
Project Title:			
Project Leader:			
Action/Milestone	**Responsible Team Member**	**Target Completion Date**	**Actual Completion Date**
Define:			
Define:			
Define:			
Define:			
Define:			

Figure 7-3. Project Plan Worksheet

4. **Set a target completion date** for each task. Target dates are important because they help a team decide what tasks are priorities and what resources to allocate to a task.

5. **As the team progresses, document the actual completion date for each milestone.** Jot down brief notes on explanations for delays or early completion.

Related Tools: A form as shown in Figure 7-3 is often used to summarize more detailed plans. You might want to use something like a Tree Diagram (see p. 312) or Gantt chart (p. 108) to identify and coordinate your work at a more detailed level.

Gantt Chart

Purpose: To help a team plan the efficient use of its time by identifying the appropriate sequence and potential overlaps in tasks.

Applications:

- Identify and organize the steps needed to complete a project or particular task.
- Plan the implementation of improvements.

Instructions:

1. **Identify the target outcome and final deadline** for the task or project.

2. **Brainstorm a list of actions that must be taken to complete the task/project.** Refine the list by combining related actions, eliminating duplicates, and so on.

3. **List the tasks in time sequence** down the left side of chart constructed to look like Figure 7-4.

4. **Label appropriate time units** across the top of the chart (days, weeks, months, etc.).

5. **Decide how long each task will take.**

6. **Identify start and end dates for each.** Work through the tasks in sequence. Look for places where you can overlap the tasks.

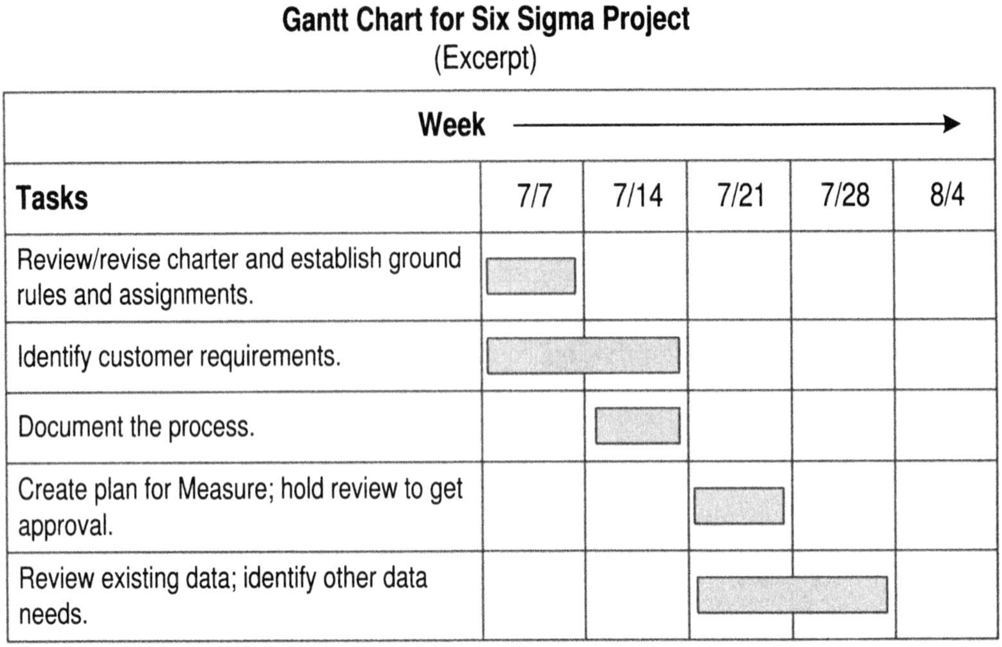

Gantt Chart for Six Sigma Project
(Excerpt)

Week →					
Tasks	7/7	7/14	7/21	7/28	8/4
Review/revise charter and establish ground rules and assignments.					
Identify customer requirements.					
Document the process.					
Create plan for Measure; hold review to get approval.					
Review existing data; identify other data needs.					

Figure 7-4. Example of Gantt Chart

7. **Do a reality check:** Do the time estimates look realistic? Can you get all this done with the time/resources allocated to the project? Revise as needed. *Tip:* Teams usually end up moving both the tasks and timelines around as they wrestle with creating a plan that is realistic yet meets the required deadlines. So prepare the initial draft in a way that is easily changed: using a flipchart and self-stick notes, a computer, a whiteboard, etc.

Related Tools:

- Use as background for the summary plan represented in your DMAIC Project Plan.

Project Stakeholder Analysis Worksheet

Purpose: To help a team stay linked with people or groups inside and outside the organization who can influence its success.

Applications:

- Early in a project, to help the team develop a communication strategy and ideas for keeping those outside the team informed of the team's progress.

Instructions:

1. **Brainstorm with all the groups or individuals who have a stake in your team's Six Sigma project.** For example, include people who can shape the project, those who will be affected by the outcome, and those who might be resources for the team. This might include:
   - Key customers, the Champion/Sponsor, Leadership Council, employees who work on the process under study.
   - Other managers/executives who can aid (or hinder) the team.
   - External groups or organizations who may affect or be affected by the project (suppliers, regulatory agencies, etc.).
2. **Sort and organize the brainstormed ideas** to get an agreed-on list, then write each person/group down the left-hand side of the Project Stakeholder Analysis Worksheet (Figure 7-5).
3. **Evaluate each stakeholder's relationship to the project** and check the appropriate box(es) on the form.
4. **Develop strategies for dealing with each stakeholder.**
   - Discuss what each stakeholder wants from the project or how they can influence it.
   - Determine a strategy for communicating with that stakeholder.
   - Check any boxes that apply to your strategy and/or write a brief summary of the team's decision.
5. **Assign responsibilities in the team for carrying through on the strategies.**

Related Tools: Use as background for completing your Team Charter.

Project Stakeholder Analysis										
	Relationship to Project					Communication/Involvement Strategy				
Stakeholder	Is affected by outcome	Can influence outcome	Has useful expertise	Provides resources	Has decision authority	Meet with regularly	Invite to team mtgs	Send copy of mtg mins	Speak with informally as needed	Other (describe)

Figure 7-5. Project Stakeholder Analysis Worksheet

B. Identifying Customer Requirements

A "Requirement Statement" is a brief but thorough description of the performance standard established for an Output or Service encounter. Composing requirement statements isn't easy. If you have sketchy or conflicting customer input, for example, it can be a big challenge to "nail down" requirements. And even *with* good data it's easy to be vague or to violate some of the guidelines of a well-stated requirement.

Service or Output Requirement Instructions

(See Figure 6-2, p. 85, for examples of these statements.)

1. **Define and re-examine the process output.**
   - What is the current Output or end product of the process?

- Is this Output still the best "thing" to fulfill the needs and objectives of the customer?
- What other alternatives—products or services—might we offer instead or how might the nature of the Output be changed?

2. **Clarify and scrutinize key requirements of the output.**
   - What features or characteristics of the Output make it usable/effective for the customer?
   - What other unmet features or characteristics are not being met?
   - What are the needs or changing requirements of the customer's customers that we can help them meet more effectively?
   - What other opportunities are there for the product/service to be more valuable, usable, convenient for the customer?
   - What lessons or other needs can we identify from understanding how the customer uses the Output?

3. **Review and retest output and requirements assumptions with customers.**
   - How can we check the validity of our or the customers' assumptions about what's required?
   - What recent data confirms these requirements? Which ones might be questioned?
   - Are there different groups within the process "customer base" that should be addressed separately?

Related tools:

- Use your SIPOC diagram to define the process outputs.
- Alternatively, use the Requirement Statement Worksheet.

Requirement Statement Worksheet

Purpose: To identify what it is about your service/product that is important to customers (see Figure 7-6).

Applications:

- In the Define stage, use it to make sure the improvement effort is targeted appropriately.

CUSTOMER COMMENT	IMAGE OR ISSUE	REQUIREMENT

Figure 7-6. Requirement Statement Worksheet

- In later stages, revisit and confirm these requirements to make sure your team is doing work that will be noticed and appreciated by customers.

Instructions:

1. **Assemble sources of data** for "Voice of the Customer" input. (Attach relevant data as needed.)
2. **Select representative statements.** Sort through any notes you have from interviews, surveys, etc., and select a statement in the customer's own words that best reflects a particular customer need.
3. **Identify the issue or image** behind the customer comment.
4. **State a specific requirement** that addresses the comment or issue.
5. **Validate the requirements** using the rest of the Define Customer Requirement process (p. 82).

Related Tools: Use the Kano Analysis to help prioritize the requirements.

Kano Analysis Instructions

Purpose: To help you understand and prioritize customer requirements.

Applications:

- In the Define stage, to help narrow your improvement targets to key customer requirements.
- In later stages, to help understand the priority of features you might want consider adding to products or services.

Instructions:

1. **Generate a list of customer requirements.**
2. **Review the Kano Analysis** (see Chapter 6, p. 90).
3. **Identify the appropriate category for each requirement.** Work through the list of requirements one by one. Mark each as to whether it is a ...
   - Dissatisfier/Basic Requirement
   - Satisfier/More Is Better
   - Delighter
4. **Determine priorities:**
   - First look at anything labeled as a Dissatisfier or Basic Requirement. These represent customers' minimal expectations, and even though doing them well doesn't win you a lot of points, doing them poorly puts you out of the running! Assess whether your organization is currently filling these requirements or not.
   - Perform the same assessment on the Satisfiers. Are you doing them at all? Can you do more?
   - If you aren't meeting the Basic Requirements and have a lot of work to do on the Satisfiers, don't worry about the Delighters yet. If you have all the Basic Requirements covered, and most of your satisfiers are done well, evaluate the Delighters and identify those that would contribute the most to customer satisfaction.

Related tools: Use one of the requirement statement tools first.

C. Developing a Process Map

In the Define stage of the DMAIC process, it is important for teams to understand the boundaries of their project. The SIPOC method discussed in Chapter

6 (pp. 94-96) is a simple tool that helps a team define those boundaries in practical terms that make it obvious where they should focus their attention.

SIPOC Analysis and Map

Purpose: To provide an "at-a-glance" perspective of the high-level process steps, in conjunction with key suppliers, inputs, outputs, and customers (see Figure 7-7).

Applications:

- Identifying boundaries (start/stop points) for process or process improvement efforts.
- Understanding the scope (magnitude) of the process or process improvement efforts.
- Identifying relationships between suppliers, inputs, and the process.
- Determining key customers (internal and external).
- Linking other SIPOC maps to understand "upstream" and "downstream" processes.

Supplier(s)	Inputs/Req'ts	*PROCESS*	Output(s)/Req'ts	Customer(s)
Enablers				

Figure 7-7. SIPOC Worksheet

Instructions:

1. **Identify the process to be mapped and name it.** For example, you may be focusing your attention on "Completing the Customer Application Form."
2. **Define the scope of the process.** Where will you start your map? With which supplier? Where will your SIPOC end? With which customer? The aim here is to prevent the team's work from trying to cover too much ground, and thus avoid "scope creep"—going further "upstream" and "downstream" into related processes instead of staying clearly focused on the process suspected of creating defects (see Figure 7-8).
3. **Name the Outputs and their Customers.**
4. **Name the Suppliers and the Inputs they provide.**
5. ***Optional:* Identify process enablers:** A process enabler is something that is not consumed or transformed in the process, but which makes the process possible. Equipment, for example, is an enabler for many processes. It is not an "input" because it is not consumed or transformed, but the process wouldn't be able to function without it. Distinguishing enablers from inputs can help you pinpoint sources of variation later in your DMAIC project.
6. **Document customer requirements for the Outputs if you already have them.** Otherwise, add them when you complete your work on translating the Voice of the Customer into measurable requirements.
   - What are key features/characteristics of the Output for each Customer?
   - What are key features/characteristics of the Input for the Process?
7. **The "P" or Process part of the SIPOC is best drawn as a block diagram, with each block standing for a major activity or sub-step in the process.** Establish five to seven "high-level" process steps to broadly describe the process in the order in which the steps actually take place.
   - Brainstorm major process activities. Use a noun-verb sequence to describe each broad step, like "Collect information," "Enter information," etc.
   - Group activities into similar categories or "major steps" in the process. (See Affinity Process and Diagram, pp. 56-57.)
   - Place major steps in most appropriate order in "block diagram."

   Tip: List action steps as they actually occur today in the process. Don't

Define Checklist

1. Define

Instructions:

If you can respond "yes" to each statement below, you're off to a good start with your project, and are ready to move into the "Measure" phase of DMAIC.

For our project we have ...

1. Confirmed that our project is a worthwhile improvement priority and is supported by the Quality Council. YES NO
2. Been given (or written) a brief business case explaining the potential impact of our project on customers and profits and its relationship on business strategies. YES NO
3. Composed and agreed to a two-to-three sentence description of the problem as we see it—the Problem Statement—focusing on symptoms only (not causes or solutions). YES NO
4. Prepared a Goal Statement defining the results we're seeking from our project, with a measurable target (or placeholder to add one). (No solutions are proposed in the Goal Statement.) YES NO
5. Prepared other key elements of a DMAIC team charter, including a list of constraints and assumptions, a review of players and roles, a preliminary plan and schedule, and (if needed) a process scope. YES NO
6. Reviewed our Charter with our Sponsor for this project and confirmed his/her support. YES NO
7. Identified the primary customer and key requirements of the process being improved and created a SIPOC diagram of the areas of concern. YES NO
8. Prepared a detailed process map of areas of the process where we expect to focus our initial measurement. YES NO

Figure 7-8. Define Checklist

use other process mapping tools to show rework loops, decision points, etc. You're not ready for that level of detail yet.

8. **Validate SIPOC map with others.**
   Here are some helpful hints for keeping SIPOC mapping nearly painless:
   - If you have more than seven or eight steps under the "P" you're probably getting into the detailed sub-process steps. Remember, the SIPOC is supposed to give us a map of the process from 30,000 feet. So stay as high as you can for as long as you can.
   - Create the SIPOC on a couple of large sheets of paper hung on the wall. Put the name of the process at the top of the sheets and under it the words *Suppliers* — *Inputs* — *Process* — *Outputs* — *Customers* running across the sheets.
   - Use sticky notes to display the information—just in case you have to rearrange it.
   - Limit "Inputs" to information or material actually used in the process. Include other physical items (such as equipment, buildings) as "enablers."
   - Figure out who supplies the inputs. Don't be surprised if it's often the customer, particularly if you're providing a service based on information and requests they provide initially.
   - Revise the Project Charter if the SIPOC map indicates you need to redefine the scope of the project.

D. Completion Checklists

Define Completion Checklist

Purpose: To bring a formal end to the Define stage of a team's project.

Applications:

- Use during the Define work to track progress.
- Use at the end of the Define stage to make sure all essential tasks have been completed.

Instructions:

1. **Walk through this checklist item by item at a team meeting.**
2. **Mark a "yes" only if everyone on the team agrees the task has been completed.** If anyone says no, ask him or her to state why they think the task is incomplete.
3. **Reach agreement as a team on each answer before marking the checklist.**
4. **If there is unfinished work, ask for volunteers, assign responsibilities, and set deadlines for completion of those tasks.**

Define Tollgate Preparation Worksheet

Purpose: To help a team prepare a presentation for a tollgate review (Figure 7-9).

Applications:

- At the end of any DMAIC stage, to help the team prepare its presentation.

Instructions:

1. **Brainstorm a list of messages** the team thinks their Sponsor/Champion, Leadership Council, or other management group should hear about the project. Sort and organize the ideas.
   - Include both what the team did in Define and what you plan to do in the Measure stage.
2. **Identify priority messages.** Use discussion and multivoting to messages that have the highest priority. You want to focus on no more than three to five during the presentation. Trying to do too much will just confuse people who did not actually participate in the work.
3. **Decide on a sequence** for the presentation, and complete the left column on the worksheet.
4. **Identify presentation methods.** For each message, identify how that information can best be presented to someone unfamiliar with the details of the project. Be creative! Look for ways to convert messages into data charts, pictures, or other high-impact visuals. Also identify what format that information will take in the presentation (such as handouts, flipcharts, slides or overheads, etc.). Complete the middle column of the worksheet.

Define Tollgate Preparation Worksheet		
Key messages to cover in the Review. (List no more than 3 to 5 in the sequence in which they will be covered in the presentation.)	**Best way to present this information.** (Be creative in finding high-impact visuals–handouts, overheads, Powerpoint slides, flipcharts, storyboards, etc.–to use in the presentation.)	**Person or persons responsible for this portion of the presentation.**
List the highlights of your plan for the Measure stage. Include estimated timeline and any additional resources needed.		

Figure 7-9. Define Tollgate Preparation Worksheet

5. **Ask for volunteers and/or assign responsibilities** for each section of the presentation. Try to involve the whole team.
6. **Prepare an agenda** for the tollgate review. Identify information that will need to be sent to the reviewers ahead of time.
7. **Do a dry run** of the presentation to make sure it can be completed in the time allotted and to help team members get more comfortable with their role.

Related Tools: Use the brainstorming, affinity diagram, and multivoting instructions provided in Chapter 5.

Advanced Define Tools

When the stakes are really high and the future of your department or organization rests in the balance, the Define stage of a DMAIC process takes on extra import. Writing down a few customer requirement statements simply won't cut it. You have to be much more rigorous in your investigation of customers and their needs, and in defining specific requirements.

If your team is in this situation, you will probably want to investigate Quality Function Deployment (QFD), a method that moves from highly specific customer requirements to what needs to happen in the process to make sure those needs are met. While the detailed work involved with QFD is demanding and exhaustive (not to mention exhausting), the basic methods themselves are based on common sense principles and tools already seen.

The QFD Cycle

QFD is an iterative process for continually refining customer requirements to ever-increasing levels of detail and specificity. The cycle has four phases:

1. Translate customer input and competitor analysis into product or service features (basic design elements).
2. Translate product/service features into specifications and measures.
3. Translate product/service specifications and measures into *process* design features (how will the process deliver the features per specifications?).

4. Translate process design features into process performance specifications and measures.

Equally importantly, the relationship between all these elements is continuously evaluated through correlation and prioritization (see Figure 7-10). The degree of relationship between features, requirements, and process capabilities are then used to inform design decisions.

Eventually, the process leads to a very detailed version of Figure 7-10 that is called the House of Quality (shown schematically in Figure 7-11).

	PRODUCT/SERVICE FEATURES							
CUSTOMER REQUIREMENTS	choice of point sizes	variety of ink colors	choice of finishes (gold, marble, etc.)	3 price levels ($12, $40, $75)	available neck chain	non-toxic ink	retailed at jewelers & specialty shops	sold direct through Web
stylish			●	△	△		●	△
multiple choices	●	●	●	●	○		○	○
safe to use			△		△	●		
value for money	△		△	●			△	●
hard to lose			△		●			

Contribution: ● Strong ○ Moderate △ Weak

Figure 7-10. Simple L Matrix

As noted, generating the information needed for each of these rooms is a complex and detailed process—but the rewards lie in generating innovative solutions to meeting customer needs. It is processes like QFD that lead to true Delighters that will amaze your customers.

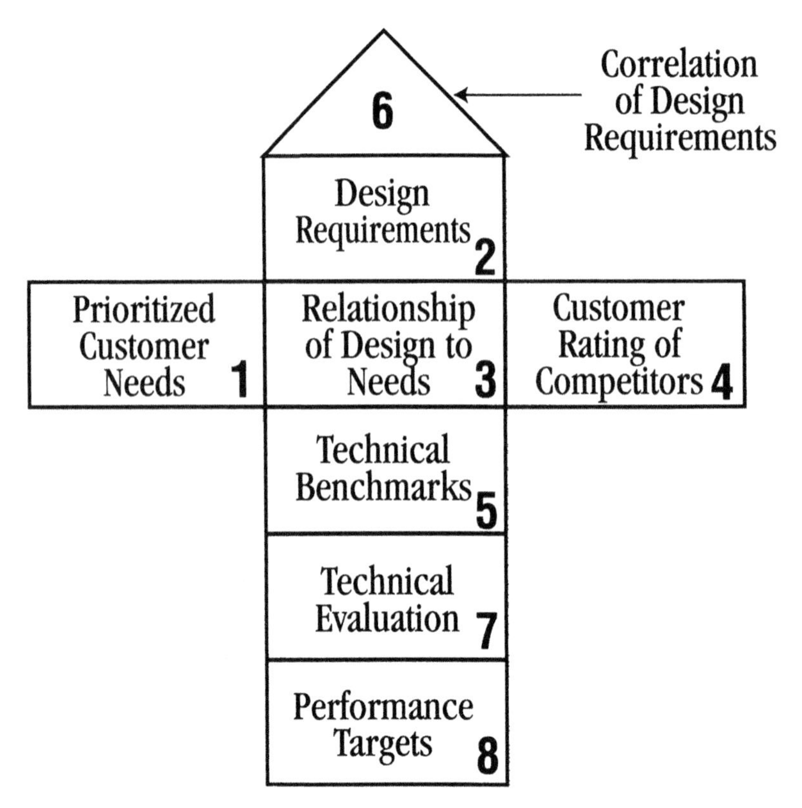

Figure 7-11. House of Quality

Chapter 8
Coming Together as a Team
Working Through the Forming Stage

In the Forming Phase of team development, Six Sigma team members gradually clarify the team's goals, their own jobs on the team, and their relationships with other team members, the Team Leader, and, perhaps, the team Champion. During this "honeymoon" phase, team members are usually excited about the chance to make some real improvements in their jobs. Though they may complain about having to do their Six Sigma team assignments plus their "regular" jobs, they secretly may be proud to have been selected as team members. They are usually anxious to get started on the project, and may start offering solutions even while the causes of problems are still unknown. Without good direction from the Team Leader and the use of DMAIC, the team might make little real progress or would jump to short-term solutions that turn out to be band-aids.

Tips for the Forming Stage

- Allow enough time for team members to get to know one another better. What may look like "socializing" now will build a reserve of trust and respect the team will need lots of later on.

- Collaborate (a fancy word, but a good one, meaning "work together") to create ground rules as soon as possible, and make sure every member expresses their understanding and support of them vocally.
- Use an agenda to give plenty of structure to meetings during Define.
- Get focused on the Project Charter and stick to it.
- The team needs to know where it's going. Create a DMAIC project plan with milestone dates at the first meeting.
- Get the project Champion to spend time with the team, but don't let the Champion take the team over! That's the Team Leader's job.
- Make sure that team members take turns leading parts of the meeting and passing around the job of facilitator, scribe, and recorder/minutes keeper. "Shared leadership," done carefully, can help everyone feel greater ownership of the project.

6σ

Many of the tools and skills described in Chapter 5 will help you implement these ideas. See also the DMAIC Project Charter worksheet in Chapter 7 (p. 103).

Troubleshooting and Problem Prevention for Define

Most of the causes of project failure can be traced to avoidable errors in the first and hardest step in DMAIC: Define. Omissions and defects in the Define stage can haunt the team throughout its work. Left uncorrected, they will prevent the team from achieving its project goal.

Here are the most common failures, their causes, and the preventive actions to be taken against them.

Failure #1: The Problem Statement Is Actually a Solution for an Implied Problem

Instead of finding the causes of the problem, the team tries one band-aid after another, usually without applying a permanent solution.

Why this happens: Champions or Black Belts working on their first DMAIC project often do not understand that *a problem is a gap in performance or high level of defects with unknown causes.* The problem can be due to unusual events (or "special" causes) or to a normal variation and performance of the process (or "common" causes). In either case, we do not know with certainty the *real* cause or causes, so the problem statement must be *only* a description of the symptoms generated by the underlying causes. "Car tire is flat" is a description of symptoms, while "Flat tire should be replaced" is a potential solution.

How to avoid it: Before they select problems or projects for teams, Champions and Black Belts should understand that they will be working on problems whose causes are unknown at the moment. This is a big change for people in most organizations, where saying you don't know the cause of a problem is usually not a milestone on the promotion fast track. The many measurement and analytic tools of DMAIC are designed specifically to probe and uncover buried causes of problems, not simply to implement pet solutions that may or may not satisfy customer requirements.

Failure #2: The Project Is Too Broad and Imprecise

The team flounders, spending its time trying to narrow the focus of the project.

Why this happens: This mistake is especially common for first-time Six Sigma teams and their Champions. Thinking that DMAIC has some "instant solution magic" in it, Champions assign teams projects that require an entire core process—say Sales or Marketing—to be overhauled. Partly, this arises from newness to the DMAIC process and ignorance of the fact that the first problems can't be huge ones—or the projects drag on forever. Also, of course, managers are accustomed to handing people big problems with the feeling that by aiming *big* they'll get more done. While that may work in some instances, it can be a bad formula for DMAIC projects. Rather than using a jackhammer to break up boulders, DMAIC is more like a laser cutting finer and finer slices until it uncovers causes of variation in processes.

How to avoid it: Experience usually reduces the frequency of this failure. On first teams, Champions and Black Belts need to know that they will be measuring and analyzing a good deal of data about the operation and variation of a process. Until their ability to use DMAIC tools has increased, they need to limit themselves to investigating key sub-steps of core processes that can be improved in a few months. Usually the availability of data or the time needed to collect it will help to keep the scope narrow. It is better for a team to have a string of small wins than a single home run, especially when you haven't even figured out the whole field yet.

Failure #3: Lack of Measurable Customer Requirements

The team *assumes* it knows what customers want and defines its project accordingly.

Why this happens: Finding out exactly what customers require and then putting measures to those requirements is not easy. Some teams think they already know what customers want when in fact the team is working from its own assumptions of what customers need. Then again, perhaps a customer survey was done a couple of years ago and the team decides that this is recent enough, and so goes for the convenient sample available to it.

How to avoid it: A large proportion of measurement and analysis in Six Sigma ultimately goes back to measurements of customer requirements. Teams must assume that what customers wanted last week may have changed. They must also find out how customers use the product or service provided to them by the team's process. Finally, they must translate what customers say they want into measurable requirements.

The Measure Stage

Chapter 9

Measuring Process Performance

Baselining and Refining the Problem Statement

WHEN A TEAM MOVES INTO MEASUREMENT, some members will comment, "But we can't measure that!" They are often referring to things that have never been measured before, such as services provided in direct customer contact—things we can see, but can't measure directly.

Of course, there *is* someone measuring what we do: our customers. Formally or informally, internal and external customers are constantly evaluating everything we do for them. With external customers, that evaluation includes "Do I want to keep giving this company my business?"

That's why it's necessary for teams to push past any initial denial around measuring what they do. Without facts and measures, the team will be lost in a sea of subjectivity and go nowhere.

Measurement is a key transitional step on the Six Sigma road, one that helps the team refine the problem and begin the search for root causes—which will be the objective of the Analyze step in DMAIC. Deciding what measures to take is often difficult, especially for teams working on their first Six Sigma project. Data collection can be difficult and time-consuming. It's easy to collect data you can't use or that doesn't tell you what you need to know. Initially, you may have no choice but to rely on educated guesses to identify what and where to measure; with experience, you will get better at knowing what kind of data to collect to help you answer specific questions such as "How is this process performing?" "What's the impact of variation on the customer?" Where are the causes of this problem?" etc. Just be clear that any data you collect should throw light on why or how your process does or does not meet customer requirements profitably.

Basic Measurement Concepts

If working with data and using measures is new to many of your team members, be sure to review the following basic concepts:

1. Observe first, then measure.
2. Know the difference between discrete and continuous measures.
3. Measure for a reason.
4. Have a measurement process.

Measurement Concept #1: Observe First, Then Measure

Even if some of your team members work on the process being studied every day, your first step should be to go and watch what happens with that process, or talk to people involved. You'll be amazed at what you learn simply by observing a process at work. You'll start to notice where people have to redo a step to correct errors. You may see the face of the customer who walks away either delighted or disappointed with the service. You might pick up on the fact that there is little consistency in how different people perform a step.

This observational experience will help you decide what and where to measure the process. It works whether you're interested in something as concrete as the dimensions of a brick or as elusive as attentiveness to a hotel guest's needs. Go stand in the hotel lobby and watch the interactions between guests and with hotel staff. Maybe you'll notice long lines at the check-in counter. If so, observe check-ins periodically for a week or two, record the time it takes for guests to check-in. Calculate the average check-in time and variation in the process, and start to interpret the data.

If we can observe an event (or even its effects) we can measure it. If we can measure it, we can improve it.

Measurement Concept #2: Continuous vs. Discrete Measures

Understanding the difference between "continuous" data and "discrete" (or, as it's sometimes called, "attribute") data is important, because the difference influences how you define your measures, how you collect your data, and what you can learn from it. The difference also affects the sampling of data and how you'll analyze it.

Sometimes the difference between these two types of measures may seem a little confusing, so we'll make the rule as clear as possible:

- **Continuous** measures are only those things that can be measured on an infinitely divisible continuum or scale. Examples: time (hours, minutes, seconds), height (feet, inches, fractions of an inch), sound level (decibels), temperature (degrees), electrical resistance (ohms), and money (dollars, yen, euros, and fractions thereof).
- **Discrete** measures are those where you can sort items into distinct, separate, non-overlapping categories. Examples: types of aircraft, categories of different types of vehicles, types of credit cards. Discrete measures include artificial scales like the ones on surveys, where people are asked to rate a product or service on a scale of 1 to 5. Discrete measures are sometimes called *attribute* measures because they count items or incidences that have a particular attribute or characteristic that sets them apart from things with a different attribute or characteristic: Is the customer male or female? Was the delivery on time or late? Was the address correct or incorrect?

The confusing part is that sometimes discrete data shows up disguised in continuous form. Say you find that 37.81% of your customers are between the ages of 66 and 70. Just because you've got decimals and numbers here doesn't make this continuous data. You're still counting people *who share one common characteristic or attribute*: they fall into the category called "age 66 to 70." The other 62.19% apparently fall into some other distinct age categories.

Here's a quick test for distinguishing between discrete and continuous measures. Think about the "unit of measure"—the *thing being measured*—and ask yourself if "half of that thing" makes sense. If the answer is yes, the measure is continuous; if no, you have a discrete measure. For example:

Unit of Measure	The "Half" Test
Customers who complain	"Half a customer" doesn't make sense. This is a discrete measure.
Hours lost to rework	"Half an hour" makes sense. This is a continuous measure.
Defects per application	"Half a defect" doesn't make sense. Discrete measure.
Impurities (ppm)	"Half a part-per-million" makes sense. Continuous measure.
Impurities (yes/no)	"Half an impurity" doesn't make sense. Discrete measure.

The second confusing issue is that some things that can be measured on a continuous scale are sometimes converted into discrete measures. For example, delivery times can be measured as "on time" or "late" (discrete categories) rather than in days, hours, minutes, and seconds (continuous data). On many automobile dashboards, oil pressure gauges showing continuous data have been replaced by warning lights that tell you when pressure is too low, versus OK. Figure 9-1 provides some examples of discrete and continuous data, and how continuous data can be converted to discrete.

The concept of discrete data is important because Six Sigma performance is based on measures of defects, which are usually discrete data, or measures that are converted to discrete items (such as "defects"). Ordinarily, continuous data is preferred because it gives a greater sense of the true variation in the process.

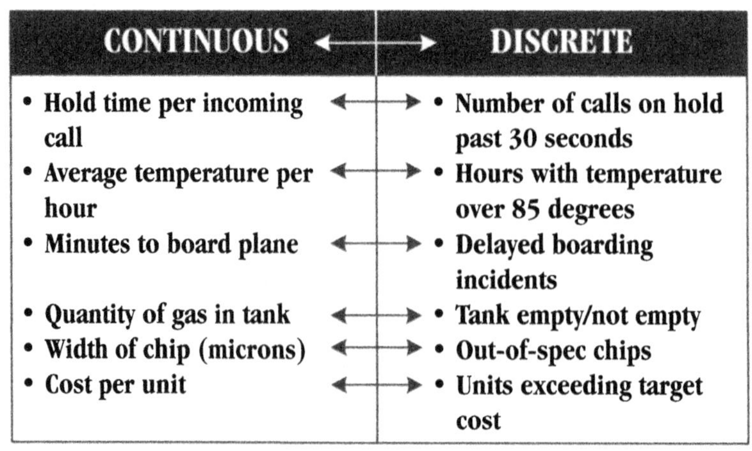

CONTINUOUS ←→	DISCRETE
• Hold time per incoming call ←→	• Number of calls on hold past 30 seconds
• Average temperature per hour ←→	• Hours with temperature over 85 degrees
• Minutes to board plane ←→	• Delayed boarding incidents
• Quantity of gas in tank ←→	• Tank empty/not empty
• Width of chip (microns) ←→	• Out-of-spec chips
• Cost per unit ←→	• Units exceeding target cost

Figure 9-1. Continuous and Discrete Measures of the same variable

Also, if you start with continuous data, you can always convert it to discrete categories by comparing the data against some threshold or criteria—"anything below 50 is a defect" or "any customer with revenues smaller than $500,000 is 'small.'" In contrast, if you start with discrete data, it's usually impossible to convert it to continuous data. However, discrete data does have some advantages. Here are some of the reasons in favor of or against discrete data.

The Pros of Discrete Data

- **Ease of collection.** Collecting discrete data is often easier and faster than collecting continuous data because you are measuring whether something meets a standard or not, like a pass/fail test in school, or people who say they liked a movie or didn't. Many business processes are set up to automatically record discrete data, such as locations (country, state, city, street), customer type (new versus repeat, home versus business user), product number, and product condition (damaged/undamaged).
- **Ease of interpretation.** Intangible factors that would be difficult to measure on a continuous scale can often be converted into discrete measures. Customer satisfaction surveys convert the intangible "customer attitude" into a scale of 1 to 5.
- **Ease of determining sigma performance level.** A sigma calculation tells you how many defects fall within customer requirements—which is an attribute measure (in or out of spec; "OK" vs. "defective").

The Cons of Discrete Data

- **A loss of precision.** Which would you rather have, a doctor who puts her hand on your forehead and says you're "feverish" or one who takes your temperature with a thermometer and says your temperature is 102.3 degrees? Continuous data offers more precision than discrete data and, if your time and resources allow it, you'll want to capture continuous data whenever you can.
- **The need to collect more data.** Interpreting discrete data is basically a question of uncovering patterns in the data categories. And you need a lot of data to accurately judge whether a pattern exists. For example, you should have 50 or 100 data points to use even simple tools like a Pareto chart or some types of bar charts. (In contrast, tools based on continuous data—such as frequency plots and run charts—can often be interpreted

with far fewer data points.) The need for more data will stretch out your data collection chores. And the closer your process is to Six Sigma—as you approach as few as 50-60 defects per million opportunities to have something go wrong—you'll have to collect lots of data just to catch the defects.

- **Increased likelihood of missing important information.** Because of its "either/or" nature, discrete data can hide important detailed information about a service or product.

In short, whenever possible, start with continuous data if time and your budget allows.

Measurement Concept #3: Measure for a Reason

Ever notice how much useless data gets collected at work? It's probably because the computer has made it easy to collect tons of numbers, however trivial. But don't let your team get sucked into that quagmire. Unless there's a clear reason to collect data—a key variable you want to track—don't bother. There are basically two reasons for collecting data:

1. **Measuring efficiency and/or effectiveness**

   - Looking at measures in terms of efficiency and effectiveness keeps your team focused on who will benefit from your improvement efforts: your organization, your customers, or (we hope) both.
   - Efficiency measures focus on the volume and cost of resources consumed in your processes, and on the improvements you've made inside the process resulting in lower costs, less time, fewer materials and staff, etc. Your own organization benefits directly from such reductions, but they will benefit your external customers only if you pass the savings along in some form.
   - Effectiveness measures reveal what your product or service looks like to the customer. How closely have you met or exceeded their requirements? What defects were delivered to them?

2. **Discovering how variables (Xs or causes) upstream in the process affect the outputs (Ys or effects) delivered to the process customer.**

   - This can be described as looking at the relationship between "Predictors" and "Results" (or "leading indicators" and "lagging indicators"). It's typical to start by measuring outputs or results delivered to customers—be it "good" or "defective." In the course of your project, you'll then work your way back into the process to discover measures that predict certain outcomes. For example, a measure of internal cycle time can be a predictor of decreased customer satisfaction if the cycle time means we'll deliver our product late to the customer.

Being clear about what you want to accomplish with data is the first step in making sure you will be measuring for the right reasons. In choosing measures for your Six Sigma project, make sure you have not focused only on output measures, but have a balance between output and process measures, predictor and results measures. Teams (and their Champions) are tempted to boost efficiency (with its quick bottom-line impact) by streamlining internal processes while forgetting the long-term impact of such measures and changes on the customer.

Measurement Concept #4: A Process for Measurement

Remember the old carpenter's saying, "Measure twice; cut once"? It should remind us of the importance of getting our measures right the first time. There is nothing more tedious and frustrating than having to collect data a second time because it wasn't done right the first time. Treating data collection as a process that can be defined, documented, studied, and improved is the best way to make sure you only have to "cut once." A detailed process is described below.

Two Components of Measure

The guidelines for data collection given above have been incorporated into the two procedures that comprise the Measure stage of DMAIC:

A. Plan and measure performance against customer requirements.
B. Develop baseline defect measures and identify improvement opportunities.

A. Plan and Measure Performance Against Customer Requirements

The following five-step measurement collection plan can help you avoid the most common problems with data collection:

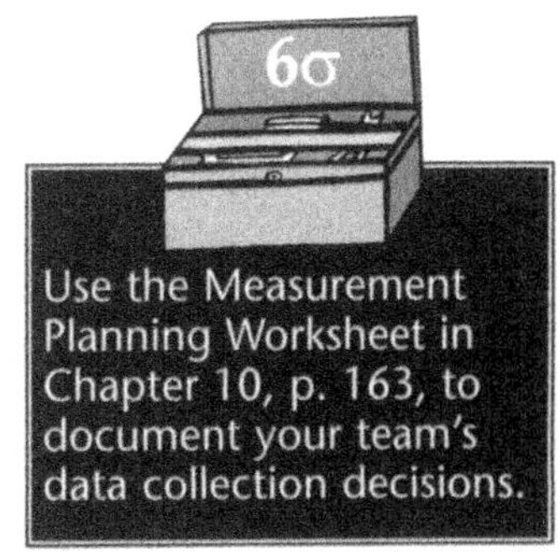

Use the Measurement Planning Worksheet in Chapter 10, p. 163, to document your team's data collection decisions.

Step A1. Select what to measure.
Step A2. Develop operational definitions.
Step A3. Identify data sources.

Step A4. Prepare a data collection and sampling plan.
Step A5. Implement and refine the measurement process.

Each of these steps is described in more detail below.

Step A1. Select What to Measure

In the Define stage, your team identified the primary problem your DMAIC project will tackle, along with critical customer requirements. Here, start by measuring to validate or refine your understanding of the size and frequency of the problem, along with how well you are meeting the customer requirements (these may be one and the same). In most cases these initial measures will amount to counting the defects that show up in your process's outputs. Your team will also want to measure the performance of those process steps that seem to contribute most to the output defects (the suspected Xs). In general, pay attention to two areas in selecting measures: what's valuable for analyzing the problem and what's feasible to collect. Figure 9-2 shows some criteria for selecting measures.

Value/Usefulness	Feasibility
• Link to high priority customer requirements • Accuracy of the data • Areas of concern or potential opportunity • Can be benchmarked to other organizations • Can be a helpful ongoing measure	• Availability of data • Lead time required • Cost of getting the data • Complexity • Likely resistance to "fear factor" associated with a particular type of measure

Figure 9-2. Criteria for useful measures

What key questions do you need to answer? What data will provide the answers? What output or Service requirements will best gauge performance against customer needs? What upstream variables in the process will predict problems downstream in the process? How will we display/analyze the data?

Since this is a DMAIC project, one set of data you gather should focus on defects, because you'll need that data for Step B, in which you refine your process baseline measure—often accompanied by determining baseline sigma levels.

Using the CTQ Tree to Identify Measures

Another approach to identifying measures that relate to customer requirements is called the CTQ Tree. This diagram is like a tree chart except here the focus is on defining measures that are "critical to quality." Figure 9-3, for example, shows where the team was interested in measuring "timely resolution of service disruptions." They decided that the measure had two broad components (disruptions fixed per day, and time to restore service), then identified specific, easily measurable data they could collect that would allow them to measure those broad components.

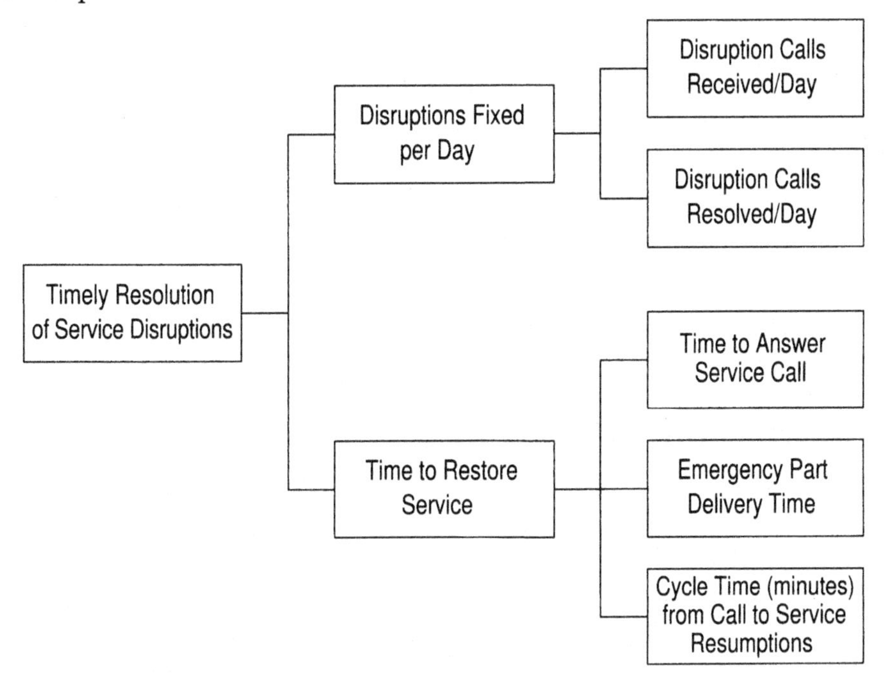

Figure 9-3. CTQ Tree

Identify Potentially Related Factors (Stratification)

Imagine that you are collecting complaint data about a product sold nationwide by your company. Now imagine you have that data in hand. What are some questions you'd like that data to answer? For example ...

- Are there differences by state or region?
- Are there differences by demographic factors such as gender, age, or income level?
- Are there differences by month of purchase?

All of these questions represent different ways you will want to "slice and dice" your data once you have it in hand. In Six Sigma terms, they are **discrete** categories (e.g., state, gender, month) that you will use to **stratify** the data.

Why is this important? Stratification of information can give you clues about where to look for the causes of problems. For example, you might discover that customers in the southwest complain twice as often as those in other parts of the country. That would raise a host of questions for your team: what is it about people in the southwest and how they obtain or use your product that is different from everyone else?

Here's another example. In one recent nationwide product recall, the cause of the problem wasn't uncovered until the investigators stratified their data first by product type, then by product size, then by manufacturing plant, and finally by product age at time of failure. Just having data on the number of products that failed couldn't lead to a solution; this company needed all of that stratification information—product type, size, manufacturing plant, and product age—to dig out the root cause.

Now here's the key lesson: **You can't stratify data unless you gather the stratification information at the same time as you gather the data.** So you need to think about what stratification data you are interested in beforehand, and build those questions into your data collection plan.

There are endless ways to stratify a set of data: knowing which ways will be most useful to your team is five parts experience and five parts guesswork. You'll find some instructions in Chapter 10 to help you get started.

See Stratification Instructions on p. 165.

Measurement Assessment

A third approach to identifying data that may be useful to your team combines the concepts of a CTQ Tree and stratification. It's called the Measurement Assessment Tree (see Figure 9-4).

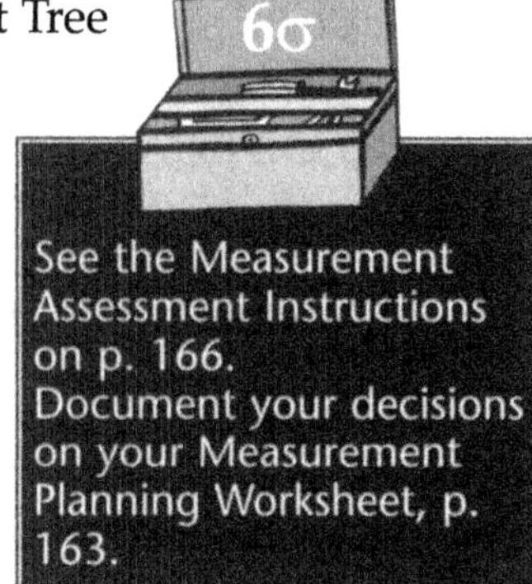

See the Measurement Assessment Instructions on p. 166.
Document your decisions on your Measurement Planning Worksheet, p. 163.

This tree started with an important output for the Six Sigma Pizza Company: number of late deliveries. The metrics—such as "# late by region"—can be directly linked back to the key defect (number of late deliveries) and to important questions about that defect ("what trends or patterns do we see?").

This kind of Measurement Assessment Tree helps your team keep a clear connection between what

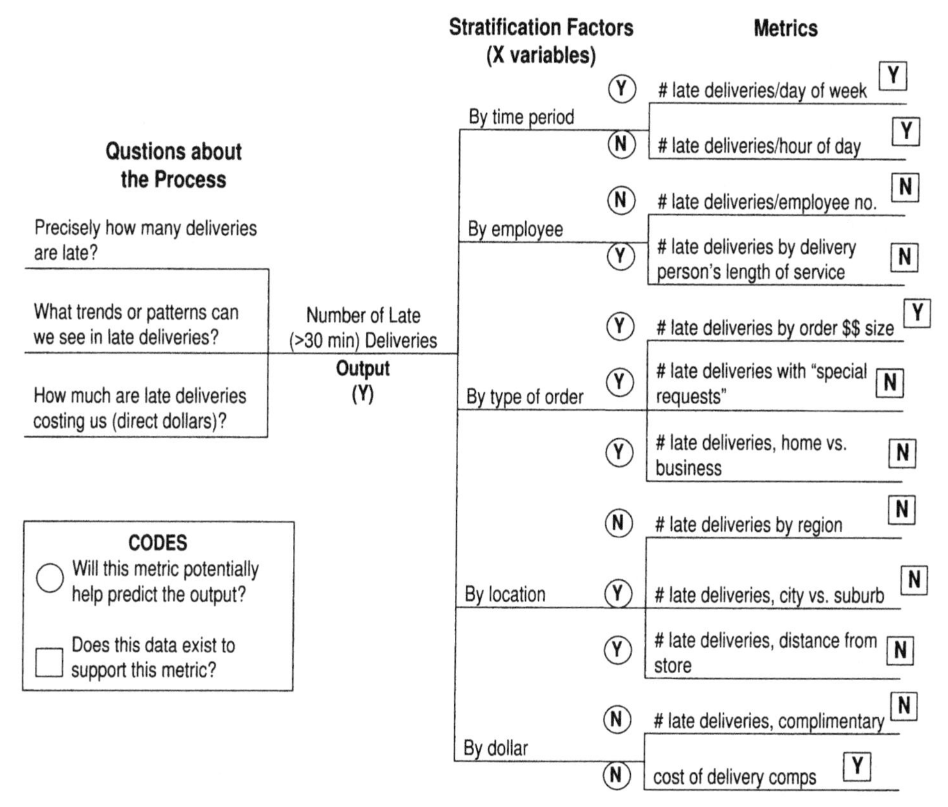

Figure 9-4. Measurement Assessment Tree

you're trying to accomplish and what data will help you get there. By linking the various levels of detail in defining a measure, you can avoid measuring the wrong things and improve the odds of measuring the right things correctly.

Before moving on...

At the end of *Measure Step A1: Select What to Measure*, you should have:

- ✔ At least one target measure linked to the project problem and goal—often called the "project Y."
- ✔ Potentially one or more target measures of *predictors* or Xs in the process or inputs that you suspect may help you find or narrow the causes of the problem.
- ✔ An idea of where you'll gather the data and a fair degree of certainty that the data is available and feasible to be collected (to be refined in coming steps).

Step A2. Develop Operational Definitions

Saying that your team will count the number of defects in a product or service is easy. But what exactly do you mean by "defect," "product," and "service"? Without having precise definitions for the things you're trying to measure, different people will count different things in different ways. To avoid this confusion, you need to have operational definitions:

> **Operational definition**: a clear, understandable description of what's to be observed and measured, such that different people taking or interpreting the data will do so consistently.

In a recent example of a failure to have such a clear definition, recall the Mars Polar Orbiter that crashed onto the planet surface because one group of engineers had written procedures in English units (pound-seconds) and the computer interpreted the data in metric units (newton-seconds). Or think back to the Florida ballot recounts in the presidential election of 2000: how consistently do you think people interpreted a "pregnant chad"?

The purpose of your operational definition is to translate what you want to know into something you can observe and measure. (As noted, nothing can be measured until it can be observed.)

You'll find instructions for writing an operational definition in Chapter 10. To make sure your operational definition is airtight, factor in ...

- Different ways people interpret the same words.
- The ability to stay focused on what needs to be observed and/or measured.
- Changes or situations that might emerge that require special interpretation.
- Events or observations that can fit under more than one grouping or that might be interpreted/measured several ways.

6σ

See the Operational Definition Worksheet on p. 169. Document your decisions on your Measurement Planning Worksheet, p. 163.

Step A3. Identify Data Sources

There are two main sources of data available for the team:

1. Data that is already being collected in your organization and has been around for some time (usually called "historical" data).
2. New data your team collects now.

Before moving on...

At the end of *Measure Step A2: Develop Operational Definitions,* you should have:

- ✔ A clear, detailed, and unambiguous description of what's being measured.
- ✔ Guidelines for data collectors on how to interpret routine and/or unusual instances or items.
- ✔ An initial plan for collecting the data (when and how)—still to be finalized in coming steps.

Historical data can be handy, when you have it—it requires fewer resources to gather, it's often computerized, and you can start using it right away. But be warned: existing data may not be suitable if...

- It was originally collected for reasons other than improving a process or detecting defects.
- It was collected using different definitions and methods than what your team has developed.
- The data is structured in a way that makes it hard to apply to your needs. For example, key "stratification factors" may not be present. Or the database may not have the capability to sort the specific data you want.

To see if historical data will meet your needs, compare your new operational definition with whatever definition was used at the time the old data was collected. You may be able to adjust your operational definition to fit what's in the data—as long as it still works in giving the answers you need. Checking the existing data will require a little research, but it'll be worth it (for example, to make sure that what the shipping department counted as "late deliveries" is the same thing that your customers who *take* deliveries count as "late").

Step A4. Prepare a Data Collection and Sampling Plan

A data collection and sampling plan covers three main issues:

A4.1: Identify or confirm the stratification factors.
A4.2: Develop a sampling scheme.
A4.3: Create data collection forms.

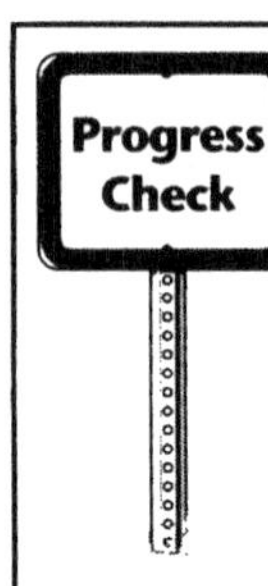

Before moving on...

At the end of *Measure Step A3: Identify Data Sources,* you should have:

- ✔ A decision on whether you need to gather new data or can rely on existing or historical data.
- ✔ Validation of the ability to access and sort existing data (if that's the choice).

Step A4.1: Identify or Confirm the Stratification Factors

Stratification was introduced earlier in this chapter (pp. 135-136) as a technique for gathering information that can help in tracking down clues to root causes. Your decisions about what stratification questions you're interested in will affect how you gather and sample data. For example, if you want to compare survey results between the Central Rockies and West Coast regions, you need to make sure that you've collected enough information from both regions to draw valid conclusions.

If you have already identified stratification information, you just need to confirm those decisions here. If not, make sure that you *don't* want to gather such information before proceeding with your sampling strategy.

Step A4.2: Develop a Sampling Scheme

Deciding how you will collect sample data involves logical thinking about how your potential data sources are structured. In the example below, pay attention to the italicized words—they reflect key concepts in sampling.

A Brief Sampling Saga: HomeHealth Products

HomeHealth Products distributes medical supplies and equipment to healthcare resellers throughout the United States. As part of its implementation of Six Sigma in its eight regional centers, HHP (as its employees call it) has taken on a chronic problem that has been an issue for the company and its reseller clients for years: the variance between what appears on the bills of lading HHP sends with its shipments and the data for the same materials that appears in the internal logistics computer system. The differences between the numbers lead to billing and inventory problems that a recent survey of HHP's top customers described as "unprofessional," "intolerable," and "likely to lead to non-renewal of existing contracts."

"The Voice of the Customer is talking loud and clear," said Bill Wrigley, Directing Manager of the Eastern Regional Center, to Ed Magos, who managed several of the processes under review, and Jane Medawar, a Six Sigma Black Belt who had been leading a team working on the problem for the last few weeks. The team had finished the "Define" stage and was about to get into "Measure."

"Just measuring the problem is a bear," said Ed Magos. "We take in nearly 800 shipments at the warehouse every day, and send out 1000 shipments of our own to the resellers every day."

"Obviously you can't check them all," said Bill Wrigley. "You'll have to do some sampling."

"Our team is putting together its data collection plan now," said Jane Medawar, "and sampling is part of the plan. But we're worried about *bias* getting in the way of a good sample."

"I don't know about bias," said Bill. "What you need is a good sampling plan."

Ed suggested that the team might sample the shipments by having the packers and shippers collect data when they're not so busy, like on the swing shift.

"We talked about that," said Jane, "but the data will be biased if we just collect data when it's *convenient* for us."

"Well," said Bill, "I'm no statistician, but why not have the shippers pick out the shipments they think are most representative of a normal day? After all, they know more about shipping than the rest of us."

"I agree," Jane replied, "and we've got three people from shipping on the team. But when I asked them which shipments they *judged* most representative of a normal day, they all had different ideas and disagreed with one another."

Bill seemed a little frustrated as he spoke next.

"So how are you going to sample the shipments, Jane?"

"We've pretty much decided to take a *systematic* approach. We'll take some measures on every tenth or twentieth shipment. We haven't figured out yet exactly how often or how many we'll look at. That's coming up at the next meeting. Once we've figured that out we'll have more confidence in our data collection plan."

"I know you're the Black Belt," said Ed, who was the project Champion and Jane's boss, "but isn't it always better to take a *random* sample?"

Jane smiled. The team had been through all this at their last meeting.

"In theory, yes. Random is best, but in practice it's not all that easy. Systematic sampling will help us spot any patterns in the data from the shipments. Of course, we'll have to make sure our systematic sample isn't based on some hidden pattern itself, like always picking samples right after break or something like that."

"Well, Jane," said Bill, "it sounds like you've thought it through. Good luck. If we can get some improvement on this problem it'll save us a ton of rework, and maybe just two or three key customers."

Jane's team still had some work to do on the sampling plan....

In data collection, sampling means measuring some of the items in a group or process to represent them all. If you're processing large numbers of things, you can't count every item. But if you only pick out a few, how can you be confident they represent the whole group?

If you've taken a statistics course, you know how big a part sampling plays. Most college statistics courses focus on "population statistics," which assume that we have a large, standing pool of water (or data), and that if you take a dipperful at any point, it will represent all the rest of the water (or data).

In the business world, however, there isn't much standing water, so your team may be more interested in "process statistics," or taking a sample from a running stream of water that may be changing minute by minute, depending on what you're measuring. Table 9-1 shows examples of both types.

Whether you're sampling a whole population of data or a moving process,

Population Sampling		Process Sampling
◆ Tallying the average loan amount from a group of applications	→	◆ Capturing the average loan amounts requested by day, week, month
◆ Recording the age of all parts of inventory currently in stock	→	◆ Tracking the average age of parts inventory by week
◆ Conducting a survey of customer perceptions	→	◆ Polling every 10th customer on his/her service experience each day
◆ Compiling all the reasons for inbound calls among all calls over the past six months	→	◆ Recording inbound call volume every quarter-hour

Table 9-1. Examples of Population and Process Sampling on similar items

you still need a valid sample—one that actually represents the whole. The next few pages cover some sampling basics to get you started. If your situation is complex and the results are critical, contact a statistical expert for specific advice.

Sampling Concept #1: Bias

Bias is the difference between the data collected in a sample and the true nature of the entire population or process flow. Bias that goes undetected will influence your interpretation and conclusions about the problem and process. Here are some examples of types of sampling bias:

- **Convenience sampling:** Collecting data simply because it's easy to collect. Example: collecting data every afternoon 20 minutes before closing time because "things are quiet then," thus ignoring data when things are busy—which might be very different data.
- **Judgment sampling:** Making educated guesses about which items or people are representative of your process. Example: surveying only those customers who scored high on your last customer satisfaction survey.

Better sampling strategies—better able to avoid bias—include:

- **Systematic sampling:** This is the method we'd recommend for most business processes. By systematic sampling of a process we mean taking data samples at certain intervals (every half-hour or every 20th item). A systematic sample from a population would be to check every 10th item in the database. But beware! Make sure your systematic sampling doesn't correspond to some hidden pattern that will bias the data. Example: Sampling every 10th insurance claim might mean that you always get claims reviewed by the same clerk on the same computer, while ignoring four other clerks and their computers.
- **Random sampling:** By random we mean that every item in a population or a process has an equal chance to be selected for counting. Selecting data randomly has its own challenges, not least of which are unconscious biases or hidden patterns in the data. Most random sampling is done by assigning computer-generated random numbers to items being surveyed.
- **Stratified sampling:** Suppose your company had a customer base of 10,000 purchasers, and your job is to survey a sample to determine customer satisfaction. Are you equally interested in what all 10,000 customers have to say? The answer is probably "no." It's likely more impor-

tant that you understand the needs and perceptions of your biggest customers or most reliable purchasers than it is that you find out what a one-time customer thinks. (The opposite would be true if you were trying to expand market share; in this case, you might want to understand how you could convert infrequent customers into regular purchasers/users.)

If there are cases like this where there is structure in the population or process flow, you can develop a *stratified sampling scheme*—either random or systematic. In the customer satisfaction example, that might mean dividing the 10,000 names into, say, four groups: large regular purchasers, small regular purchasers, infrequent but recent purchasers, and lapsed customers. You could then use different schemes to *randomly* sample from each of these groups, perhaps choosing one out of every five large regular purchases but only a handful of lapsed customers.

See the Sampling Definitions and Worksheets on pp. 170-173.
Attach your sampling plan to your Measurement Planning Worksheet, p. 163.

A stratified sample helps avoid the gaps that can come up if data are collected over a large population where key subgroups of the population are underrepresented.

Sampling Concept #2: Confidence Level or Interval

This is your level of confidence that the data you sample actually represents the entire population or process under study, provided the sample was gathered randomly. In the business world, a confidence level of 95% is standard. It means that you have five chances out of 100 of drawing the wrong conclusion from your data.

There's a "Catch 22" to developing a good sampling plan: You *already have to know something* about the data you're collecting *before* you collect it. As a result, your first measures won't be as reliable as you'd like because they're based on educated "guesstimates." The longer you take measures, however, the better you'll know your process and the better your sampling plan will be.

Instructions for Developing a Sampling Plan

The Sampling Worksheets in Chapter 10 will help you develop a sampling scheme appropriate for your project. As a preview, you will need to know whether you are measuring from a *process* or *population*; the form itself will walk you through the decisions necessary to determine an appropriate sampling scheme.

If you have trouble making the decisions required to complete the sampling worksheets, contact a statistician familiar with your processes and operations.

Step A4.3: Create Data Collection Forms

Now that you've made decisions about what data and stratification information you want to collect and what sample sizes are appropriate, you need to document those decisions on a Data Collection Form. Spreadsheets or checksheets are the workhorses of data collection. While each checksheet will vary depending on the data collected, the following guidelines will help you avoid some common pitfalls in collection forms:

See Checksheet Development Instructions on pp. 175-176.

- **Keep it simple**. If the form is cluttered, hard to read, or confusing, there's a risk of errors or nonconformance.
- **Label it well**. Make sure there is no question about where data should go on the form.
- **Include space for date, time, and collector's name**. These obvious details are often omitted, causing headaches later.
- **Organize the data collection form and compiling sheet** (the spreadsheet you'll use to compile all the data) **consistently**.
- **Include key factors to stratify the data**.

Common types of checksheets include …

- **Defect or Cause Checksheet**. Used to record types of defects or causes of defects. *Examples:* reasons for field repair calls, types of operating log discrepancies, causes of late shipments.
- **Data Sheet**. Captures readings, measures or counts. *Examples:* transmitter power level, number of people in line, temperature readings.
- **Frequency Plot Checksheet**. Records a measure of an item along a scale or continuum. *Examples:* gross income of loan applicants, cycle time for shipped orders, weight of packages.
- **Concentration Diagram Checksheet**. Shows a picture of an object or document being observed on which collectors mark where defects actually occur. *Examples:* damage done to rental cars, noting errors on application forms.

- **Traveler Checksheet**. Any checksheet that actually travels through the process along with the product or service being produced. The checksheet lists the process steps down one column, then has additional columns for documenting process data. Figure 9-5, for example, shows a simple traveler checksheet where the team monitoring how long it took loan applications to complete each step of the process (*time* information should almost always be collected) and the number of defects found. Some examples of traveler checksheet uses are capturing cycle time data for each step in an engineering change order, noting time or number of people working on a part as it is assembled, tracking rework on an insurance claim form.

Before moving on...

At the end of *Measure Step A4: Prepare a Collection and Sampling Plan,* you should have:

✔ A list of stratification factors that are potentially important to your project.
✔ A completed sampling plan.
✔ Data collection forms.

Step A5. Implement and Refine the Measurement Process

There are five steps in implementing and refining the measurement process:

A5.1: Review and finalize your data collection plans.
A5.2: Prepare the workplace.
A5.3: Test your data collection procedures.
A5.4: Collect the data.
A5.5: Monitor accuracy and refine procedures as appropriate.

Step A5.1: Review and Finalize Your Data Collection Plans

Complete your Measurement Planning Worksheet. Determine how you will assess the accuracy and reliability of the measurements. Measurement practices and measuring devices themselves are subject to variation. No matter how well you train people, it's likely they will vary slightly in how they collect data; and instruments are known to degrade in precision over time. Six Sigma teams need

Traveler Checksheet
Loan Application - Underwriting

Loan # 3256-879

Loan Type: ☒ Conventional ☐ Jumbo ☐ VA/FHA

Amount Requested 194,000

Customer Location ☐ NW ☐ W ☒ SW ☐ E

Process Step	Date/Hour Received	Defects Found
Application Completion	0623/13:42	|||
Packet Preparation	0626/09:00	||||
Underwriting	0715/16:30	~~||||~~ |

Figure 9-5. Example of a Traveler Checksheet (one that is attached to a document or product as it goes through a process)

to take this variation into account. Especially if you repeat measurements over time, you'll have to keep an eye on several factors:

- **Accuracy:** How precise is the measurement: hours, minutes, seconds, millimeters, two decimal places?
- **Repeatability:** If the same person measures the same unit with the same measuring device, will they repeat the same results every time they do it? How much variation is there between measurements?
- **Reproducibility:** If two or more people or devices measure the same thing, will they produce the same results? What is the variation?
- **Stability:** How much do accuracy, repeatability, and reproducibility change over time? Do we get the same variation in measures that we did a week ago? A month ago?

Part of your plan should include procedures to ensure that your data continue to be valid throughout the data collection process.

Step A5.2: Prepare the Workplace

Explain clearly why you're gathering the data. Describe what you plan to do with the data—including your plan to share the results with the data collectors, keeping identities confidential, and the like.

Step A5.3: Test Your Data Collection Procedures

Be careful whom you choose as data collectors; avoid making data collection a reward or punishment. When you collect new data using your operational definitions, it's best to experiment a little at first, gathering data manually from people in the process. (There's a temptation for many teams to create an elaborate IT computerized system right away. But you'll learn more and be able to better refine your system if you do some initial data collection with pencil and paper or your own laptop.)

In manufacturing, accuracy of gathering continuous data is ensured through the calibration of measuring devices. With discrete measures—either in service or manufacturing—one method of testing repeatability and reproducibility is to have one or more people count the number of defects on documents that have been carefully inspected before by an expert who has counted the precise type and number of defects. The results of the counters are then compared to those of the expert, and the variation measured.

Step A5.4: Collect the Data

Implement your plans. Remember that part of your plan includes the "sample size," that is, the number of data points you have to collect. Your data collection should stop when you've reached the appropriate sample size, unless there were problems with some of that data. Do not continue to collect data unless there are plans to make it a standard part of the process.

Step A5.5: Monitor Accuracy and Refine Procedures as Appropriate

Throughout the data collection, be sure to monitor both the procedures and devices (if any) used to collect the data.

Before moving on...

At the end of *Measure Step A5: Implement and Refine the Measurement Process*, you should have data in hand that:

- ✔ Meet your data collection priorities.
- ✔ Were sampled according to your plan.
- ✔ Reflect accurate, repeatable, reproducible, and reliable measurement practices.

B. Develop Baseline Defect Measures and Identify Improvement Opportunities

The measurement and sampling methods described above are important any time your team is gathering data. At this point on the road to Six Sigma, however, your team needs to baseline the performance of the process under investigation—determine how the process and product/service are working today, before you start making changes. You'll gauge improvements against this baseline later.

Instructions for completing this process appear later in this chapter (starting on p. 150). Before you begin, however, you need to understand what it means to measure the sigma performance of a process. Typically, people start by looking at measures for process outputs, then work their way upstream to look for measures of how the process itself is performing.

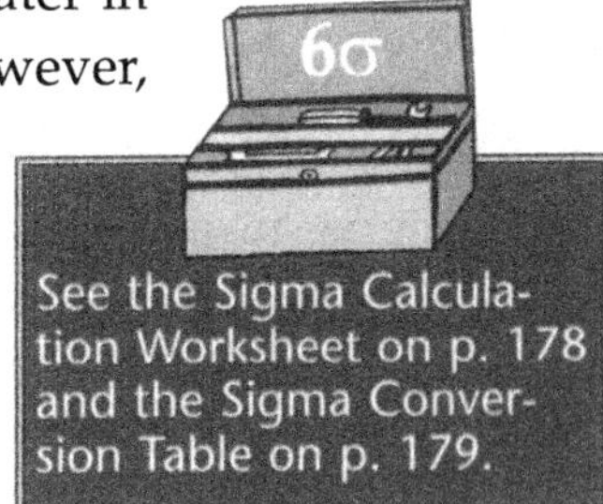

See the Sigma Calculation Worksheet on p. 178 and the Sigma Conversion Table on p. 179.

Output Performance Measures

Six Sigma performance measures are most often based on defects produced by the process. There are several advantages to basing measurements on defects, including:

- **Simplicity:** Anyone who can understand "good" and "bad" can understand "good" and "defective."
- **Consistency:** Defect measures apply to any process for which there are customer requirements, whether we are measuring manufacturing or services, using continuous or discrete data.
- **Comparability:** Motorola and other Six Sigma companies use defects to track and compare performance in very different areas across the business.

The same measure allows teams to measure their improvements over the course of their projects and beyond.

A drawback of looking only at good versus defective is that defect counts may hide important variations in the numbers—especially with continuous data like time. The aim here is not to burden you with lots of qualifiers, but to give you a good foundation for measuring what's actually going on in your processes and determine the causes of problems and unwanted variation.

Here are three steps that will help translate the concepts about sigma capability into concrete numbers useful to your team:

Step B1. Calculate baseline sigma levels for the process as a whole.
Step B2. Calculate final and first-pass yield.
Step B3. Determine the "Cost of Poor Quality."

Step B1. Calculate Baseline Sigma Levels for the Process as a Whole

Calculating baseline sigma for process, product, or service is a simple four-step process. You'll need to do some simple math and consult one conversion table. A worksheet for the sigma calculation is in Chapter 10 (p. 178); you'll need to be familiar with the following terms and concepts in order to complete that worksheet.

The Key Definitions: Units, Defects, and Defect Opportunities

The Six Sigma team needs to understand a few key terms both to collect and analyze data used to determine the capability of its process:

- **Unit:** An item being processed, or the final product or service being delivered either to internal customers (other employees working for the same company as the team) or external customers (the paying customers). *Examples*: a car, a mortgage loan, a computer platform, a medical diagnosis, a hotel stay, or a credit card invoice.
- **Defect:** Any failure to meet a customer requirement or performance standard. *Examples:* a poor paint job, a delay in closing a mortgage loan, the wrong prescription, a lost reservation, or a statement error.
- **Defect Opportunity:** A chance that a product or service might fail to meet a customer requirement or performance standard.

Of these terms, defect opportunity is the trickiest to implement and most critical for calculating a reliable sigma capability figure. The defect opportunity component of a Six Sigma calculation is what enables us to compare processes of different complexity. As a simple example, consider two people who are both making phone calls. Rita is calling local numbers and only has to dial seven numbers per customer. Gordie has a long-distance calling card (with a PIN) and is making international calls. He often has to dial two or three times as many numbers as Rita just to complete one call.

The data show that both Rita and Gordie have one wrong number for each 100 calls they make. But is that an accurate comparison? In Six Sigma terms, Gordie's process has many more "opportunities" for making a mistake when dialing a phone number because he has to dial so many more numbers. Looked at another way, Rita is making one mistake in every 700 digits she dials; Gordie is making one mistake in every 1,400 to 2,500 numbers he dials. So whose process has fewer defects?

In a world where Murphy's Law is widely understood, you might think that there are thousands and thousands of chances for things to go wrong. In reality, that is seldom the case, though in some very complex systems it could actually be more than that.

Two guidelines prevent the use of defect opportunities from becoming a nightmare:

- First, you need to focus on **defects that are important to the customer**. Consider a bank that regularly makes two kinds of mistakes: (1) mailing out monthly statements a day late, and (2) entering interest payments a day late. Which of these defects do you think is really important to most customers? So we link opportunities in most cases to a CTQ Tree (see p. 135).
- Second, defect opportunities reflect the **number of places where something in a process can go wrong, not all the ways it can go wrong**. So, for example, you would define "wrong address" as one opportunity for a defect on a database record rather than describing all the ways in which that address could be wrong (incorrect street number, street name, wrong ZIP code, etc.). The more opportunities we add, the more things look fishy—because more opportunities mean better-looking performance.

Here's an example. Any time a clerk types a form, application, report, etc., every key stroke could theoretically be counted as a *defect opportunity*. However, that is not only impractical, but it would also clump important defects along with lots of unimportant defects. In many cases, reports and forms like applications have standard templates that are filled automatically with identical text. The key in these kinds of situations would be to focus on defects that are important to and would be noticed by either the customer or the next step in your process.

For example, here's a list of 16 opportunities for defects on a generic invoice:

Customer name	Total price
Contact name	Tax
Customer address: street and number, city, ZIP, mail stop	Shipping costs
	Payment due date
Account number	Remittance address
Purchase order number	Printing errors
Items ordered	Folding/stuffing errors
Quantity of items ordered	On-time delivery of invoice to customer
Discounts	

Your team might feel that list of defect opportunities is still too long: it might be difficult to measure all these defects, and some of them are less important to customers than others. So another option would be to combine the various individual defects into four categories of opportunities, as follows:

1. Customer data (name, billing address, purchase order number)
2. Order information (items, quantity, shipping address)
3. Pricing (unit price, tax, discounts)
4. Production (print quality, mailers included)

The shorter list of categories still reflects the complexity of the invoice and the importance of the information to the customer, but makes it easier to capture the data. In fact, as long as the team is consistent in the counting of defect opportunities, and your reasoning is sound, you can make a case for either 16 or four as the number to use.

For the sake of convenience and practicality, it makes sense to define defect opportunities in a way that keeps the number fairly low. On the other hand, you can artificially inflate your sigma level by making the number high.

Consider this example. You know your order entry process generally results in three typos per form. If, like the invoice example above, you define four defect opportunities on the form, you're looking at a defect rate of three out of four—and a sigma level near zero! On the other hand, if you count every key stroke as an "opportunity," you're looking at something closer to three defects per 200 opportunities, and a sigma level of about 3.7. Which sigma level would you rather report? Which would you rather *have*?

Tips for Defining Defect Opportunities

Here are a few more tips on using defect opportunities for your products and services:

- **Focus on "routine" defects:** Defects that are extremely rare shouldn't be considered as opportunities.
- **Group closely related defects into one opportunity category:** This will simplify the work of gathering data.
- **Be consistent:** As Six Sigma spreads throughout your company, you should consider using standard definitions of defect opportunities across the board.
- **Change definitions only when necessary:** The team will use the number of defect opportunities to calculate a baseline sigma measure at the beginning of the project, and then compare that number to the improved sigma number near the end of the project. So stick with the same defect opportunities throughout the project.

Example of a Six Sigma Calculation

Figure 9-6 shows some examples of Six Sigma levels calculated using the worksheet in Chapter 10 and applying the definitions given above (DPMO = defects per million opportunities).

If the data is accurate and the defect opportunities are consistently applied, the microchip manufacturing process is functioning most effectively and the advertising contract process is worst. Another way to say this is that the manufacturing process does a better job of meeting customer requirements and has the least amount of rework to fix defects.

Assuming that your team did a good job identifying customer requirements and carrying out your data collection plan, you should have the correct data you need to calculate the baseline sigma number for your output measures. Outputs, or Ys, you'll recall, are the products and services you deliver to external, paying customers or to other internal customers within your own organization.

See Sigma Calculation Worksheet on p. 178.

Once your team has completed the sigma level, keep that number handy; you'll want to recalculate the sigma level after you've made improvements and compare it to this original baseline.

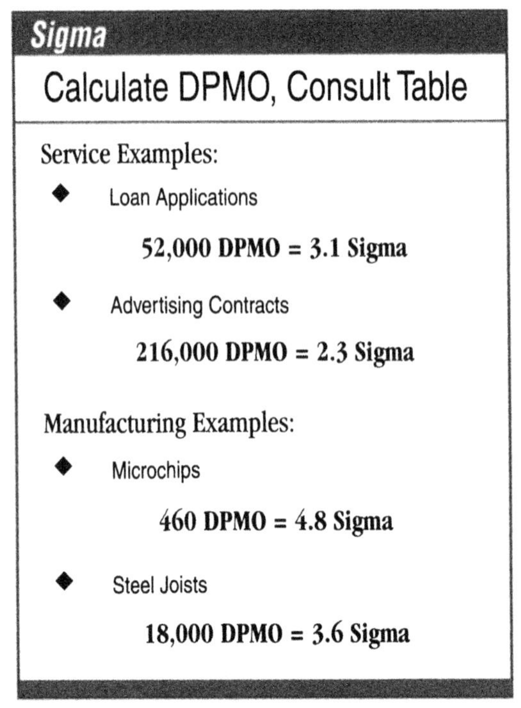

Figure 9-6. Sample Six Sigma calculation

Before moving on...

At the end of *Measure Step B1: Calculate Baseline Sigma,* you should have:

✔ Defined units, defects, and defect opportunities for your process.

✔ Calculated baseline sigma.

Step B2. Calculate Final and First-Pass Yield

The previous discussion has focused on determining sigma capability at the *end* of a process, based on the results (output). That's fine if all you want to do is focus on the capability of our process to meet external customer requirements. But a low output sigma number obviously means that the "innards" of our process are not working very well.

Imagine a process in either services or manufacturing. As Figure 9-7 shows, data collected at the output of the process showed a final yield of .985 (98.5%) and

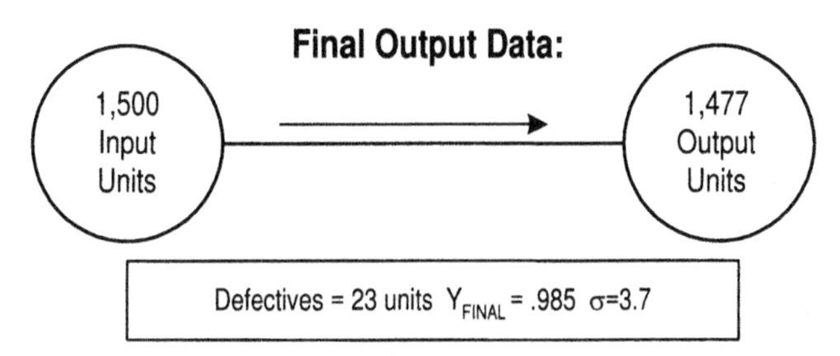

Figure 9-7. Overall process sigma based on final output

a sigma level of 3.7. Of the original 1,500 units (orders, parts, etc.) that entered the process, only 1,477 emerged "defect-free" as outputs at the end of the process.

Now look inside this process. It has three major subprocesses, each of which operates with a yield of good product in the upper 90th percentile range. The company catches and reworks defects, and over the course of the whole process, 89 units have to be reworked before delivery to the paying customer (see Figure 9-8). So of those 1,500 units, only 1,411 remained defect-free throughout the whole process; the other 89 needed some rework. (Apparently, some of that rework was beneficial since customers received 1,477 "defect-free" units!)

The comparison of these two types of yield introduces two important Six Sigma terms:

- The figure of 1,477 is called the **final yield**, because it measures how many units finally came through the process without defects.
- The figure of 1,411 is called **first-pass yield**, because it measures the number of units that made it through the first time without needed rework.

As Figure 9-8 shows, once you take into account all the rework that has to take place, the percentage of "defect-free" items falls to 94%.

The comparison of these two measures of yield points out the difference between focusing only on outputs (final yield) versus looking at what happens inside a process (first-pass yield). Yields that are measured only as outputs hide defects *and the costs associated with them.* In some service businesses the costs associated with a low first-pass yield can reach 20% or more of total sales revenues.

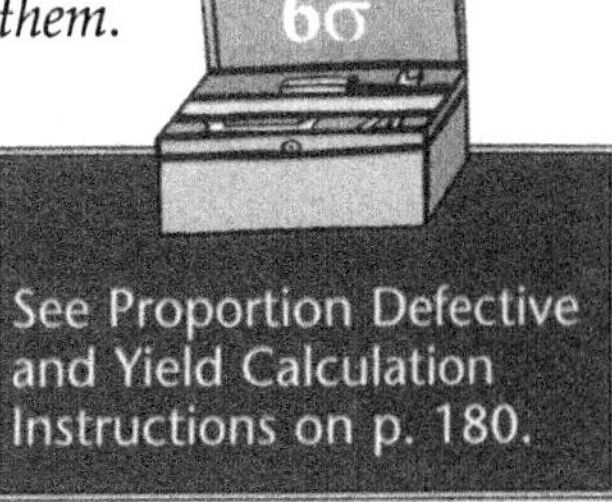

See Proportion Defective and Yield Calculation Instructions on p. 180.

Your Six Sigma team will have to decide whether to focus on output results only, or look at the various

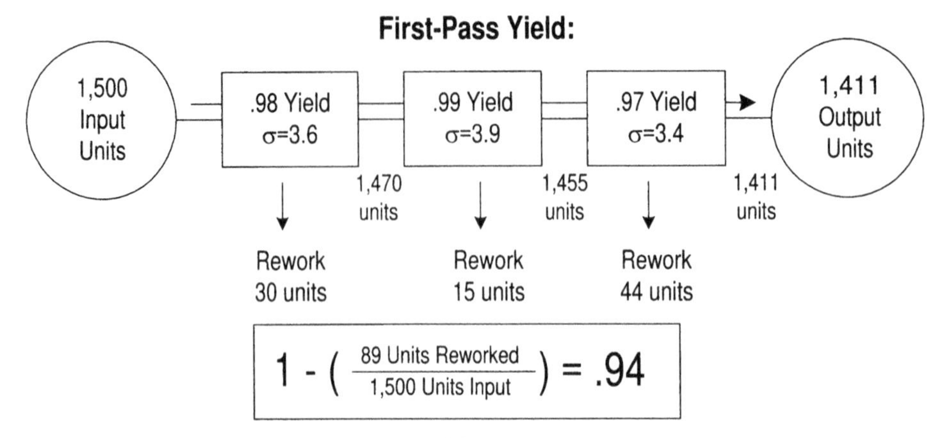

Figure 9-8. Calculating first-pass yield

components of a process. Looking at *internal* subprocesses can help you target your improvement efforts. Figure 9-9 shows the first-pass yield for the three component steps in our imaginary process. As you can see, the third step has the lowest yield, and therefore might be ripest for improvement.

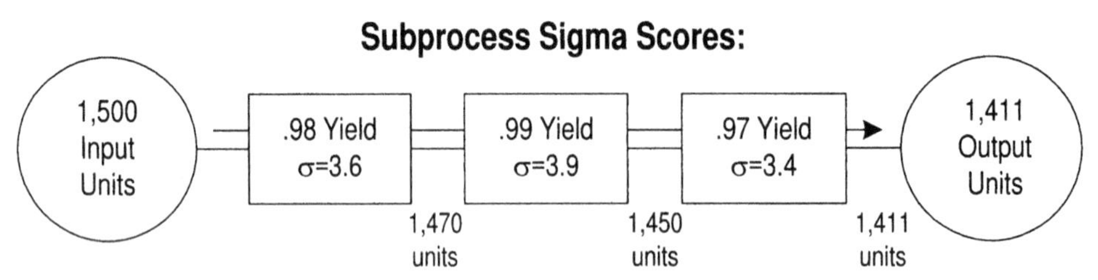

Figure 9-9. Sigma calculation based on subprocess capability

If you choose to take the latter approach, remember that just as the outputs for the entire process are measured against the requirements of the *external* customer, the outputs of key internal process steps are measured against the requirements of the *internal* customer. These internal customers are other people who work for the same company as you and are your customer in the sense that they are closer to the external customer and their requirements help determine what you do in your own process.

Step B3. Measuring the "Cost of Poor Quality"

Neither defect counts nor sigma measures directly capture the costs associated with poor quality. Two different processes may both measure in at 3.5 sigma,

Before moving on...

At the end of *Measure Step B2: Calculate Final and First-Pass Yield,* you should have:

✔ Calculation of final yield.
✔ Calculation of first-pass yield.

meaning that their capabilities are roughly equal, but the dollars lost due to the defects in the processes can be very different. For example, a malpractice suit costs more than retaking an x-ray picture.

For this reason, you should measure the Cost of Poor Quality (COPQ) as soon as you have collected defect data. This means *translating problems or defects into dollar costs per defect—including labor and materials costs for rework.* Measuring COPQ can help get support for improvements the team creates, and gets the attention of managers who may find the language of sigma measures a little strange at first, but who recognize the value of increased revenues or savings when they see them.

See Cost of Poor Quality instructions on p. 181 and Measure Checklist on p. 183.

Before moving on...

At the end of *Measure Step B3: Measuring the "Cost of Poor Quality,"* you should have:

✔ Identified labor and materials costs for rework.
✔ Translated defects into dollar costs per defect.

Getting Ready for Analyze

By the end of the Measure phase of DMAIC, your team has data that will inform your improvement decisions, and a baseline against which your progress will be measured. The measurement of sigma, yield, defects, and cost of poor quality lay the foundation for what every Six Sigma company needs: a reliable and thorough *system* of measuring both processes and outputs for customers.

Three of the last steps for completing Measure are the same as for Define, but there is an additional first step:

1. Revisit your problem statement.
2. Create a plan for Analyze.
3. Update your Project Storyboard.
4. Prepare for your tollgate review by your Sponsor or Leadership Council.
5. Celebrate.

1. Revisit Your Problem Statement

Even before completing a thorough analysis of your data, it's likely your team has learned more about the initial problem that sparked this project. Refine the Problem Statement as appropriate, perhaps by providing more specificity about what the problem is or how it impacts the organization.

2. Create a Plan for Analyze

By the end of Define, you know how much data you've collected, and probably have a gut feeling for how easy or difficult it's going to be to analyze that data. Think ahead and create a plan for the Analyze work. As before, if you are a novice team leader, you can keep the plan simple:

- Assign responsibilities within the team for completing the data analysis (often, the initial work is completed by a subteam, who presents the results to the team for discussion).
- Set or confirm a target date for completion of the data analysis.
- Have your team meetings scheduled.
- If most people on the team are new to data analysis, you may want to check with your Coach/Black Belt or a Master Black Belt to arrange for extra support.

3. Update Your Storyboard

The Measure section of your storyboard should display your data collection plan, a few data collection forms, and a revised Problem Statement (if you changed it). You might also consider including a process diagram that highlights problem areas. There may be measures of the sigma level of the process here, as well as rough estimates of the cost of defects and their rework in the process.

4. Prepare for the Tollgate Review

Before you prepare for the Measure Tollgate Review, do a debrief with the team on what happened last time:

- What did you do well as a team in the Define review?
- What could have been done better?
- What did the reviewers (your customers!) say?
- Which of the support materials (slides, overheads, handouts, flipcharts, etc.) worked and which didn't?
- How did you do on time? If it was too long, what could you do this time to make sure you keep it brief? If it was too short, do you need to add in more detail? Speak more slowly?

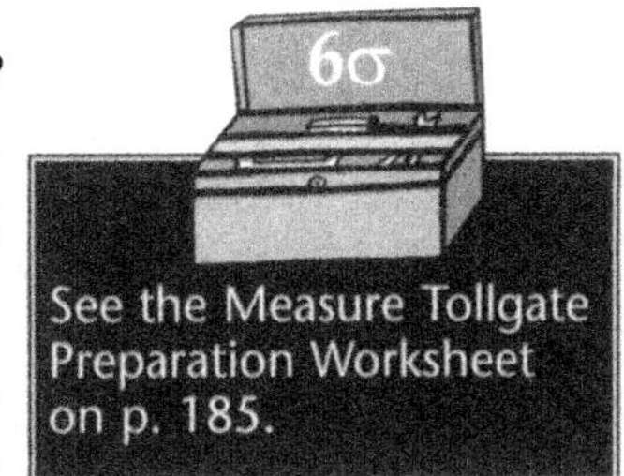

See the Measure Tollgate Preparation Worksheet on p. 185.

After this general review, start your preparation for the Measure Tollgate Review. The main difference between this review and that for Define is that you need to clearly link the work you did here with *what came before* and *what you expect to come after.*

Review the tips given in the Measure Tollgate Instructions (p. 182).

5. Celebrate

Once again, take time to celebrate the *work* and *progress* on your Six Sigma projects. Be sure to point out particular challenges that the team handled well in its Measure work; for example:

- Coming up with innovative data collection ideas.
- People maintaining their level of commitment, carrying through on assignments.
- Finally having a sigma level for the process (which might be the first time an objective measure has been used!).

Once these steps are complete, your team is ready for *Analyze.*

Chapter 10

Power Tools for "Measure"

Collecting and Using Data

If You're Gonna Do It, Do It Right

"Whew!" said Jake, as he finished writing. "At last we've finished our data collection, Clare. I can't wait to get this analyzed so we can start making some real improvements around here."

Clare was excited, too. "I'm with you there, Jake. Three weeks with a stop-watch in my hand measuring how long it takes the warehouse guys to package the orders is enough for me! I think my thumb is going numb!"

Jake glanced down at the stack of papers on the table. "I was getting pretty sick of it, too," he agreed. "I hope we don't have to go through this again. Now let's get to work organizing this data so we'll have something to present at the team meeting tomorrow. I'll bet everyone is as anxious as we are to look at the results."

Clare and Jake showed up at the team meeting with a handful of charts they had created from the data. They could tell their team-mates how long it took on average to pack an order, how many steps were involved, and where they thought there was some rework in the process. At the end of their presentation, they invited their team-mates to ask them questions.

Moira was the first to speak up: "I know that the Southwest Sales team had a special promotion two weeks ago where customers could get their items shipped free if they bought at least $500 worth of product. That's about double our normal order size. Did any of those orders from Southwest come through when you were collecting data?"

Jake and Clare glanced at each other, then finally Jake spoke up, "Um, we didn't ask whether the orders were special or not. And nobody mentioned it to us. Uh, and I guess we never thought to mark down where an order originated."

Moira wasn't sure what to say. "Well, maybe I can find out if any were processed, though I'm not sure how to tell which data points that might be. I just think they'd take longer than our typical orders to pack."

Marcus, the team leader, spoke up then, trying to rescue Jake and Clare from an embarrassing situation. "That's a good idea, Moira. Anyone else have questions for Jake and Clare, or ideas about how to get more from this data?"

Ben, the packaging supervisor, raised his hand. "My main question is what contributes the most to the time it takes to complete a package. Is it the number of items, or the way that different items have to be wrapped? And I've always thought that processing the paperwork took longer than the actual assembly. Is that true?"

This time it was a slightly red-faced Clare who answered. "Well, we didn't really break things out by steps. We just timed the whole thing from start to finish. And I guess we didn't really count the number of items per package, either." She looked at Jake. "Guess we need to go back to the drawing board on this."

Marcus stood up, grabbed a marker, and went to the flipchart. "I think we all dropped the ball on this one, Clare. Obviously, none of us thought much ahead of time of what we really wanted to know so we could collect the kind of information that would answer our questions. Let's say we brainstorm now all the questions we have about the time it takes to package orders. Then we can rethink what data we need to collect."

Has is happened to you yet? Have you collected the wrong data on the wrong things? Or not enough data on the right things? If not, it probably will. It's natural for Six Sigma teams to be anxious to get to work, to get some data quickly so they can do the "real work" of making improvements. Most of us have fallen victim to the temptation to just go collect data.

Unfortunately, succumbing to that temptation usually ends up putting your team behind schedule rather than ahead of it. Taking the time to create a data collection plan isn't a luxury; it's a necessity if you want to use your team's time effectively and efficiently.

The tools in this chapter help a team with its two key Measure tasks:

A. Deciding what data to collect and collecting it.
B. Using data on defects to determine a baseline sigma level.

In additional, two further items for the toolkit:

C. Measure completion checklists.
D. Advanced sigma tools.

A. Collecting Data/Taking Measurements

The example that opened this chapter illustrated some of the common problems teams face when collecting data, such as not collecting enough information about each data point. Jake and Clare got plenty of measurements of their most important indicator, time, they didn't collect any other information associated with each measurement, such as the size of the order, or different packaging techniques used for different types of products included in each order, or where the order originated.

All of the data collection tools and methods described in this section will help you avoid the most common errors associated with data collection. Much of your work will be summarized in the very first tool, the Measurement Planning Worksheet.

The length of time needed to complete a data collection plan varies greatly from team to team based on team members' prior experience, the types of data that will be collected, and so on. The best advice, especially to novice teams, is not to rush through it. As Jake and Clare's experience showed, that may just end up costing you more time in the long run.

Measurement Planning Worksheet

Purpose: To capture a team's plan for collecting useful, meaningful data.

Application: Should be completed any time a team is going to collect data, which can occur in any phase of DMAIC.

Instructions: Read through the appropriate sections of Chapter 9, and use the other tools in this section to generate the information you need to complete this form.

Measurement Planning Worksheet

1. What are we trying to learn, track, or evaluate? ______________________________

____

____

2. a) What will we count or measure (the "unit")? ______________________________

   b) How will the measure be expressed? (e.g., number, percent, weight, time, dollars)

   c) Is this measure _____ Continuous? _____ Discrete? (Use Continuous if possible.)

3. What is the Operational Definition for the measure? (If you are examining multiple factors, you may need more than one Operational Definition.) ______________________________

____

____

____

4. Will new data need to be collected for this measure? _____ Yes _____ No
(If "No," skip to 6, but first make sure you aren't settling for convenient data.)

5. In gathering data, will you be tracking changes over time? _____ Yes _____ No
(If "Yes," use the Process Sampling worksheet. If "No," use the Continuous or Discrete Population Sampling worksheet.)

6. How do you plan to use/display the measure or data? ______________________________

____

____

7. What is the plan for ensuring the measure's accuracy, repeatability, and reproducibility?

____

____

Attach additional pages as needed, one page per measure.

Figure 10-1. Measurement Planning Worksheet (for other worksheets referred to in Item 5, see Figures 10-6, 10-7, and 10-8)

Related tools: This worksheet summarizes data collection decisions your team will make using other tools in this chapter.

CTQ Tree

Purpose: To link measure to an important outcome.

Application:

- Use in data collection to make sure you collect data that is meaningful to your project.

Instructions: The CTQ Tree in Figure 10-2 shows the basic structure. The exact number of branches on your tree will be determined by your team.

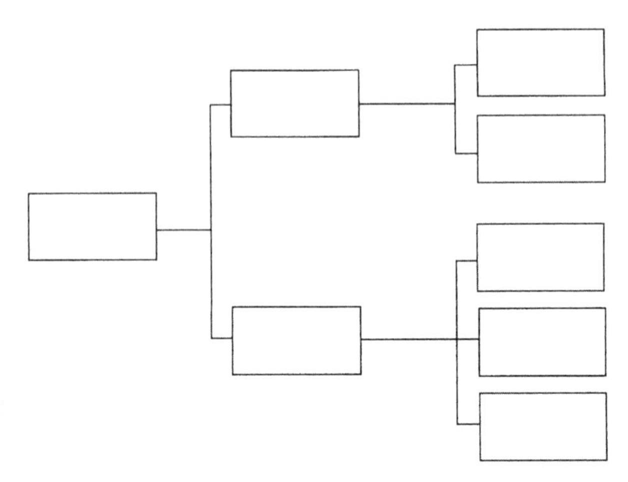

Figure 10-2. CTQ Tree Structure

1. **Identify an output that is important to customers.** (Use your SIPOC diagram as a starting point.)
2. **Identify a characteristic of that output that is "critical to quality"** and write it in a box on the left-hand side of a sheet of flipchart paper, whiteboard, etc.
   - If appropriate, have your team brainstorm a list of potential characteristics, then use multivoting or other decision-making tools to select the *one* that is most critical.
3. **Brainstorm specific kinds of data** associated with the critical-to-quality characteristic and arrange them logically on branching limbs of the diagram.
   - You may want to use an affinity process to identify related sets of measures. This can help you determine logical groupings for the CTQ Tree.
4. **Do a reality check** on the final diagram. Is it feasible and desirable to collect all the data identified?
5. **Confirm which of the data** you will collect.

Stratification Factors

Purpose: To collect information that will help you pinpoint the patterns and causes of problems.

Application:

- You should consider collecting stratification information any time you collect data.

Instructions:

1. **Identify questions you might want to investigate once you have the data in hand.** (Use Figure 10-3 as a starting point.)
2. **Decide which stratification factors are most important to your team** (that is, are most pertinent to questions that are key to being able to solve the problem under study).
3. **Document those decisions.** (You will incorporate these decisions into your data collection form, described later in this chapter.)

Data Stratification	
Factors	**Examples (Slice the data by...)**
Who	◆ Department ◆ Individual ◆ Customer type
What	◆ Type of complaint ◆ Defect category ◆ Reason for incoming call
When	◆ Month, quarter ◆ Day of week ◆ Time of day
Where	◆ Region ◆ City ◆ Specific location on product (top right corner, on/off switch, etc.)

Figure 10-3. Common Stratification Factors

Measurement Assessment Tree

Purpose: To link data collection to key issues in a project.

Application:

- Used primarily in the Measure phase of DMAIC to help identify measures (metrics) that will produce useful, meaningful data for the team.

Instructions: Use Figure 10-4 as a model for drawing your own Measurement Assessment Tree.

1. **Identify a customer-related defect in a key output,** and write it above the designated line on the chart. (Use your SIPOC diagram as a starting point.)
2. **Brainstorm a list of questions** that relate to that defect, and write them on the left side of the tree.
   - What patterns do you suspect you might find?
   - What factors do you think might influence the type or amount of that defect?

   Note: If your team generates a lot of questions, use an affinity process (pp. 56-57) or multivoting (p. 56) to develop a short list of critical questions.
3. **Identify stratification factors** (p. 165) that will help you answer the questions about the output. Write these on the branches to the right of the output.
4. **Identify specific types of data** (metrics) you could collect that would answer the question of how the stratification factor did or did not affect the output.
5. **When the diagram is complete**, have your team review each of the metrics and rate them as follows:
   - Put a square with Y (for yes) by any of the metrics for which you think there is existing data.
   - Put a circle with a Y (for yes) by any of the metrics that you think will help you predict Y (that is, the status or change in this metric is a clue linked to the level of defects in the output).
6. **Use this analysis** to help your team decide which of the metrics will be most useful for your project.

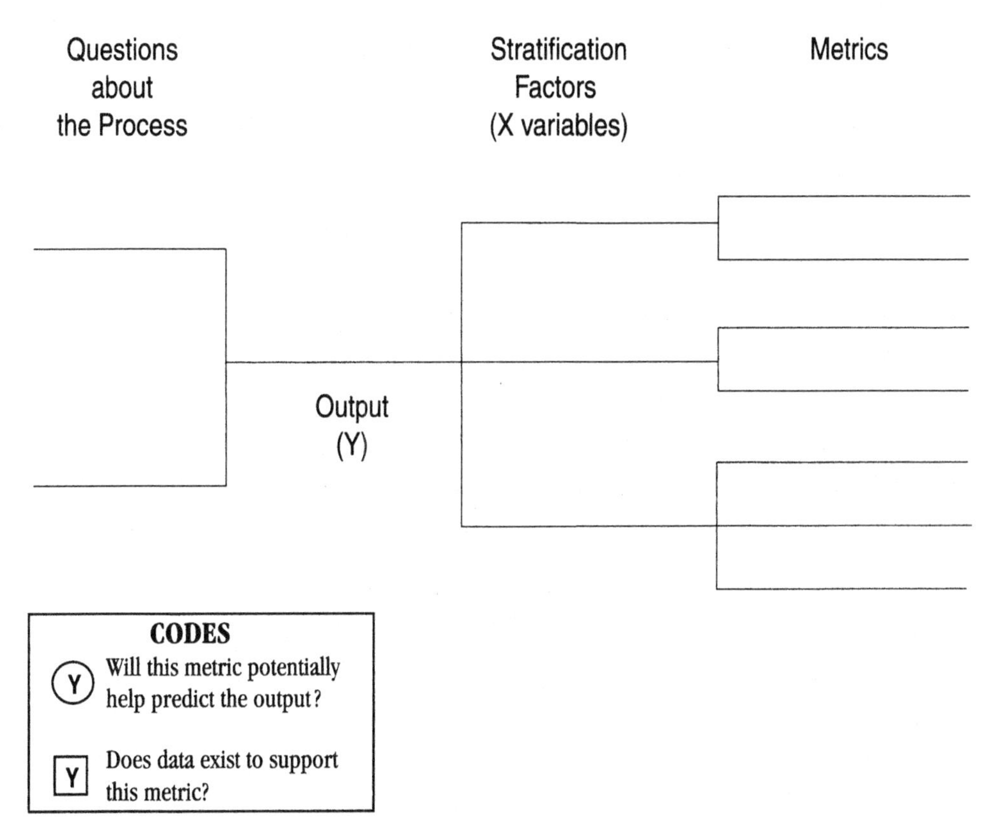

Figure 10-4. Measurement Assessment Tree

Operational Definition Worksheet

Purpose: To ensure that all persons collecting data collect it the same way.

Application: Should be completed any time the team collects data.

Instructions:

1. **Ask one or two team members to write a draft definition of the data and how it will be collected**. The definition should include specifically the issues listed in Table 10-1.
2. **Ask different team members to read the definition and try to shoot holes in it.** Is every word understandable? Make revisions as appropriate.
3. **If relevant, check the definition with customers**. Is the definition of a defect exactly the same as theirs?

4. **Develop job aids if appropriate.** For example, show a range of color swatches from a cloth to show unacceptable and acceptable shades. Compile a set of photos depicting what is and isn't a "surface defect."
5. **Have people who were not involved in developing your definition apply it in collecting data.** Plan to observe the testers to watch for problem areas and sources of confusion.
6. **Finalize the definition and train all data collectors in its use.** Use the worksheet shown in Figure 10-5 to develop operational definitions.

Operational Definition	
Elements	**Examples**
What you are trying to measure.	◆ Satisfaction of customers in the Northeast region with telephone support services. ◆ Number of surface defects on the rear panel. ◆ On-time delivery for Product X.
What the measure isn't.	◆ Are "customer comments" included under "complaints"? ◆ Does "surface defects" include smears or only scratches and dents?
Basic definition of the measure.	◆ Satisfaction = X% of customers giving us a score of 80 or above. ◆ Surface defect = any dent or scratch visible from a distance of 3 feet under normal light.
How to take the measurement (in detail).	◆ "Start the stopwatch when the customer steps into the line, and stop it when the customer leaves the front desk." ◆ "Use the standard calipers placed at the X-junction to measure width in centimeters."

Table 10-1. Elements of an operational definition

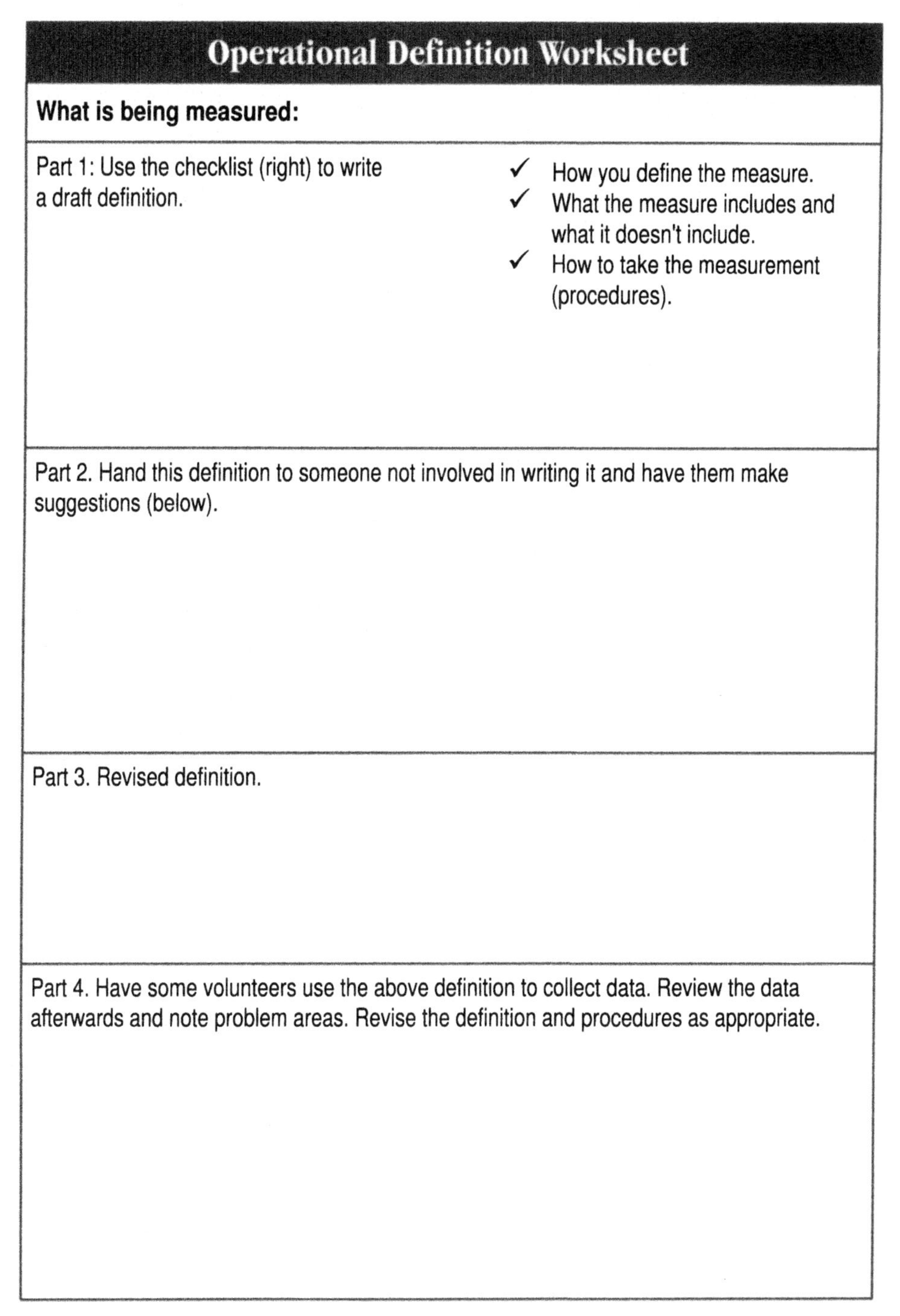

Operational Definition Worksheet

What is being measured:	
Part 1: Use the checklist (right) to write a draft definition.	✓ How you define the measure. ✓ What the measure includes and what it doesn't include. ✓ How to take the measurement (procedures).
Part 2. Hand this definition to someone not involved in writing it and have them make suggestions (below).	
Part 3. Revised definition.	
Part 4. Have some volunteers use the above definition to collect data. Review the data afterwards and note problem areas. Revise the definition and procedures as appropriate.	

Figure 10-5. Operational Definition Worksheet

Process and Population Sampling

Purpose: To help a team decide when to collect data from a process or population and how much data it will need to draw valid conclusions.

Application: Any time it is impractical or simply unnecessary for a team to measure everything produced by a process (within a given time frame) or all the items in a population.

Instructions:

1. **Review the distinction between population and process sampling** in Chapter 9 (pp. 140-143). Decide which type of sampling your team will be doing.
2. **Select the appropriate sampling worksheet** (Figure 10-6, 10-7, or 10-8).
3. **Review the important sampling definitions** (Table 10-2).
4. **Divide your team in half. Have each subgroup complete the selected worksheet independently of the other subgroup.**
5. **Compare answers.** Did both subgroups come up with the same sampling size, frequency, etc.? If not, where do they differ? Were they interpreting the terms in the same way? Did they make different judgment calls along the way?
6. **Reach agreement on a single sampling plan for your team.**

Term	Definition
Sampling event	The act of extracting items from a process or population to be measured.
Subgroup	The number of consecutive units extracted for measurement at each sampling event. A subgroup can be just one item or several items.
Sampling frequency	The number of times per day or week a sample is taken; sampling events per period of time. Sampling frequency tends to increase as the number of cycles or changes in a process increases (i.e., if you process 50 requests per day, you would measure more frequently than someone who processes only five requests per day).
Example	Sampling plan: Measure call length on the first five calls of each hour from 8 AM to 5 PM for two consecutive weeks, M-F. ◆ This plan has 90 sampling events–nine samples per day for 10 working days. ◆ Subgroup size is five. ◆ Sampling frequency is hourly for two weeks.

Table 10-2. Definitions of Key Sampling Terms

Sampling Worksheet #1: Discrete data from a population

Population Sampling Worksheet, Discrete Data

1. Develop Initial Data Profile

A. What's being counted ("the unit")? ______________________

B. What is the size of the population? N = __________
(consider each stratum as a unique population)

C. What's the measure (e.g., defects)?

D. What proportion of the population do you estimate contains this defect/characteristic?
(express in decimals [percent/100]) p = __________

E. Within what percentage precision?
(express in decimals [percent/100]) +/- d = __________

2. Select a Sampling Strategy

A. Selected strategy: ☐ Random ☐ Systematic

B. Means of identifying sample units? ______________________
(e.g., Are they numbered? How will you generate the random numbers?)

3. Determine the Minimum Sample Size

$$n = (2/d)^2 \times p\,(1-p)$$

n = __________

4. Adjust for Finite Population

A. Check proportion of sample to population. n/N =

B. If n/N is greater than .05, adjust using formula:

$$n_{finite} = \frac{n}{(1+n/N)}$$

n_{finite} = _______

Figure 10-6. Population sampling when you have discrete data

Sampling Worksheet #2: Continuous data from a population

Population Sampling Worksheet, Continuous Data

1. Develop Initial Data Profile

A. What's being counted ("the unit")? ______________________

B. What is the size of the population? N = ________
(consider each stratum as a unique population)

C. What's the measure (e.g., defects, time, minutes)?

D. For this population, what do you estimate one standard deviation of this variable to be?
(express in the same units as C) p = ________

E. Within what precision?
(express in the same units as C) +/- d = ________

2. Select a Sampling Strategy

A. Selected strategy: ☐ Random ☐ Systematic

B. Means of identifying sample units? ______________________
(e.g., Are they numbered? How will you generate the random numbers?)

3. Determine the Minimum Sample Size

$$n = (2s/d)^2$$

n = ________

4. Adjust for Finite Population

A. Check proportion of sample to population. n/N = ________

B. If n/N is greater than .05, adjust using formula:

$$n_{finite} = \frac{n}{(1+n/N)}$$

n_{finite} = ________

Figure 10-7. Population sampling when you have continuous data

Sampling Worksheet #3: Continuous or discrete data from a process

Population Sample Worksheet, Continuous or Discrete Data

1. Develop Initial Data Profile

A. What's being counted ("the unit")? ____________________

B. How many units are processes? Per Week? _______ Per Day? _______

C. What's the measure (e.g., defects, time, volume)? ________

D. Is this ☐ Continuous? ☐ Discrete?

E. (Discrete Data only) What estimated proportion of units have the characteristic being measured? (express in decimals [percent/100]): _________

F. How many "cycles" of work are there per day or week (e.g., rush periods, shifts)? _________

2. Determine Strategy and Minimum Sample Size

A. Is quantity processed > 10 per day? (see 1B)

☐ Yes, Use Daily Sampling Strategy ☐ No, Use Weekly Sampling Strategy

B. Use appropriate Sample Size Selection Chart (Daily or Weekly) to determine the minimum sample size. Record Minimum Sample Size: _________

C. Note sampling strategy. Use Systematic unless Random selection is feasible: _________

3. Select Subgroup Size and Sampling Frequency

A. Is your measure continuous? (see 1D)

☐ Yes, Use subgroups of 1. Sample at appropriate intervals (every hour, every nth item). Move to 3D

☐ No, Use following steps to identify subgroup quantity and sampling frequency.

B. 5/_______ (Occurence rate (see 1E) = _______ Subgroup Quantity

C. Is this subgroup quantity ≥ number processed per day/week? (see 1B)

☐ Yes, Sorry you need to "sample" *all* the items or reconsider the measure.

☐ No, Apply the following formula to identify frequency:

_______________ ÷ _______________ = _______________

(Min. Sample Size (see 2B) (Subgroup Quantity (see 3B) Base Frequency (round to nearest whole number >0)

D. If there is a cycle to the process (see 1F), increase the frequency to *twice* the number of cycles per day or week. (E.g., 3 cycles, take sample 6 times per day.)

Figure 10-8. Process sampling

Daily and Weekly Sampling Charts

Instructions: Locate the appropriate daily or weekly quantity on the vertical (Y) axis on Figure 10-9 or 10-10. Now move horizontally directly across the chart until you hit the diagonal line. Move from that point down to the horizontal (X) axis and determine the minimum sample size appropriate for the quantity of items. Note that these are *minimums*; you can collect more data if you like, but not less.

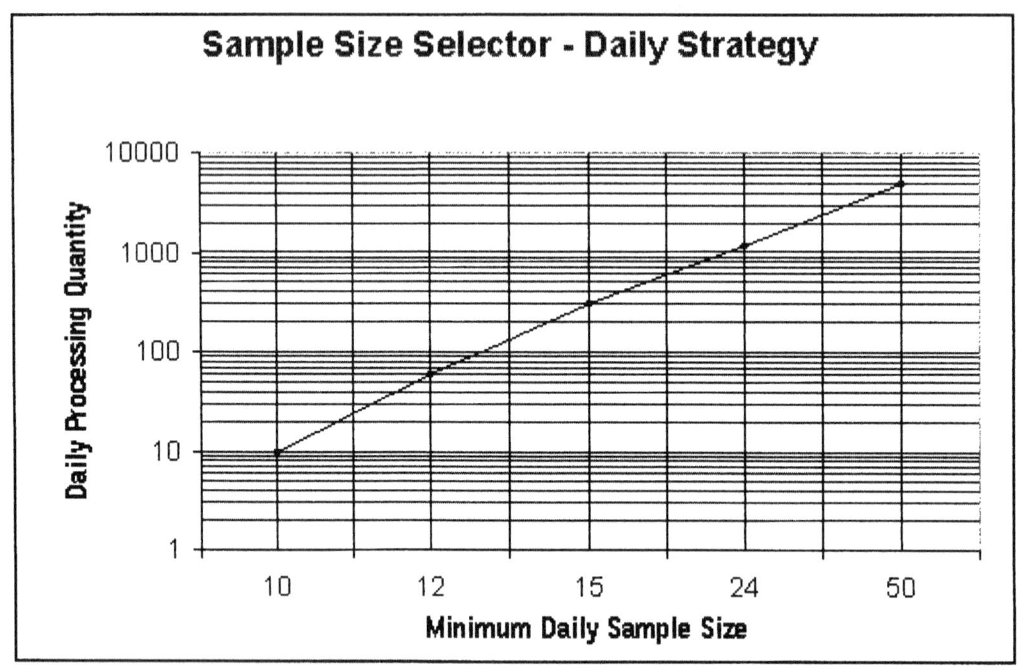

Figure 10-9. Sample Size Selector Chart—daily strategy

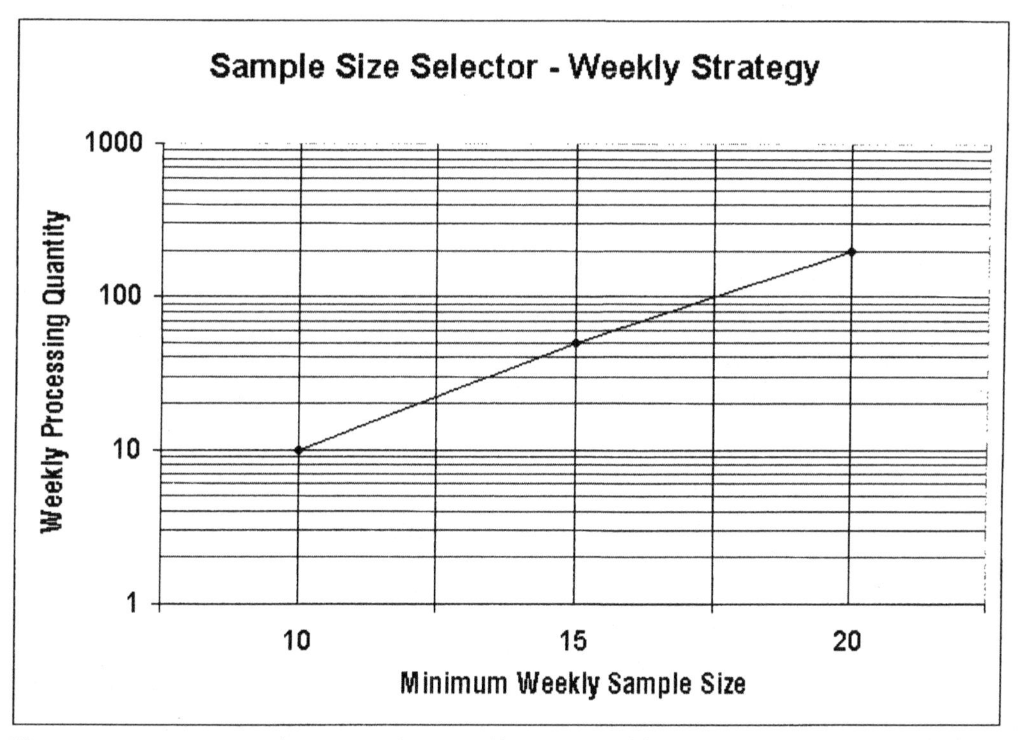

Figure 10-10. Sample Size Selector Chart—weekly strategy

Checksheet Development Instructions

Purpose: Used in data collection to categorize and measure/track frequency of process problems, causes, or other performance factors. Serves as "starting point" for Pareto charts and other display tools (e.g., pie chart, run chart).

Application:

- Providing consistent data collection.
- Identifying and defining problems/opportunities.
- Setting priorities.
- Identifying root cause(s).
- Following up and verifying results.

Instructions:

1. **Determine the data to be gathered.** Options include:
   - Types of problems in a process.
   - Possible causes of one or more problems.

- Satisfaction of a customer requirement.
- Other process measures.

Note: Item to be noted should be objective and yes/no.

2. **Decide on frequency** to be covered by checksheet (hour, shift, day, week, month).
3. **Design a checksheet** matrix or form:
   - One axis for categories, the other for frequency.
   - Use complete dates (e.g., 10/10/02 or week of 10/10-10/17/02).
   - Include space for data collectors' full name and for stratification factors as needed.

Designing the form in the same software in which you'll compile data saves time and work. (Using Excel, for example, helps in design, compilation, and graphing of data.)

4. **Ensure effectiveness** of data gathering or observation. (Train users.)
5. **Place checkmarks or hashmarks in appropriate boxes** when occurrences are observed.
6. **Compile results.**
7. **Chart and analyze data as needed.**

B. Counting Defects and Calculating Sigma

The following tools are what distinguishes Six Sigma methods from all other improvement and management philosophies. A key reason that organizations are increasingly turning to Six Sigma is the ability to develop comparable measures of performance across a wide range of processes, products, and services—and put numbers to issues that were formerly thought to be too fuzzy to withstand a rigorous business analysis. That "comparable measure" is the sigma level, which is based on defining and counting defects—the tasks that the following tools will help you complete.

Sigma Calculation Worksheet

Purpose: To help a team determine the sigma capability of a process.

Application:

- Determining baseline process capability.
- Assessing whether process changes have improved capability.

Instructions: Complete the worksheet shown in Figure 10-11. If you are confused about what the terms mean, refer to Chapter 9 (pp. 150-157). Here are some additional hints:

1. **Select the process.** Your team should have already identified the process under study, its customers, and their requirements when you completed your Define work.
2. **Define "defect" and "number of opportunities" (i.e., defining "defect opportunities" for your product, process, or service).**
   - *Develop a preliminary list of defect types.* For example, a coffee mug might have the following types of defects:
     – Leaks – Glazing/finish blemishes
     – Misshapen container – Misshapen handle – Broken
   - *Re-evaluate the list based on which opportunities are realistic, customer-critical, and specific. Combine or reorganize items.* Usually, some defects realistically never happen or might reflect two types of defects. So it's a good idea to scrutinize your first draft list. For the mug example, using common sense leads to three opportunities for error on a mug, as follows:
     – Glazing/finish blemishes
     – Misshapen (container or handle)
     – Broken

     Leaks no longer appear on the list because they are so rare that it's not a realistic consideration for day-to-day measures of performance. Also, it's simple and also realistic to consider all malformed mugs to fall under one opportunity.

     Note: There is no single right answer to what a "defect opportunity" is. Use your team's judgment to come up with a list that seems reasonable, realistic, practical, and, most importantly, consistent with other such measures in your organization.

Sigma Calculation Worksheet

Sigma levels of a process can be determined several ways. The steps below use the simplest method, based on number of defects at the ***end*** of a process (usually called, appropriately, "process sigma").

STEP 1. Select the Process, Unit, and Requirements

- Identify the process you want to evaluate: _______(process)
- What is the primary "thing" produced by the process? _______(unit)
- What are key customer requirements for the unit? _______

____

____(requirements)

STEP 2. Define the "Defect" and "Number of Opportunities"

- Based on the requirements noted above, list all the possible ***defects*** in a single unit (e.g., late or missing data, wrong size, delivered to wrong address, etc.). Be sure the defects described can be identified ***objectively***. _______________________

____

____

____(defects)

- How many defects could be found on a ***single*** unit? _______ (opportunities)

STEP 3. Gather Data and Calculate DPMO

- Collect end-of-process data _______(units counted) _______(*total* defects counted)
- Determine total opportunities in data gathered:

  # Units Counted x Opportunities = _______(total opportunities)
- Calculate defects per million opportunities:

  (# Defects Counted ÷ Total Opportunities) x 10^6 = _______ (DPMO)

STEP 4. Convert DPMO to Sigma

- Use simplified conversion table and note estimated sigma here: _______

NOTES: 1. The table will give you a very rough range of your sigma level. 2. Your sigma figure can vary significantly based on the accuracy of your data and the number of opportunities you identify on a unit.

Figure 10-11. Sigma Calculation Worksheet (for the conversion table referred to in Step 4, see Figure 10-12)

- *Check proposed number of opportunities against other standards.* Over time there would likely be guidelines or conventions for numbers of opportunities for certain products. (For additional discussion of these terms, refer to Chapter 9, pp. 150-157.)

3. **Gather data and calculate DPMO.**
   - If your team has not collected defect data already, follow the guidelines discussed in the first half of Chapter 9 to do so.
   - **Calculate defects per unit:** Baseline sigma reflects the number of defects that might occur if we had a million opportunities for defects (defects per million 0pportunities = DPMO). Start by dividing the number of observed defects by all the defect opportunities across all the units (products/services) included in the defect count. Here's the formula: $\frac{D}{N = O}$

     (D = defects, N = number of units produced, O = opportunities for defects)
   - **Calculate DPMO:** Take the number from Step 1 and multiply by one million. That gives you DPMO.

4. **Convert DPMO into sigma.** Use the sigma conversion table (right) to determine the sigma level for your process.

YIELD (%)	DPMO	Sigma
6.68	933200	0
8.455	915450	0.125
10.56	894400	0.25
13.03	869700	0.375
15.87	841300	0.5
19.08	809200	0.625
22.66	773400	0.75
26.595	734050	0.875
30.85	**691500**	**1**
35.435	645650	1.125
40.13	598700	1.25
45.025	549750	1.375
50	500000	1.5
54.975	450250	1.625
59.87	401300	1.75
64.565	354350	1.875
69.15	**308500**	**2**
73.405	265950	2.125
77.34	226600	2.25
80.92	190800	2.375
84.13	158700	2.5
86.97	130300	2.625
89.44	105600	2.75
91.545	84550	2.875
93.32	**66800**	**3**
94.79	52100	3.125
95.99	40100	3.25
96.96	30400	3.375
97.73	22700	3.5
98.32	16800	3.625
98.78	12200	3.75
99.12	8800	3.875
99.38	**6200**	**4**
99.565	4350	4.125
99.7	3000	4.25
99.795	2050	4.375
99.87	1300	4.5
99.91	900	4.625
99.94	600	4.75
99.96	400	4.875
99.977	**230**	**5**
99.982	180	5.125
99.987	130	5.25
99.992	80	5.375
99.997	30	5.5
99.99767	23.35	5.625
99.99833	16.7	5.75
99.999	10.05	5.875
99.99966	**3.4**	**6**

Figure 10-12. Sigma Conversion Table

Optional: Teams who are familiar with sigma calculations may also want to calculate the "Z shift," which tells you what your process's short-term capability is compared to its long-term capability. (See pp. 186-188 later in this chapter.)

Proportion Defective and Yield Calculation Instructions

1. **Calculate the proportion defective for the process.** This is the fraction or percentage of the item sampled that had one or more defects.
2. **Calculate final yield** for your process using the simple formula shown at the top of Figure 10-13. (Several examples are included to illustrate the basic concept.)
3. **Calculate first-pass yield.**
   - Construct a simple diagram like that shown in Figure 10-14 (showing the initial units input, boxes reflecting the number of steps in your process, and the final output).
   - For each step, determine the number of units that make it through without requiring rework. Use this figure to calculate a yield for each step.
   - Determine first-pass yield (Figure 10-14) for the process as a whole. (That is, the proportion of units that make it through the entire process requiring rework.)

Final Yield

Formula: 1 - Proportion Defective

Service Examples:

- 43 of 250 loan applications contain defects

  1 - .172 = .828 or 82.8% Yield

- 66 of 186 advertising contracts contain defects

  1 - .354 = .646 or 64.6% Yield

Manufacturing Examples:

- 97 of 750 microchips contain defects

  1 - .129 = .871 or 87.1% Yield

- 99 of 1150 steel joists contain defects

  1 - .086 = .914 or 91.4% Yield

Figure 10-13. Final yield calculation

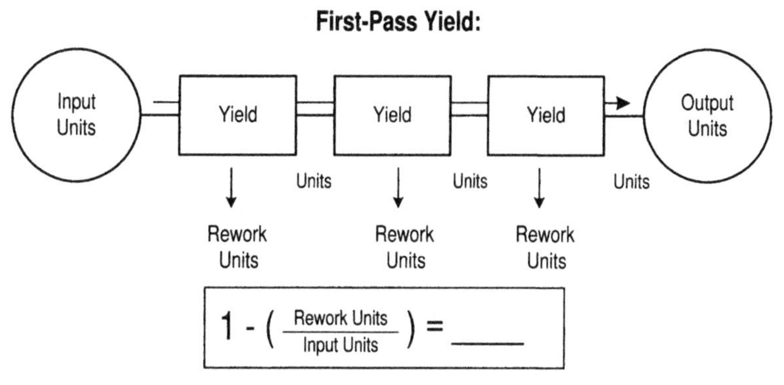

Figure 10-14. First-pass yield

Cost of Poor Quality (COPQ) Calculations

Purpose: To assign a dollar figure to the amount of defects produced by a process.

Application: After collecting defect data to gauge the impact of those defects on profitability.

Instructions: For any type of error, defect, or mistake…

1. **Count the number of incidents** over a period of time (once, per day, per week).

2. **Determine the labor cost** associated with fixing those incidents (reworking, storing, retrieving, etc.).

________	x	________	x	________	x	________	=	$ ________
# defective per day/ # of good per day		# of people who work in the area		Hours worked per day		Hourly pay for the time period		

3. **Determine the material cost** for the defects.

________	x	________	=	$ ________
Cost per item used		Quantity defective per day		

4. **Add the totals from Steps 2 and 3.**

C. Measure Completion Checklists

Measure Checklist

Purpose: To bring a formal end to the Measure stage of a team's project.

Applications:

- Use during the Measure work to track progress.
- Use at the end of the Measure stage to make sure all essential tasks have been completed.

Instructions:

1. Walk through this checklist (Figure 10-15) item by item at a team meeting.
2. Mark a "yes" only if *everyone* on the team agrees the task has been completed. If anyone says no, ask him or her to state why they think the task is incomplete, and to offer specific actions needed to complete it.
3. Reach agreement as a team on each answer before marking the checklist.
4. If there is unfinished work, ask for volunteers, assign responsibilities, and set deadlines for completion of those tasks.

Measure Checklist

2. Measure

Instructions:

If you can respond "yes" to each statement below, you're doing well with measurement and are ready to move into the Analyze phase of DMAIC.

For our project we have ...

1. Determined what we want to learn about our problem and process and where in the process we can go to get the answer. YES NO
2. Identified the types of measures we want to collect and have a balance between effectiveness/efficiency and input/process/output. YES NO
3. Developed clear, unambiguous operational definitions of the things or attributes we want to measure. YES NO
4. Tested our operational definitions with others to ensure their clarity and consistent interpretation. YES NO
5. Made a clear, reasonable choice between gathering new data and taking advantage of existing data collected in the organization. YES NO
6. Clarified the stratification factors we need to identify to facilitate analysis of our data. YES NO
7. Developed and tested data collection forms or checksheets which are easy to use and provide consistent, complete data. YES NO
8. Identified an appropriate sample size, subgroup quantity, and sampling frequency to ensure valid representation of the process we're measuring. YES NO
9. Prepared and tested our measurement system, including training of collectors and assessment of data collection stability. YES NO
10. Used data to prepare baseline process performance measures, including proportion defective and yield. YES NO

Figure 10-15. Measure Completion Checklist

Tollgate Preparation Worksheet

Purpose: To help a team prepare a presentation for the tollgate review at the end of the Measure stage.

Application: Use at the end of the Measure stage to help the team prepare its presentation.

Instructions (Figure 10-16): Follow the general tollgate guidelines given for the Define review (pp. 118 and 120) to identify priority message to include in your presentation. In addition:

1. **Document any commitments or promises** the team made to the Sponsor/Champion, Leadership Council, etc., during the Define tollgate review.
2. **Compile information about progress** on meeting those commitments.
3. **Decide on a sequence for the presentation,** and complete the left column on the worksheet (key messages).
4. **For each message, identify how that information can best be presented** to someone unfamiliar with the details of the project. Be creative! Look for ways to convert messages into data charts, pictures, or other high-impact visuals. Also identify what format that information will take in the presentation (such as handouts, flipcharts, slides or overheads, etc.). Complete the middle column of the worksheet (vehicles for key messages).
5. **Ask for volunteers and/or assign responsibilities** for each section of the presentation. Try to involve the whole team.
6. **Prepare an agenda** for the tollgate review. Identify information that will need to be sent to the reviewers ahead of time.
7. **Do a dry run** of the presentation to make sure it can be completed in the time allotted and to help team members get more comfortable with their roles.

MEASURE Tollgate Preparation Worksheet		
Key messages to cover in the Review (list no more than 3 to 5, in the sequence in which they will be covered in the presentation)	**Best way to present this information** (be creative in finding high-impact visuals–handouts, overheads, flipcharts, storyboard, etc.–to use in the presentation)	**Person or persons responsible for this portion of the presentation**
List promises or commitments made in the previous review.	Identify ways to show progress on the issues raised in the previous review.	
List the highlights of your plan for the Analyze stage. Include estimated timeline and any additional resources needed.		

Figure 10-16. Measure Tollgate Preparation Worksheet

D. Advanced Sigma Tools: Understanding Long-Term Variation

Tracking Long-Term Variation and Process Shifts

Two Six Sigma teams are studying the same manufacturing process and measuring the same quality characteristic or defect. Team Rabbit collects 50 data points all on the same day. Team Hare collects one data point a day for 50 days. If you plotted these data on a frequency plot, which do you think would show more variation?

The answer, of course, is that you'd expect to see more variation in Team Hare's 50 points than in Team Rabbit's. Why? Because a process will change a lot more over the course of 50 days than it will in one day. A 50-day span would cover changes in batch materials received from a supplier, changes in physical conditions, possible deterioration of equipment, etc.—all of which would be less likely to happen (or would not be as pronounced) in the course of a single day.

The concept of "variation over time" is just as relevant to administrative processes, as they, too, will change over time. In addition, administrative processes are very prone to variation between different individuals, groups, locations, etc. For example, suppose our two Six Sigma teams were studying "application processing time." Team Rifle collects data on processing time for 50 applications processed by one office during the course of one week. Team Shotgun collects data on processing time from 50 different offices on the same day. Which of these sets of data do you think would show more variation? The answer is that it's more likely that the 50 different offices will each have their own individual application processes, and therefore you'd expect to see more variation among different offices than within a single office.

The lesson to learn from these examples is that over the long term, or in widest application, processes experience much more variation than they do in the short term or in limited applications. This concept is captured in Figure 10-17. The smaller distributions at the top of the figure all reflect what can happen in the short term to any process. If you compiled all this data together, you would get the long-term distribution shown at the bottom.

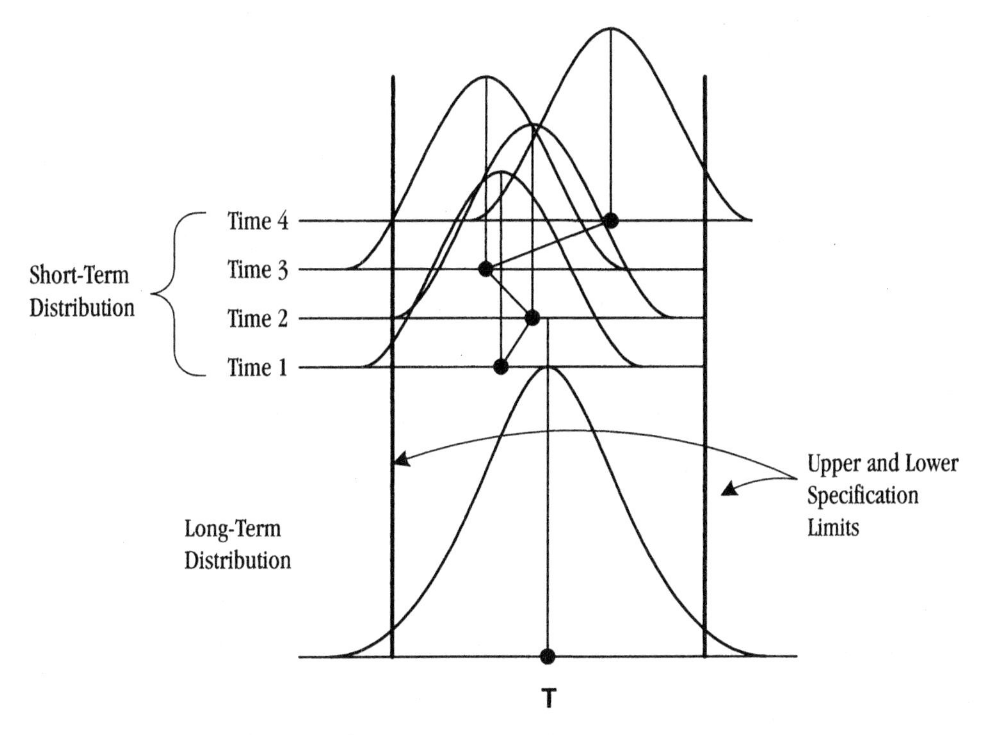

Figure 10-17. Long- and short-term variation

Long-Term Variation and Sigma Capability

This difference between short- and long-term variation has a direct relationship to process capability as well. Look again at Figure 10-17 and notice the relationship between the various distribution curves and the specification limits drawn on the chart. As you can see, in the short term, the process can drift closer to one of the specification limits, then back in the other direction. This leads to two key concepts:

- **Short-term capability:** the best the process can be if centered.
- **Long-term capability:** sustained reproducibility of the process.

Let's say you have a process that has a short-term process capability of 3.2σ. You know that over time, the process will likely shift in one direction or the other. **Experience has shown that this shift often reduces capability by 1.5σ:** that means your 3.2σ process is really only "1.7σ capable" in the long run (see Figure 10-18).

Now here's the curious part: the sigma conversion table used with the sigma calculations described earlier in this chapter has the 1.5σ built into it—oddly, the

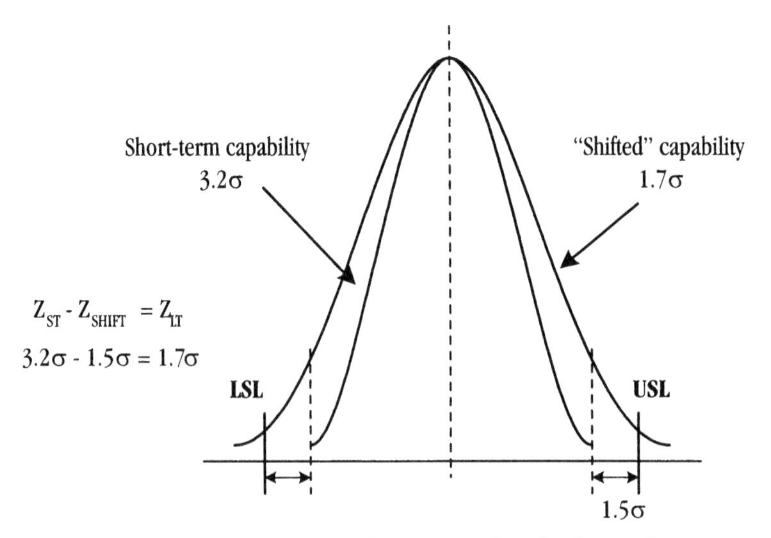

Figure 10-18. Shifts in process capability over time

table assumes you are using long-term data to calculate short-term capability. However, you can use a separate **Z conversion table** to calculate both the long- and short-term sigma capability for your processes as follows. Table 10-3 shows examples of these calculations; the Z conversion table (normal distribution) is in Table 10-4.

Discrete Data	Continuous Data (where there is only one specification limit)	Continuous Data (where there are two specfication limits)
1. **Determine p, the proportion defective.** p = # of defective units (p = d/n) 2. **Determine yield.** Yield = 1-p 3. **Look up Z_{LT} (the yield in the Z conversion table).** 4. **Add the shift.** $Z_{ST} = Z_{LT} + 1.5$ **Example:** d = 5 n = 25 d/n = p = 0.2 Yield = 1 - 0.2 = .80 Z_{ST} = .84 + 1.5 = 2.34	1. **Determine the average (X-bar) and standard deviation (s)** for your data. 2. **Identify a specification limit**–either upper (USL) or lower (LSL). 3. **Determine Z** by calculating the difference between the specification limit and the average. Use the appropriate equation below depending on whether you have a USL or LSL. $Z_{LT} = (USL - \bar{X})/s$ $Z_{LT} = (\bar{X} - LSL)/s$ 4. **Add the shift.** $Z_{ST} = Z_{LT} + 1.5$ **Example:** $\bar{X}$ = 120, s = 25 USL = 160 Z_{LT} = (160 - 120)/25 Z_{LT} = 40/25 Z_{LT} = 1.6 Z_{ST} = 1.6 + 1.5 = 3.1	1. **Calculate the Z values.** $Z_{USL} = (USL - \bar{X})/s$ $Z_{LSL} = (\bar{X} - LSL)/s$ 2. **Use the Z values to look up yields** (Y values) for both the specification limits (Y_{LSL} and Y_{USL}). 3. **Determine the total yield.** $Y_{TOTAL} = Y_{USL} - (1 - Y_{LSL})$ The Y value for the USL is the "proportion good" from that end of the curve–that is, it represents everything BELOW the upper specification limit. To complete the calculation, we need to "chop off the other end of the tail" by eliminating output that is below the LOWER specification limit–that's why there is a subtraction at the end of the equation. 4. **Add the shift.** $Z_{ST} = Z_{LT} + 1.5$ **Example:** $\bar{X}$ = 120, s = 25 USL = 160, LSL = 90 Z_{USL} = (160 - 120)/25 = 1.6 Z_{LSL} = (120 - 90)/25 = 1.2 Y_{USL} = .9543 Y_{TOTAL} = .9543 - (1 - .8849) = .9543 - .1151 = .8301 Z_{LT} = .95 Z_{ST} = .95 + 1.5 = 2.45

Table 10-3. Examples of calculating shift

Minor increments of Z

Major increments of Z

Z	.00	.01	.02	.03	.04	.05	.06	.07	.08	.09
-3.80	.0001	.0001	.0001	.0001	.0001	.0001	.0001	.0001	.0001	.0001
-3.50	.0002	.0002	.0002	.0002	.0002	.0002	.0002	.0002	.0002	.0002
-3.00	.0013	.0013	.0013	.0012	.0012	.0011	.0011	.0011	.0010	.0010
-2.50	.0062	.0060	.0059	.0057	.0055	.0054	.0052	.0051	.0049	.0048
-2.00	.0228	.0222	.0217	.0212	.0207	.0202	.0197	.0192	.0188	.0183
-1.50	.0668	.0655	.0643	.0630	.0618	.0606	.0594	.0582	.0571	.0559
-1.00	.1587	.1562	.1539	.1515	.1492	.1469	.1446	.1423	.1401	.1379
-0.50	.3085	.3050	.3015	.2981	.2946	.2912	.2877	.2843	.2810	.2776
0.00	.5000	.4960	.4920	.4880	.4840	.4801	.4761	.4721	.4681	.4641
0.00	.5000	.5040	.5080	.5120	.5160	.5199	.5239	.5279	.5319	.5359
0.50	.6915	.6950	.6985	.7019	.7054	.7088	.7123	.7157	.7190	.7224
1.00	.8413	.8438	.8461	.8485	.8508	.8531	.8554	.8577	.8599	.8621
1.50	.9332	.9345	.9357	.9370	.9382	.9394	.9406	.9418	.9429	.9441
2.00	.9772	.9778	.9783	.9788	.9793	.9798	.9803	.9808	.9812	.9817
2.50	.9938	.9940	.9941	.9943	.9945	.9946	.9948	.9949	.9951	.9952
3.00	.9987	.9987	.9987	.9988	.9988	.9989	.9989	.9989	.9990	.9990
3.50	.9998	.9998	.9998	.9998	.9998	.9998	.9998	.9998	.9998	.9998
3.80	.9999	.9999	.9999	.9999	.9999	.9999	.9999	.9999	.9999	.9999

This is an abbreviated version of the Z or "normal" table, in which you can find in most basic statistics texts.

If you know the Z value, first find the major increment that is closest to that value, then move across to the appropriate minor increment. For example, if your Z value is 2.07, start at the "2.00" row, the move across to the ".07" column. (Since this is an abbreviated table, you will need to interpolate for values that are not represented, locate a statistics text, or check a statistics computer program.)

If you know the yield of your process, find the closest value in the four-digit numbers in the body of the table, then read across and up to find the closest Z value. For example, if your yield is 85%, the Z value is somewhere between 1.03 and 1.04. As noted in the text, if your process yield is less than 50%, don't bother calculating shift! Go improve your processes!

Table 10-4. Z conversion table

Chapter 11

Guiding the Six Sigma Team in the Measure Stage

Storm Clouds Ahead

As one of our Texas colleagues likes to say, when Six Sigma teams move from the Define to the Measure stage, they often go from "discussin'" to "cussin' and fussin.'" Team members start arguing about what should be measured and who should measure it. The original Project Charter comes under fire from some team members who openly wonder what the Champion was thinking when the project was conceived. People complain that collecting data is interfering with their "real" jobs. Team assignments may go unfinished and project milestones slip by uncompleted. Some team members may stop coming to meetings.

For the team and the team leader this "storming" phase can be a nightmare. Fortunately, it's a natural part of the team's evolution, and—if managed properly—can strengthen the team in the long run. Knowing about the natural stages of team development can prepare the team leader for the storms of the Measurement phase. Forewarned, as they say, is forearmed.

The Anatomy of Team Storming

Storming and the behaviors around it appear as the team begins to realize that the problem they've been asked to solve is tougher than they thought at first, and will not be solved in a few meetings. Impatience and touchiness are typical emotions at team meetings in this stage, especially if the collection of data runs into snags or dry holes and the expected data fails to appear.

Often uncertain about DMAIC, and unfamiliar with its painstaking methods, team members complain that the process is taking too long, and fall back on their own problem-solving methods, or simply propose solutions. They complain about being forced to follow a process that really doesn't apply to the problem before the team. Cliques may appear with two or three members regularly siding with themselves against another team member. Far from looking after one another, members of the group actually attack one another.

There may, of course, be some grounds for these complaints. If the Champion has not done a good job of scoping the project, the team may waste a lot of time forcing the project into a manageable frame. Worse yet, the Champion may already have decided on a solution, and is using the team as a rubber stamp for preconceived ideas. The cause of these problems, of course, usually lies in the poor training of Champions themselves.

Team leaders themselves may contribute to storming problems during the Measure phase if they try to play too many roles: facilitator, scribe, timekeeper. The team members are reduced to passive critics focused on the team leader instead of the project at hand.

Overall, storming is not a pretty picture. Nevertheless, most teams will experience something like it as they finally accept the sobering thought that they are indeed trying to track down the elusive and unknown causes of problems. This requires a level of humility and willingness to rely on other people that are often absent in the work lives of many successful people.

Tips for the Storming Stage

Don't Panic! Team members will start showing signs of impatience, touchiness, and downright anger as they begin to realize the problem they're working on simply won't disappear with a wave of a magic wand. Don't ignore these emotions or try to suppress them by simply referring to ground rules. Acknowledge that storming is natural at this stage of the team's work. In fact, the wise Black Belt will lead a discussion of the forming-storming-norming-performing cycle at one of the first team meetings.

Recognize that storming reflects a high-energy state. The strong emotions associated with storming come from wanting to do a good job, to make improvements, to do *something*, and not just talk. The team will need this energy later on to collect and analyze data, and to make and install improvements. The Team Leader should reassure the team that DMAIC actually works when applied correctly. Review the original plan and note any progress, however slight.

Solve obvious problems and eliminate any process steps that clearly don't add value for customers. Teams need to be careful that they don't merely tinker with the problem, and thereby create more problems. But on the other hand, there's no need to wait for weeks or months if there is an obvious problem that can be fixed easily. (It doesn't take a control chart to tell you to remove a nail from a flat tire!) Such early improvements are risky because you don't have a complete picture yet, but teams are usually motivated more by small wins than big talk from Team Leaders or Project Champions.

Make sure everyone on the team has an assignment for every meeting and in between meetings, too. Idle hands add fuel to the storming blaze. In the Define stage, team members need to interview customers and translate their voice into measurable requirements. They also need to confirm that their high-level process map (SIPOC) is actually what happens in the process. Even the ground rules need discussion and refinement. There's plenty to do in these early meetings. Make sure no one team member has nothing to do but critique the work of others.

Be sure to educate/train team members in what is expected of them. Some storming comes from mere uncertainty and ignorance of what to do. Team members need to be trained to do their assignments. Even then they may feel unsure of themselves.

Work in pairs. Exploit "partner power" at every opportunity. Pair together people who have complementary skills. Working with a partner helps keep team members honest when they promise to do something by a certain time.

If this talk about storming makes you nervous about being a team leader or team member, it's worth remembering that practically every team that has the basics in place survives and gains strength from the storming phase. And you should have these basics in place already:

1. Clear goals, an action plan, and well-defined roles in your Project Charter.
2. Good guidelines on communication in your ground rules.
3. Tools for reaching decisions in the Improve stage.
4. Structured processes to attack problems, aka DMAIC.
5. Awareness of the natural evolution of teams from forming to performing.

Because you have these established already, don't worry when the storm clouds appear. As the Wise Old Sigma Trainer says: *"Inside every dark cloud, there's a darker cloud. Keep punching away, because there's a silver lining in there someplace."*

Troubleshooting and Problem Prevention for Measure

The Measure stage provides the raw material for analysis and has its own fail points. On the whole, it tends to s-t-r-e-t-c-h out longer than it should. The reasons for this are related to the following failures.

Failure #1: The Team Measures the Wrong Things

This becomes obvious for many teams only when they try to analyze the data they've collected, find that it's not what they need, and have to go back to "Start."

Why this happens: There are many causes for this failure, but the most common one is ignorance of the key process variables (the key Xs) that impact the output variables (the key Ys) that must meet customer requirements. Hoping to save themselves the time and trouble to collect requirements from customers and then identify which process variables affect those requirements, teams collect data that has "conveniently" been collected already. Unfortunately, most of the data clogging the arteries of corporate America was not collected to throw light on external customers and the processes that supply them.

How to avoid it: Obviously, having measures for customer requirements is the basic prerequisite. Knowing which process variables to measure is also crucial. Then the team needs a thorough data collection plan with operational definitions of the key variables it is measuring, sampling plan, etc. The team needs to be on the lookout for biases of one kind or another that creep in to support pet theories. Another good preventive measure is to test out the data collection worksheets on a small scale to make sure they actually work. Why all this fuss around measurement? Because once they have collected the data, teams hate having to collect it again. More important, if they use incorrect data, their analysis will be faulty, and their solution won't work.

Failure #2: Measurement Systems "Drift"

Why this happens: Teams that do not monitor data collection procedures usually have a rude awakening. When they examine their data, they find that something isn't quite right—data from one period of time or one person doesn't quite match that collected at another time or by another person. Without constant monitoring, measurement processes (like all processes) start to drift. People forget exactly how to take a measurement or how to record the data or how a defect is defined ... and they unintentionally introduce variation into the process.

How to avoid it: Follow the instructions in Chapters 9 and 10 for identifying and testing the operational definitions that your team and any other data collectors will use. If your data collection will extend for more than a week or so (and usually it does), identify ways to monitor the repeatability, reproducibility, and stability of your measurement procedures.

Measure Do's and Don'ts

Do

- **Balance output with process/input measures:** Make sure that you're tracking impact on the customer and end product/service (Output), even if your focus is on reducing costs and efficiency (Input/Process).
- **Stop taking measures that are not needed or useful:** When a bean-counting "measure everything because we can measure everything" mentality takes hold, it's time to recall that the primary reason for measuring anything is to make sure we're meeting external customer requirements with our products and services.

Don't

- **Try to do too much:** Even though you want to get a jump on making improvements, don't be greedy and try to measure too many things at once. Focus on those measures that you're pretty sure you can use to find the time—a week to a month being a good rule of thumb.

- **Expect the data you collect to confirm your assumptions:** It's easy to accept data that confirms your pet theories. Be prepared—and open-minded—when the data refutes what you expected to see.

The Analyze Stage

Chapter 12

Analyzing Data and Investigating Causes

"Call for Sherlock Holmes!"

Problems at Gemini

Gemini Computer Sales specializes in selling and delivering customized computer systems to individual users, small businesses, and large corporations. Most sales are made by Telephone Sales Consultants (TSCs) in the Des Moines call center, but more experienced individual and corporate customers buy direct from the Gemini web site.

The good news at Gemini is that sales are increasing 30% this year. The bad news is that a recent survey revealed a sharp drop in customer satisfaction. That's why a Black Belt and Six Sigma team have been meeting every Thursday afternoon for the past month.

The team started with a problem statement that "customers are unhappy with Gemini's service." Once they gathered more information, they were able to refine this to read "New customers and small businesses complain that their orders are incomplete, inaccurate, and different from what they ordered." The goal was to increase customer satisfaction around these issues, and their Champion told them to focus their attention on the sales process in Des Moines.

The SIPOC map that the team put together at an early meeting outlined the basic process in seven steps:

1. Customers place orders with the TSCs at the call center.
2. TSCs send billing information to Accounting and forward customer order information to the Order Verification Dept. (OVD).
3. OVD verifies the order and sends it to Order & Picking (O & P).
4. O & P picks the parts and components and sends them to Assembly.
5. Assembly assembles the orders and sends them to Shipping.
6. Shipping sends the filled orders to Customers.
7. Orders with wrong or missing components are corrected in the Returned Materials Area (RMA).

The team, with the help of their Champion, quickly decided they needed to limit the scope of its study to the first three steps of the process, because that seemed manageable. Those three steps describe the "sales" portion of the process: salespeople (TSCs) take the customer orders over the phone and send the information on to OVD clerks, who check every phone order for systems compatibility and completeness. When the OVD clerks find a problem, they contact the TSC or the customer, or they check a database used by the TSCs, called "Jimmy Gemini." Sometimes they check the information in the Gemini Web site, which has all the latest information on it, often before "Jimmy Gemini" does.

Delving into the history of this process, the team discovered several interesting facts. For example, salespeople receive a bonus on the volume of business they handle. When sales grew at a faster pace than expected in the third quarter, management moved some of the order verifiers from OVD into direct phone sales jobs. Those who made the transition into sales say they like the bonuses and they don't miss the often tedious and time-consuming work of checking every order.

When the team moved into the Measure stage of its improvement process, it discovered there was already plenty of data recorded in the automatic call directing equipment used by the TSCs. The team collected data on daily sales, number of orders received each day, and how many returns were authorized. Their data is in Figure 12-1.

Seeing the dramatic rise in returns inspired the team to name itself the *Point of No Returns* team. Along with this historical data, they decided to attach a traveler checksheet to orders returned by customers to find out if there was a connection between customer type (individual, small business, corporate) and order type (complete system or components only).

While gathering data, the team also found out the following:

- The TSCs don't always verify all customer shipping data or systems compatibility because it takes too long, and they think Order Verification clerks will make any corrections needed.
- In Order Verification, orders tend to pile up, and OV clerks do not always verify them in the order they arrive, preferring sometimes to do the ones that can be verified quickly, leaving the tougher ones for later.
- When they do check orders, OV clerks often go to the web site, where they find more current information than in Jimmy Gemini.

Sales Analysis Quick Report

Month	Average Daily Sales	Average Daily Order Count	Average Daily RMAs* Count
J	$2,039,369	756	59
A	$2,110,445	784	62
S	$2,077,845	764	61
O	$2,667,202	969	78
N	$2,766,989	993	81
D	$2,581,598	951	75
J	$2,111,429	765	62
F	$2,241,443	832	65
M	$1,906,967	705	56
A	$2,284,616	841	71
M	$2,341,695	882	74
J	$2,359,981	879	77
J	$2,496,605	907	86
A	$2,969,272	1085	102
S	$3,199,099	1183	117
O	$3,294,299	1188	120
N	$3,568,812	1333	157
D	$3,419,532	1245	169

* RMA = Returned materials authorizations

Figure 12-1. Order and return data from Gemini

If you were on the Gemini team, what would you do next? How would you use the data to help improve the process? Obviously, all the figures are increasing—but is it a significant increase? How would you analyze the process information gathered with the traveler checksheet to identify problems?

The answer: you would use data analysis tools and process analysis techniques to identify and verify root causes of the problem. That is the goal of the Analyze stage of DMAIC.

root cause of problem

Remember the old Sherlock Holmes stories? Given a few clues, the English detective could solve the most baffling mysteries. Unlike the inspectors from Scotland Yard, who often jumped to conclusions and arrested the wrong person, Holmes paid attention to subtle clues and took his time to identify the real culprit. In the novel, *A Study in Scarlet,* Holmes warned that "it is a capital mistake to theorize before you have all the evidence. It biases the judgement." Holmes would have made a great Six Sigma Black Belt!

A Six Sigma team should have plenty of evidence in the data they've collected at the "scene of the crime." Now it is time to generate theories about the cause of defects. But they can't accept the hypothesis at face value; they need to verify that it really does contribute to the problem under study. Any hypothesis or hypotheses proposed this week could get shot down by the data; new data collected next week may suggest another a new suspected case, and so on, until the team is able to confirm a hypothesis with data and can "arrest" the guilty causes.

Becoming a Defect Detective

If Sherlock Holmes were living in today's business world, he'd have a lot of powerful tools at his disposal for investigating the causes of defects. These tools fall into two main categories:

- **Data Analysis**: Using data collected to find patterns, trends, and other differences that can suggest, support, or reject theories about the causes of defects.
- **Process Analysis**: A detailed look at the existing key processes that supply customer requirements in order to identify cycle time, rework, downtime, and other steps that don't add value for the customer.

Most teams will use both types of tools in their projects. Data analysis is addressed first below, but your team may decide to use the process analysis tools first (or divide the team so you can conduct both data and process analysis simultaneously).

No matter which of these paths your team follows, there are three phases of root cause analysis:

1. **Exploring**: Investigating the data and/or process with an open mind, just to see what you can learn.
2. **Generating hypotheses about causes:** Using your new-found knowledge to identify the most likely causes of defects.
3. **Verifying or eliminating causes:** Using data, experimentation, or further process analysis to verify which of the potential causes significantly contribute to the problem.

While it's easy to lay out these stages in a nice sequence in Table 12-1, reality is not always that clean. Your team may have theories before it begins its

	Data Analysis	Process Analysis
Exploring	**Approach:** Examine the data gathered in the Measure phase in many ways to discover clues to the underlying cause of problems. **Tools:** Pareto charts Run charts (time plots) Histograms (frequency plots)	**Approach:** Generate process maps that capture the reality of what actually happens in the process. **Tools:** Basic flowchart Deployment flowcharts
Generating Hypotheses	**Approach:** Use the lessons gleaned from the exploration to generate ideas about the cause of defects. **Tools:** Brainstorming Cause-and-effect diagrams	**Approach:** Use the process maps to identify areas where the process steps, responsibilities, or outcomes are unclear or produce no-value-added work. Analyze the process for where steps add value or just add cost. **Tools:** Brainstorming Value analysis
Verifying Causes	**Approach:** Gather additional data or use pilot testing/experimentation to see if the suspects are guilty. **Tools:** Scatter diagrams Coded or stratified versions of the "exploring" tools	**Approach:** Gather data to quantify delays/lost time in various process steps. Make deliberate changes in the process to see if the identified problems disappear. Try the changes out on a small scale in case they do not work. **Tools:** Data collection tools Process maps and documentation

Table 12-1. The Defect Detective's Toolkit

exploration; if you have trouble generating theories, you may need to return to the exploration stage, and so on. **In fact, most teams can expect to go through several rounds of exploration-hypothesis-verification before pinpointing the root causes of problems.**

Even if you go through these stages out-of-sequence or need to revisit a stage, being clear about your purpose will help your team keep its meetings focused: "We're here to generate theories about causes" ... "We're here to talk about how we can verify the three causes" ... "Today's meeting will focus on how we'll analyze our data."

Data Analysis: Exploring

What's the purpose of data analysis? To turn numbers into meaning. Unfortunately, that's not as simple as it sounds, especially for new Six Sigma teams. There are three principles that can help you decide how to analyze data:

1. **Know what you need to know.** As we all know, there are a lot of numbers floating around out in the workplace, and it's easy to get drowned. Revisit your project charter and problem statement regularly to keep in mind what the team is trying to accomplish.

2. **Have a hypothesis.** There are dozens if not hundreds of ways to analyze data, and a Six Sigma team can waste a lot of time following blind alleys if they aren't careful. Now that you have all the evidence (à la Sherlock!), having a hypothesis can help you decide how to analyze that data. For example:

   Hypothesis: The rise in complaints at Chez Chic Restaurant is the results of having newer, inexperienced waitstaff.

   Analysis Approach: Divide the customer complaint data into two sets—data from customers served by new staff and data from people served by experienced staff. Look for systematic differences in the two sets.

   The caveat here is that you have to be open to the possibility (even probability) that your hypothesis is wrong! You can't ignore data that contradicts your suspicions. Experienced data collectors know that, in fact, most of their theories are wrong. But learning what *isn't* true helps them ultimately pinpoint what *is.*

3. **Ask *lots* of questions about frequency, impact, and type of symptoms associated with a problem or defect.** If you limit your investigation to one hypothesis or one question, you won't ever know if you've asked the *right* question. A better approach is to ask lots of questions about your data, and find out through analysis which of those questions are important and which aren't. Here are just a few questions the restaurant team above might ask:
   - Do customers served by new waitstaff complain *more often* than other customers? *(frequency of the problem)*
   - What do customers of new waitstaff complain about? How does that

compare to complaints received from customers of experienced waitstaff? *(type of problem observed)*

- If customers of new waitstaff complain more, does it mean they are less likely to return? *(impact of the problem)*

Your team needs to have a deep understanding of the problem in order to make sound choices about where to spend its time and where to implement solutions. Otherwise you can end up wasting three months fixing a problem that occurs infrequently or that has no impact on customers.

Applying these principles will put your team's data collection efforts to the test; here's where you get to use the stratification information (see pp. 135-136 in Chapter 9 and p. 165 in Chapter 10) you identified and gathered in the Measure phase. The restaurant team described above, for example, would need to have coded the complaint data by "new" vs. "experienced" waitstaff in order to perform its analysis.

Logical Cause Analysis

The principles described above are captured in a process called Logical Cause Analysis, the basics of which will likely be familiar to anyone who has tried to troubleshoot a car that won't start or a toaster that won't work.

In fact, let's say you've had a toaster that worked fine until this morning: all of a sudden, it won't even let you put the bread in, let alone toast it. You theorize that maybe it's not getting any electricity, but your "data" shoots that theory down in a hurry: the toaster's plugged in and the little red light on the side of the toaster is on, suggesting that the juice is getting through. Then you speculate that it's a loose wire inside, so you grab the old screwdriver, and ... oops! Don't forget to unplug it! And so on. It's pretty simple: when your data doesn't support your hypothesis, your hypothesis is probably wrong.

Although the Analysis process can get complex on complicated equipment or an insurance process with 50 steps, the basic concept is simple: **Ask questions that help you identify differences or changes between the process or material or methods or people that have the defects and similar ones that don't have them.** How does my toaster differ from a new one? It's older, loaded with breadcrumbs, off the counter last night when the cat dove behind it after an invisible mouse, etc. Here are some more useful logical questions to start analyzing for cause:

- **Do the defects clump up in categories?** What's different about the people/methods/process steps with the defects?
- **Do the problems appear more in one place than another?** What's different about those places compared to similar units? If sales of your new product are significantly lower per store in Los Angeles than in San Diego, you need to determine what's different in or about the City of Angels. And don't just look at differences in the product being sold. What's different about the sales force, store design, processes, customers, etc.? Get the idea? You're looking for some hidden difference in the problem or process.
- **Are there times when the defects really are prevalent?** What's unusual about those times compared to others? What if my toaster worked at night but not in the morning?
- **Are there any things or variables that change as the problem or defects change?** We're looking for a correlation here, some relationship that's not simply due to chance. For example, is there any correlation between the size of accounts in Los Angeles and variations in monthly sales? Is the correlation the same in San Diego?

Asking these questions will help you test your theories, knocking them down one after another until only one or two possible causes are left standing. What then? Well, Sherlock Holmes had a line on this, too. In a story called *The Sign of the Four*, Holmes told his sidekick, Dr. Watson, that "when you have eliminated the impossible, whatever remains, *however improbable*, must be the truth."

Introduction to the Power Tools

As noted above, there are many different ways to analyze data and many tools a team could use. But often, the following three tools will help the team pinpoint causes:

a. **Pareto Charts:** a special type of bar chart that helps a team focus on the components of the problem that have the biggest impact. Used with discrete or attribute data.
b. **Run (Trend) Charts:** a very important tool that helps a team look at whether there are patterns over time in the problem. Used with continuous data.
c. **Histograms (Frequency Plots):** Used with either continuous data or counts of attributes (discrete data).

To find specific instructions on how to create and interpret the power tools described below, look in Chapter 13. The information in this chapter focuses on how these tools can help a team.

Using Pareto Charts to Find the "Vital Few"

The Case of the Defective Invoices

Staff in the Accounts Receivables Department at Work World Office Supplies have formed a team to reduce the number of incorrect invoices that customers have received. They have identified 16 defect opportunities in their process (the same 16 defects that appeared in Chapter 9):

(1) Customer name
(2) Contact name
(3) Customer address: street and number, city, ZIP, mail stop
(4) Account number
(5) Purchase order number
(6) Items ordered
(7) Quantity of items ordered
(8) Discounts
(9) Total Price
(10) Tax
(11) Shipping costs
(12) Payment due date
(13) Remittance address
(14) Printing errors
(15) Folding/stuffing errors
(16) O-time delivery of invoice to customer

The team has collected data on the number of defects for all 16 opportunities. Work World is a high-volume business, and in just two months the team found 1267 defects overall. To help them see if any of the types of defects were significantly more important than others, they created a special type of bar chart called a Pareto chart (Figure 12-2).

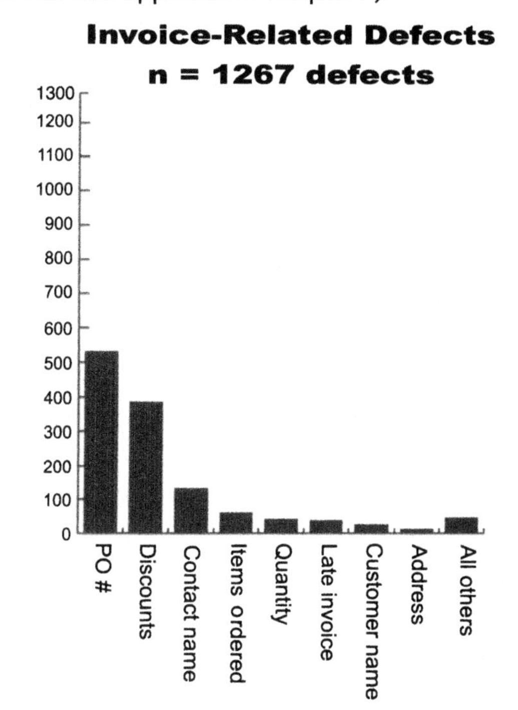

Figure 12-2. Invoice Pareto chart

Pareto charts are based on the "Pareto Principle" that *80% of the effects we see are due to 20% of the causes*. The split isn't always exactly 80 and 20 in real data, but the effect is often the same.

Two things to note about Pareto charts:

1. The vertical axis needs to be at least as tall as the *total count number.* That's the only way you can visually judge how much any given type of defect contributes *relative* to the problem as a whole.
2. The categories of data (individual bars) are arranged in descending frequency.

When looking at a Pareto chart, you need to determine whether the top "vital few" bars account for most of the problem. If so, then your next course of action is to focus on the potential causes for those contributors. (The Work World team who created the chart above, for example, would want to look at potential causes of defects in purchase order numbers and discount rates.) If you don't find a Pareto effect, then you need to look for potential causes that affect all the types of defects equally (or nearly so), or look for other patterns in your data.

Here's another example. The Gemini Point of No Returns team that opened this chapter wanted to look into the reasons for returned materials authorizations (RMAs), which is what happens when customers call wanting to return products they received. At Gemini, RMAs are a key quality measure: if customers are getting the products they want, when they want them, and in the right quantity, there should be few RMAs. So the Gemini team created the Pareto chart shown in Figure 12-3 based on the reasons why customers wanted to return products.

Looking at this chart, you can see that based on *counts* of the various types of returns, there isn't really a "Pareto effect"—there are no "vital few" causes contributing the most to the problem.

What to do? Look at the data another way. This time, the team looked at what accounted for the largest *costs* associated with the returned products (RMAs), as shown in Figure 12-4.

Looking at the data this way, the team has found that while "system incomplete" errors do not occur that much more often than other types of defects, they do account for a big majority of the *impact* (costs) associated with RMAs. However, knowing that "system incomplete" accounted for 68% of the RMA costs isn't enough to help a team move on to the "solution" stage. The team would now have to investigate possible causes for having incomplete systems in their shipments.

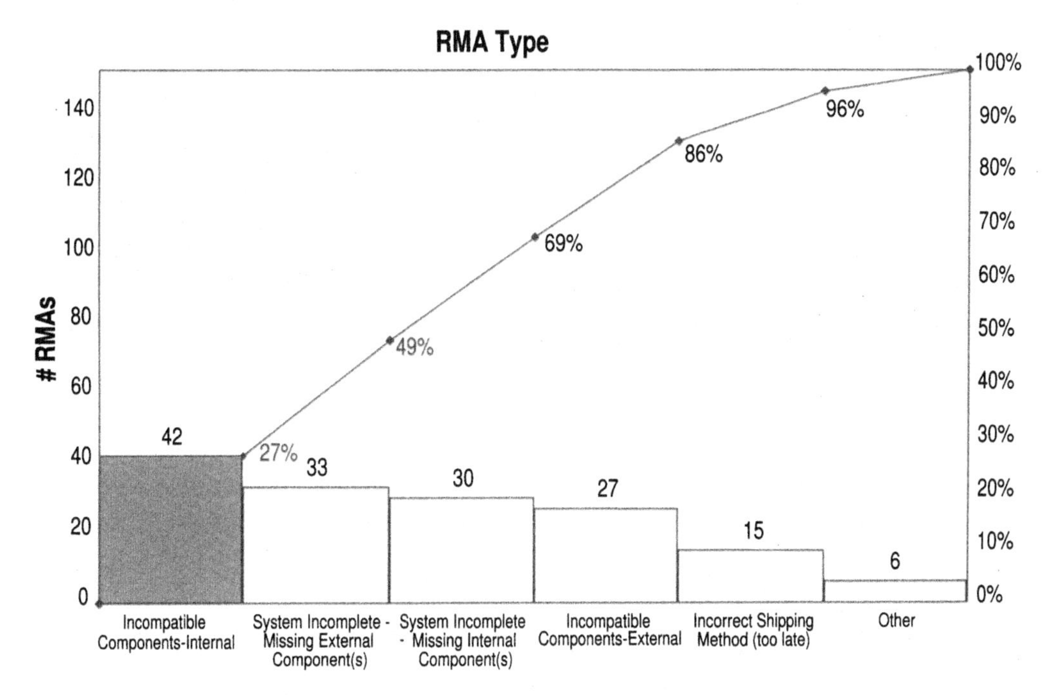

Figure 12-3. Gemini Pareto for types (reasons for returns)

Using Pareto Charts

Pareto charts are popular with teams for a number of reasons:

- The data is easy to gather (usually *counts* of different parts of a problem, which is discrete data, or some measure of impact, which might be continuous data).
- They are easy to construct.
- They are easy to interpret.

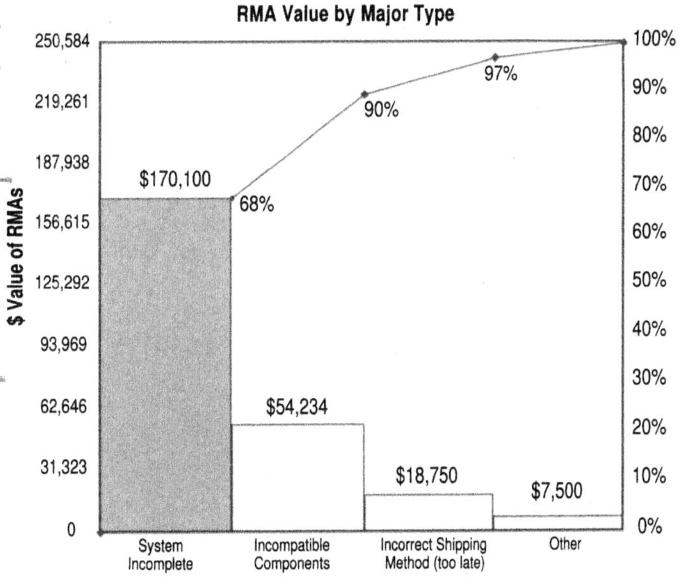

Figure 12-4. Gemini Pareto for impact (cost of types of returns)

- They help the team leverage their efforts by focusing on the parts of the problem that have the biggest impact.

6σ

For more examples of Pareto charts and instructions for creating them, see Chapter 13, p. 236-239 for instructions.

To use a Pareto chart, you need to make sure you have discrete or category data—it won't work with continuous measures (like weight or temperature).

The only limitation of Pareto charts (and bar charts in general) is that they represent a static snapshot of the data at a certain point in time. But processes are constantly operating, generating new data, and possibly changing. So a team will almost always want to look for patterns over time: is the problem increasing? decreasing? remaining steady? did it rise then fall back again? The tool to answer these questions is a run chart (also known as a trend chart or time plot).

Using Run/Trend Charts to Find Patterns over Time

Panic Time at the High School

A Six Sigma team at the local high school was in a panic. The school's ACT scores—one of the main standardized tests given each year to junior students—had been dropping recently. From a high of 22.5 just two years ago, the schoolwide score was just 21 (see Figure 12-5).

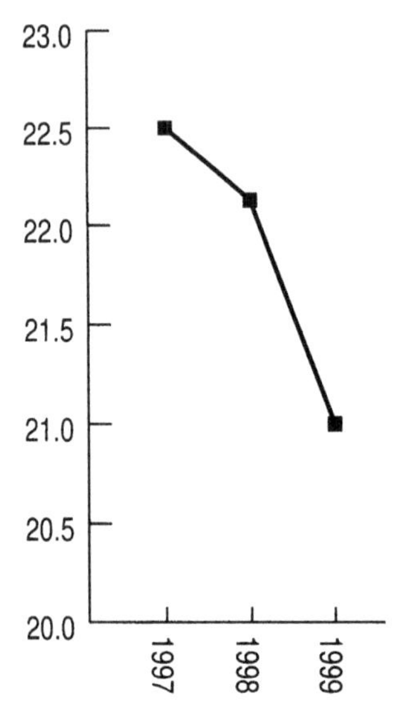

Figure 12-5. Panic time at the high school

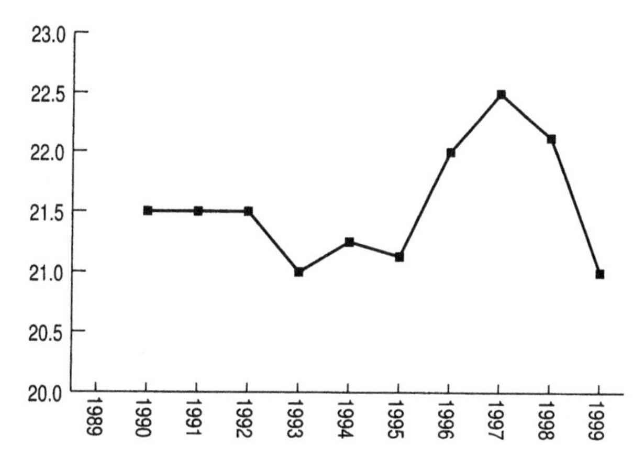

Figure 12-6. More data from the high school

The team wasn't sure what to do, but their Coach advised them to first look at more data. A three-year trend is not really enough to base improvement decisions on! So one member of the team dug through the records and came up with seven more points. The Coach then helped the team plot these data points over time (Figure 12-6).

Seeing the bigger picture didn't exactly relieve the team of its stress, but it did provide them with a better perspective on the problem: the educational and test-taking processes at their school had returned to a more normal level after a few years of unexpectedly high numbers. Their path forward: examine the curriculum and teaching practices from 1996 to 1998 and see what was different during those years compared to the other years.

Like this school team, all Six Sigma teams need to gain perspective on what their process is doing now versus what it has been capable of doing in the past. One way to do that is to create run charts, which are also known as time plots or trend charts. Run charts are constructed from a measurement that has been gathered over time (usually at regular intervals, such as hourly, daily, weekly...) and then plotted in time order.

To interpret a run chart, you have to understand something about variation. As hard as it may be to believe, every cause of variation can be put into one of two categories: *special causes* and *common causes*. The difference is important because teams need to eliminate special causes **first—before** they work on common causes.

Remember the toaster described earlier in this chapter? It worked OK until the cat knocked it off the counter, and now it doesn't toast at all. Unless you have an especially rambunctious cat, getting knocked off the counter is likely an unusual event, not part of this toaster's everyday life. It is, in fact, a special cause, impossible to predict precisely. (Cats sometimes knock things off shelves; *if* and *when* are up to the cat!)

But what if the problem wasn't that the toaster didn't work, but that it toasted with a lot of variation—though always set on "medium," sometimes it burns the bread and sometimes it barely toasts it? There are probably a number of causes for this variation, ranging from the setting mechanism on the toaster, to the type and shape of the bread inserted, to the thickness of the bread, the moisture content of the bread, etc. These factors are present in the process all the time, though their influence varies from day to day. They are common causes that are part of the normal toast-making process: it's hard to say what combination of variables contributes most of the variations in the "toastiness" of the finished product.

Though special causes are hard to predict, most of them are relatively easy to identify if you know what to look for. There are, in fact, a set of **special cause signals** you can apply to the data points on a run chart to see if there are any special causes present. Figure 12-7 shows an example of one of these signals, six or more consecutive points increasing. (Other signals are described in Chapter 13, along with instructions for constructing a run chart).

Note that run charts don't identify exactly *what* the cause is; they just help you decide whether to look for something special or common in the process. And, since they are usually time-based, they can tell you *when* special causes seem to be in effect. These special events point a finger at different types of problems in a process and can help your team decide what to do next in its analysis.

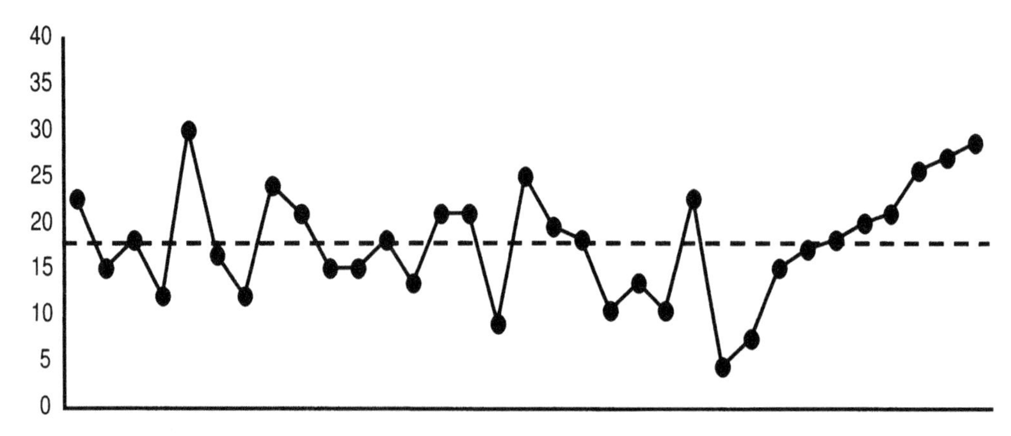

Figure 12-7. Run chart showing a special cause

Using Run Charts

See the run chart instructions on pp. 238-241.

Run charts (time plots) are also popular with teams because, like Pareto charts, they are easy to construct and easy to interpret, and can be used to plot either continuous measures or count (discrete) data. The tricky point can sometimes be gathering sufficient data. Interpretation of a run chart will be more reliable if you can gather at least 25 data points. However, if you are plotting monthly or yearly data—like the school team described above—you may have to make do with fewer points; just be sure to take your interpretations with a grain of salt!

The rule of thumb is that if you have data *collected* over time, you should *plot* it over time.

Trends at Gemini

The Gemini team displayed some of its data on two trend charts, showing average daily sales (by count) and average daily counts of returned computer components. Both charts show some normal variation until about March, when sales and returns both trended upwards until December, when they both fell slightly.

Using the special cause tests taught to them by their Master Black Belt, the Point of No Returns team concluded that (1) the rise in average sales most likely indicated a significant trend upwards, (2) it looked like the returns (RMAs) might be increasing slightly faster than sales.

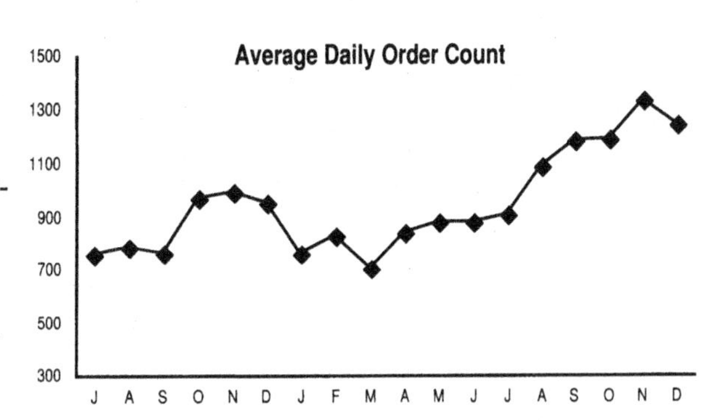

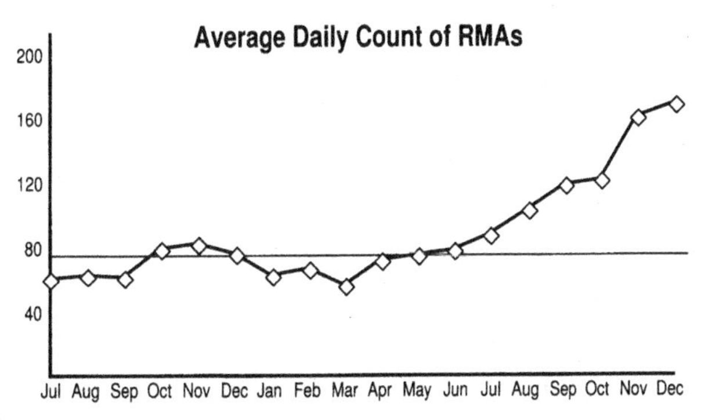

Figures 12-8 and 12-9. Gemini run charts of orders (top) and returns (bottom)

Using Histograms to Understand the Process

Because Six Sigma teams are interested in understanding and reducing variation in their processes (since variation is often at the core of defects), they will usually make a histogram about the same time as they create a run chart, from the same data. A histogram, such as that shown in Figure 12-10, gives them a better view of the center, distribution, and shape of the data.

The histogram displays continuous data, such as time, measures of length or weight, dollar amounts, and any other measures that can be sub-divided into fractions. The data is displayed on a chart on which the horizontal axis is marked off in increasing values (from right to left), and the vertical axis shows the frequency. You can use either bars or individual symbols (such as a stack of dots) to reflect the counts (frequency) of each data value.

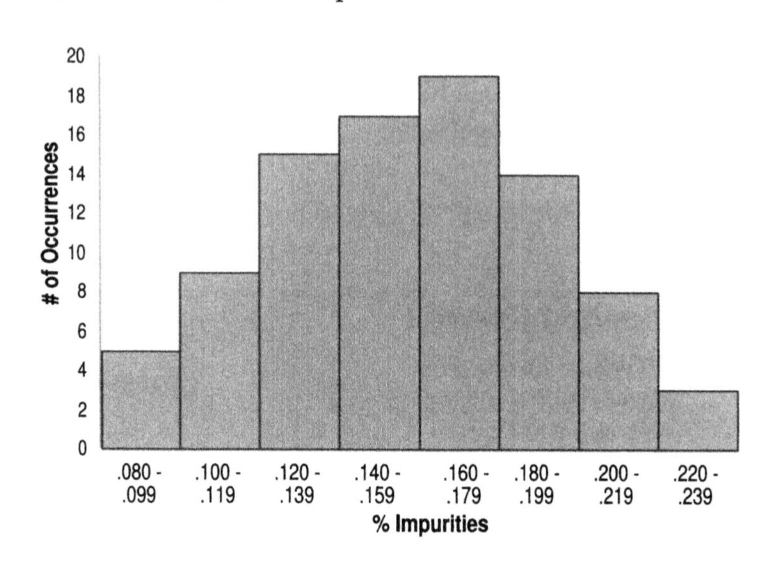

Figure 12-10. Histogram of batch impurities

A histogram allows the team to see how much variation there is in the item being measured; the center of the data and its shape. This may not sound like much, but it helps the team answer many important questions:

- Is the process centered on the required customer requirement, or is it off one way or another?
- Is the process so variable that it frequently misses the customer requirements and causes defects?
- Is the process skewed somehow, with data piling up at unexpected points?

If you have continuous data, or count data that can be plotted on a scale, you should always construct a histogram when your team starts its project to establish a baseline. You can then reconstruct the chart after you make changes, to see if those changes had the desired impact. Histograms are often used by teams to show a "before" and "after" snapshot of the process under review.

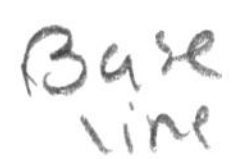

Before moving on...

At the end of the *Exploring Phase of Data Analysis,* you should have:

- ✔ Pinpointed the time, place, and description of the defect (when it occurs and does not occur, what happens when it occurs and doesn't occur, whether it is present all the time or only at specific times).
- ✔ Data charts (Pareto charts, run charts, and/or frequency plots) that demonstrate a logical probing beyond surface symptoms. (For example, if you discovered a Pareto effect in your first Pareto chart, you should have a second that breaks down the "vital few" causes from the first chart. If a run chart gave a signal of a special cause, you should have detailed information, and perhaps additional data, on what was happening at that place and time.)

Data Analysis: Generating Hypotheses About Causes

Do you know the secret to finding effective solutions?

You have to know specifically what problem you're trying to solve ... a task that's harder than it sounds! Just think about all the problems in your organization that have been "solved" over and over and over again. Obviously, whatever solutions were put in place had little effect on the underlying cause of those problems.

Do you know the secret to uncovering the underlying causes of problems?

You have to dig deep beneath the surface symptoms of that problem. By the time you finish the exploration phase of your analysis, your team should have both knowledge about when, where, and how the problem manifests itself, and lots of ideas about possible causes. The trick at this stage is keeping focused on the problem definition and organizing your detective work to make sure the causes you choose to study address that problem.

The two most common tools used at this point are the **cause-and-effect diagram** and the **relations diagram.** They provide a critical link that will help your team make sure you've isolated the underlying or root causes of a problem. Two important notes about these tools:

- First, they help you think logically about *potential* causes of a problem; you will still need to gather data to verify which are the *real* causes of a problem.
- Second, their effectiveness is directly related to the creativity and depth of the thinking that goes into creating them. That's why these tools are best used with your team as a whole—you want many minds brainstorming ideas so you have a broad and deep list of potential causes.

Cause-and-Effect Diagram (Fishbone Diagram)

Up to this point, your team has been collecting data about process outputs (the Ys, as we called them in the Measure phase). In order to make sure you're not overlooking possible causes (and measures) of defects, the team can use a tool called a fishbone or cause-and-effect diagram, which is really a structured brainstorming tool (see Figure 12-11).

A cause-and-effect diagram summarizes the team's problem statement in the "head" of the fish, with potential causes arranged sets of "bones" linked to the head. The smallest bones are the most specific types of cause that contributes to the next larger level of bone, and so on.

Cause-and-effect analysis lets a group start with an "effect"—a problem or, in some cases, a desired effect or result—and create a structured list of possible causes for it.

Benefits of cause-and-effect diagrams include:

- It's a great tool for gathering group ideas and input, being basically a "structured brainstorming" method.
- By establishing categories of potential causes, it helps ensure a group thinks of many possibilities, rather than focusing on a few typical areas (e.g., people, bad materials).
- Using a cause-and-effect diagram to identify some "prime suspect" causes gives focus to help begin process and data analysis.
- They help get the Analyze phase started, or keep the thought processes moving after an initial exploration of data and the process.

Cause-and-effect diagrams do not tell you which of the potential causes is the culprit. But creating the diagram is an extremely useful team exercise because it forces team members to consider theories in addition to their own "pet ideas." The resulting diagram is also a good way for the team to document which theories it has considered, which have been targeted for further investigation, and, ultimately, which have been verified.

To see how the Gemini team used a cause-and-effect diagram to brainstorm causes for its returned orders, look at Figure 12-11.

Obviously, whatever theories the teams brainstormed about the likely causes of defects will have to be tested against the data collected. Think the problems are caused by the use of temps on Fridays? The data shows the same problems appearing Monday through Thursday when there's nary a temp on site! Out goes the "blame the temps" suspected cause. Or, as Holmes might say, "Elementary, my dear Team Leader."

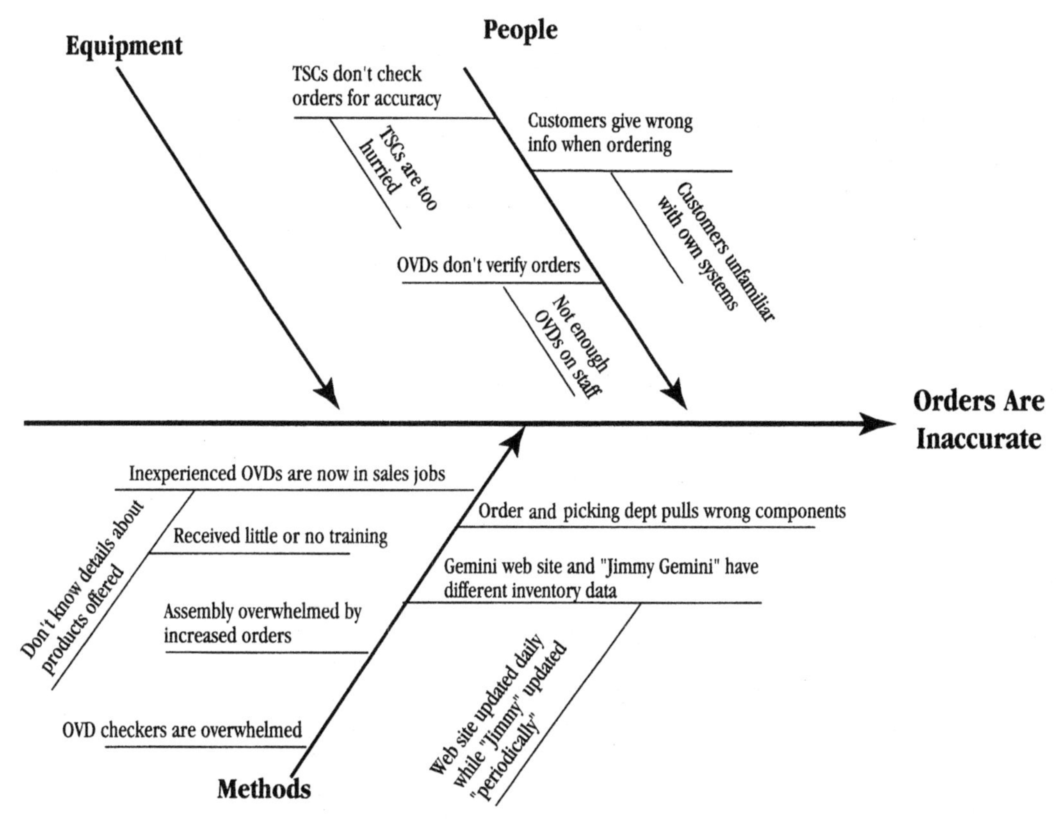

Figure 12-11. Gemini cause-and-effect showing potential causes of inaccurate orders

Analyzing Complex Systems: The Relations Diagram

Cause-and-effect diagrams are extremely adaptable tools that will help you in most situations. But there are times when a problem is so complex that that linear relationships depicted by a cause-and-effect diagram can't really capture the situation. In that case, a **relations diagram** (sometimes called an **interrelations diagram** or **digraph**) may be a more appropriate cause analysis tool.

Banking on Success

Nation's Federal Bank, a 10-state banking conglomerate, was in trouble. A high rate of employee turnover in the past years had led to a marked decrease in customer satisfaction and similar loss in organizational effectiveness. Since the problem was a "fuzzy" people issue, and not linked to any single process, the Leadership Council assigned its most experienced team members and Six Sigma support staff to study and improve the issues. The team defined a defect as "any employee who leaves Nation's Federal with less than a year of service." They later refined that definition to exclude employees whose resignation was caused by a family crisis.

The data analysis revealed that turnover was highest among younger employees (under 30) and mid-level, middle-aged staff (ages 40 to 55, bank manager level and above). They suspected that the causes of turnover would be different for the two groups, so they addressed each group separately in their cause analysis.

Here are some reasons they suspected that the younger employees might leave:

Low pay scales   Uncompetitive benefits   Lack of training

Poor job satisfaction   Stressful working environment

Difficulty in hiring appropriate people

Increase in competition for good employees

Lack of opportunities for advancement

To help them decide which potential causes to investigate further, the team selected on pair of causes and asked, "Does one of these factors cause or influence the other?" They did this with each pair of factors and summarized their results in a relations diagram (Figure 12-12).

Interpreting a relations diagram is a matter of counting the number of "in" and "out" arrows for each potential cause: those with the most "out" arrows are

Potential Causes for High Turnover in Younger Employees

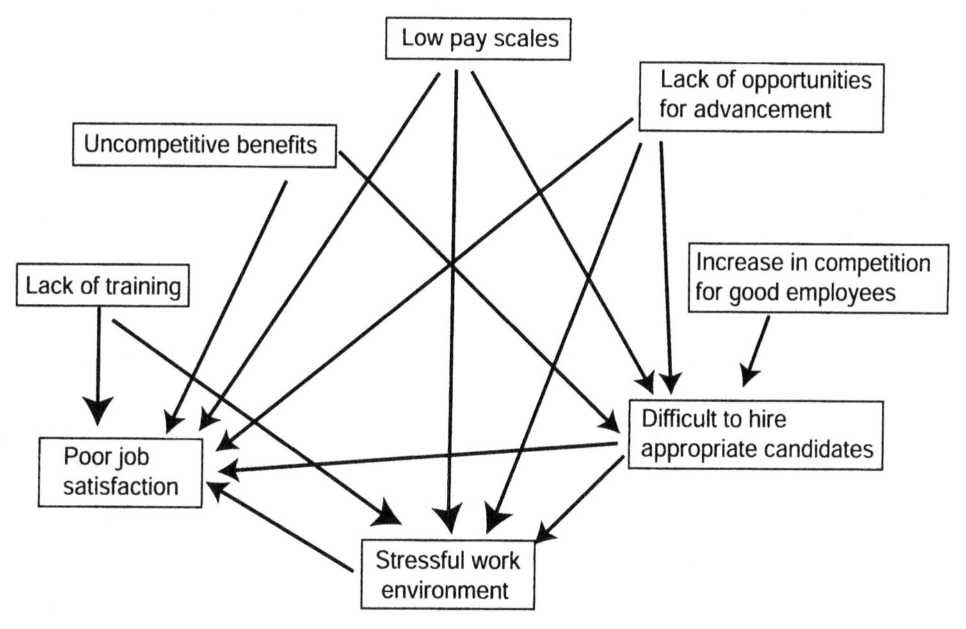

Figure 12-12. Relations diagram on high turnover

underlying or potential root causes. In this banking team's example, two of the top boxes—low pay scale and lack of opportunities for advancement—have the most "out" arrows, and therefore warrant further investigation. (In contrast, "poor job satisfaction" and "stressful work environment" have a lot of "in" arrows—that means they are the *effect* of other underlying *causes*.)

6σ See the Relations Diagram instructions on p. 252-253.

Before moving on...

At the end of the *Hypothesis Phase of Data Analysis*, you should have:

- ✔ Brainstormed ideas that represent diverse thinking about potential causes.
- ✔ A cause-and-effect or relations diagram showing potential causes that are clearly linked to the lessons learned from your data exploration.

Data Analysis: Verifying Causes

There are three ways to verify causes: one logical, one statistical, and one experimental. Let's start ... well, logically.

Causal Logic

Say that the team at Gemini decides that the cause of many of the wrong orders are glitches in the order management systems used by the salespeople. This might account for some of the wrong orders, but it does not explain why Gemini has more problems appearing with less-sophisticated customers (small businesses and new computer users). If the order management systems were at fault, you'd expect problems to affect *all* customers more or less the same. So the systems cause hypothesis might explain what we see (wrong orders), **but it does not explain what we do not see:** wrong orders for both small business orders and new individual orders.

For a causal hypothesis to really stand up, it has to pass the logical test of explaining ***both what we do see in the data*** and ***what we might expect to see, but do not.*** Here's a causal logic example from Gemini:

The team might theorize that the transfer of order checkers into the sales force led to more wrong orders being taken. In other words, the new staff are inexperienced and make errors when taking orders from customers. But inexperienced people make lots of different kinds of errors. If the data showed that the new salespeople had no problem with pricing or other parts of the order, it would weaken the hypothesis that their inexperience alone is at the bottom of the problem. And what if the number of returned orders was already moving up well before the order checkers moved into sales positions?

Whichever hypothesis best explains what the team sees and does not see in the data probably explains the most likely causes of the problem. It's pretty much the same method the great Sherlock used a century ago; the power of logic to test causal theories remains strong today. Of course, an armload of data on the computer helps, too!

Statistical Verification of Causes

There are two basic approaches to using statistics to determine cause-and-effect relationships:

- Judging the degree to which a cause (X) and an output (Y) are correlated. This can be an approximate assessment using the patterns seen in a scatter plot (see below), or an actual calculation using regression and correlation formulas (see the end of this chapter).
- Codifying or stratifying the data to expose patterns (or the lack thereof).

A. Correlation: Using Scatter Diagrams to Understand Potential Cause-and-Effect Relationships

Back to Gemini

The team suspected that there was a correlation between the number of sales and the number of items that customers returned. The hypothesis was that the more orders taken, the larger the number of items returned. They collected the data and displayed it on a scatter diagram, where the number of sales was plotted along the horizontal or X axis, and the number of returns was plotted on the vertical or Y axis (Figure 12-13).

The scatter diagram showed a positive correlation, and so supported the hypothesis that there was some relationship, but as the team's Black Belt pointed out, increased sales by themselves should not cause the number of returns reported. There had to be some other, related causes. The team needed to dig deeper.

Scatter diagrams provide another way for teams to test hypotheses about the causes of problems. These diagrams used "paired data" to test the correlation between X variables and Y outcomes. Paired data, as the name suggests, are data of two variables gathered at the same time on each item being measured. The "pairs" typically reflect a potential cause and its effect: like the complexity of a form that needs completion (an X variable or

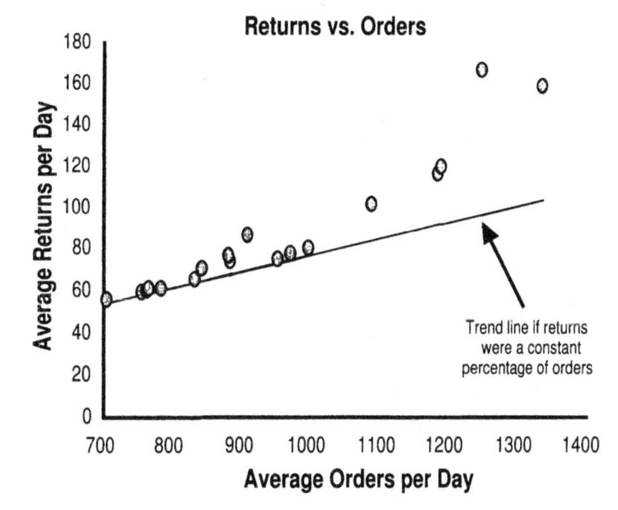

Figure 12-13. Gemini scatter plot

potential cause) and the actual time it takes to complete the form (a Y outcome or effect). The paired data are plotted along the X and Y axes of a chart, and then analyzed for signs of correlation, as shown in Figure 12-14.

If the data points cluster in a pattern running up from left to right on an angle (Figure 12-14a), that suggests a *positive* correlation between the X variables and the Y outputs. Simply, such a pattern would suggest that "the more X variable you have, the more Y variable you have." Other patterns might indicate a *negative* correlation (Figure 12-14b), no correlation (Figure 12-14c), or more complex relationships (in this case a "curvilinear" pattern) (Figure 12-14d).

Scatter diagrams showing strong correlations do not prove that the variable tracked on the X axis is actually **causing** the output measured on the Y axis, but they do suggest that they are related to one another in some way and do not occur together simply by chance. So a team can use a scatter diagram as one test of causal theories, but not necessarily the only or conclusive test. In more sophisticated correlation analyses, many different scatter plots will be developed to compare a number of suspected factors.

Figure 12-14. Basic scatter plot patterns

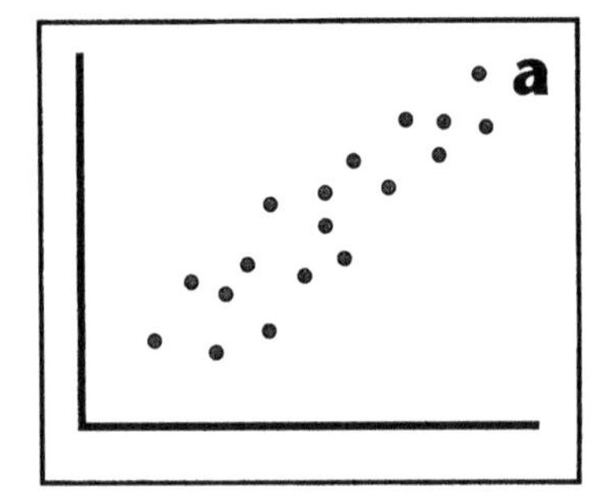

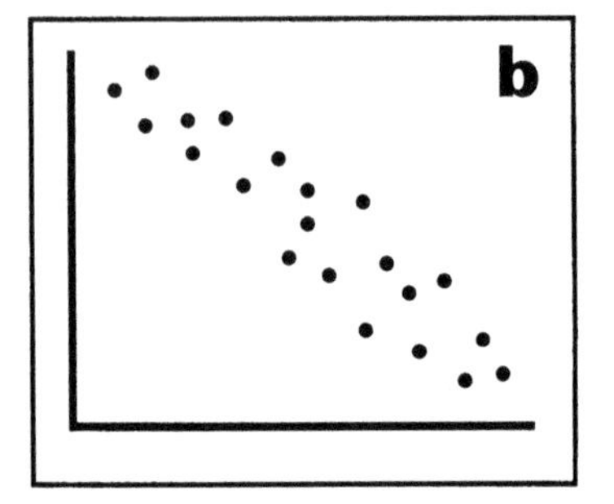

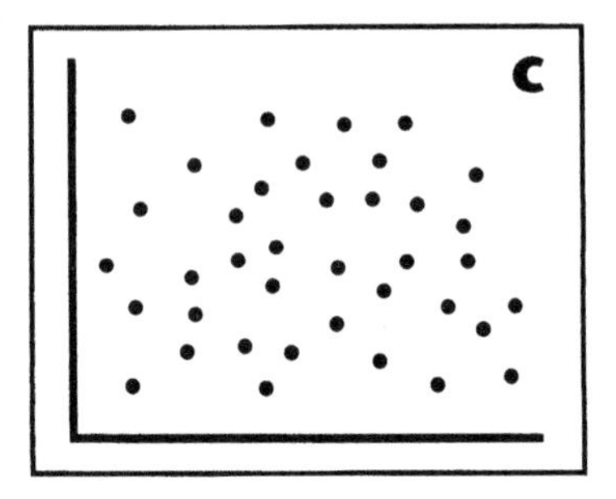

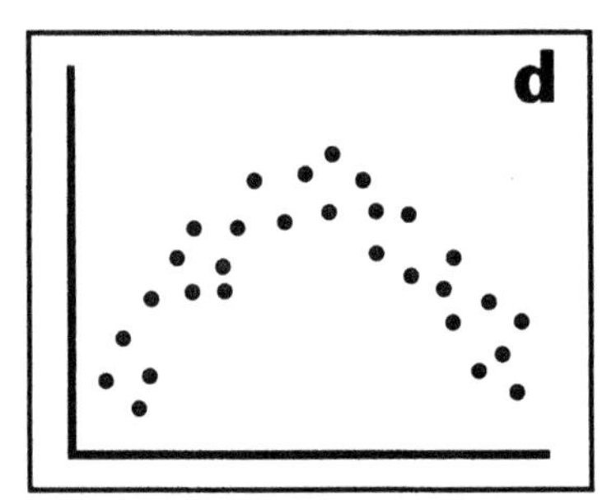

B. Codifying/Stratifying Data to Verify Causes

Pizza Time!

Everybody working at Six Sigma Pizza just knew that the most "late deliveries" occurred right at dinner time, when the number of orders was the highest. But they wanted to be able to show the data that would prove their point. As a truly customer-focused organization, they had data going back about a year on late deliveries.

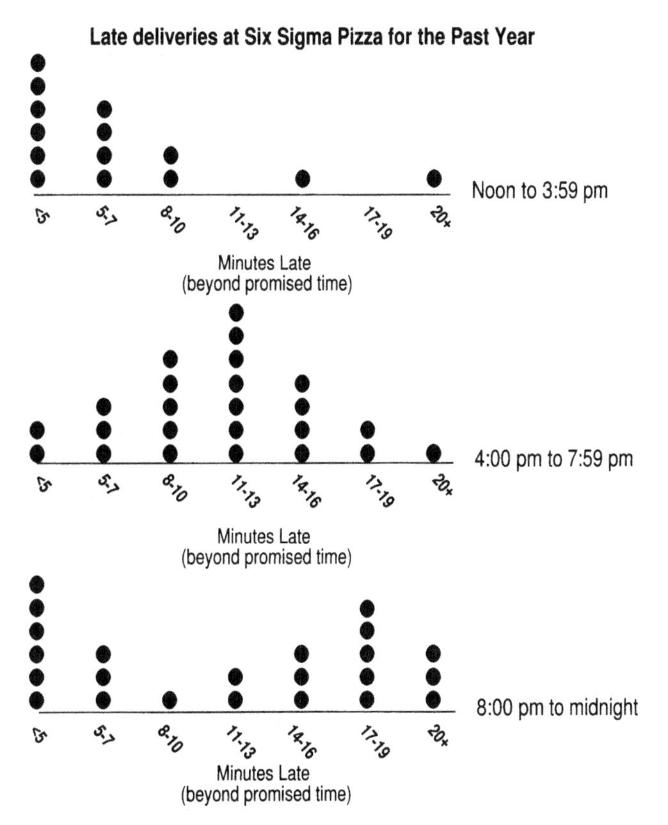

Figure 12-15. Stratified histogram

The team divided them into three categories, based on the time of day the late delivery was made. Figure 12-15 shows their charts.

After charting the data, the team saw that late deliveries were almost as common in the late shift as over the dinner hour—and what was worse, the late night deliveries were more often *really* late. The team had to rethink their main premise—that the dinner hour rush would account for most of the late deliveries. Based on these patterns, they had some more detective work cut out for them.

The Six Sigma Pizza team learned a valuable lesson by stratifying their data into different categories: not all "late deliveries" were made the same! All of the stratification factors you identified in Measure can come in handy here when you're trying to confirm cause-and-effect patterns.

6σ

See Chapter 13, pp. 254-255, for more on scatter plot interpretation.

Whenever your team comes up with a hypothesis about what's happening with the data you've gathered, look to see if there is stratification information that can help you verify or disprove that hypothesis.

Pilot Testing/Experimentation to Confirm Causes

In some cases, it is difficult, impractical, or simply too time-consuming to collect data that will verify whether a particular cause is the culprit behind high defect levels. In these cases, a viable option is to simply change the process to eliminate the suspected cause, and monitor the effects. But don't just go about changing things willy-nilly! This approach works only if you treat the change as a formal test or pilot experiment:

6σ
See the Stratified Charts Instructions on pp. 255-257.

- You know exactly what it is you will change and how you will change it.
- You know what you're going to look for.
- You try the change on a small scale first.
- You think about and make plans for mitigating potential side-effects.

For example, suppose your team believed that the "script" used by telephone service reps led to unsuspected confusion among customers and a number of incorrect orders being submitted. There isn't really any good way to measure script effectiveness because the customers don't *know* they were confused. Instead, experiment with new scripts to see if the problem disappears—but try out the new script with just a few operators, rather than disrupt the whole process.

Before moving on...

At the end of the *Verification Phase of Data Analysis*, you should have:

✔ Data that demonstrates which of the potential causes you identified are, in fact, responsible for the observed effect (defect).

Process Analysis: Exploring

In all likelihood, some if not all of the people on your team work with the process being studied in your DMAIC project. They probably think they know it well. That's why they'll be surprised at how much they'll learn by exploring that process! The tools and methods described here will help your team take an objective look at their process and expose places where confusion and variation in procedures contribute to defects.

Process Mapping

During the Define stage of its project, your team likely created a high-level process map called SIPOC. You may have even used it to drill down to a little more detail in order to identify a place to start taking measures.

But there is a lot more that happens in a process than can be captured in a SIPOC diagram. During the Analyze stage, therefore, your team will likely want to map in more detail the portions of the process where data and experience lead you to suspect some buried Xs (causes).

How much detail will you need? Enough so that daily decisions and ordinary activities are included, as well as "loop-backs," when the process has to go in reverse to get missing information or parts before the unit recycles into the mainstream process. Action steps should be detailed enough to capture everyday delays and inspection points. It's hard to say how much detail is too much, but if you're at the level of writing "Pick up phone" and "Speak into phone," you're getting too deep into the woods. Come up a level and put all these sub-sub-substeps under one step: "Answer Phone."

Putting together a detailed process map is always a lively and rewarding experience for team members. They come to respect the amount of work other people do in the process, and they also begin to realize just how much variation there is in the methods that people use, particularly in service processes.

Another effective tool at this stage is a deployment flowchart, which adds a unique element to a normal flowchart: *who* is responsible for which activities. Deployment flowcharts are particularly helpful when working in a process that has a lot of handoffs between individuals or groups (which is true of many service processes).

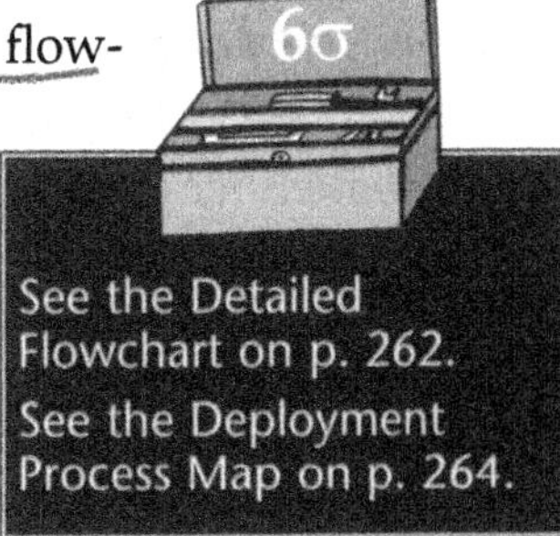

See the Detailed Flowchart on p. 262.
See the Deployment Process Map on p. 264.

Verifying a Flowchart

Once your team completes a detailed or deployment flowchart, you should take time to verify that what you've captured on paper is, in fact, what happens. To confirm this, track some typical units of product or services through the process for a few days. Use a traveler checksheet like that described in Chapter 9 (Figure 9-5) to do this, and be sure to describe the process in sufficient detail to capture each important step.

Before moving on...

At the end of the *Exploration Phase of Process Analysis,* you should have:

- ✔ Notes from team discussions about what really happens in a process.
- ✔ Either a detailed flowchart that depicts rework/complexity or a deployment flowchart that shows critical handoffs.

Process Analysis: Generating Hypotheses About Causes

Some ideas about potential causes of process problems will arise naturally as your team creates its process maps. If the three people who perform a process step can't agree on how that step should be done, you've identified a potential cause of problems. (You still need to verify that this confusion is an *important* cause of defects in the process; if someone performs a procedure as A-B-C and another does B-A-C, you should look to see whether that switch in order has an impact on the product/service.)

First-Level Analysis: Identifying Obvious Process Problems

Once confirmed, a process map can be the focus of a team meeting in which the members analyze it with marker in hand to highlight the following trouble spots. (Remember that you're looking for "typical" problems that occur often enough that they're part of the routine process.)

- **Disconnects:** Steps in the process where there are breakdowns in communications between shifts, between customers and suppliers, or manager and employees.

- **Bottlenecks:** Points in the process where volume of work often overwhelms capacity, slowing the entire process downstream. (If work has to wait for someone to return from vacation, you're probably looking at a bottleneck.)
- **Redundancies:** Steps in the process that duplicate activities or results elsewhere in the process; for example, the same information coming from two steps and going to the same place with the same information.
- **Rework loops:** Points where units with missing parts or information have to be sent back upstream or delayed at one step until the necessary work is done. Inspection steps often trigger rework loops.
- **Decision/Inspection points:** Process steps where a variety of checks and appraisals are made, creating delays and rework loops. In organizations plagued by high defect levels, these points abound.

Each of these types of problems is a potential cause of defects in a process. More importantly, from the customer's point of view, such steps do not add value because they either lead to inaccuracies in the output, or stretch out the time it takes to deliver product and service to the customer. Also the costs of all these snags usually get passed on to the customer in some form.

Second-Level Analysis: Quantifying Value-Added Steps

As processes get more complex, they tend to insulate people from the real reason customers patronize the business. The *value* provided to markets and society can seem almost incidental in the midst of a large organization. "Value Analysis" is a way of re-emphasizing the key reason-for-being of a business or process by looking at work from the external customer's point of view. In the analysis, we assign each process step into one of three categories:

1. **Value-Adding.** These are tasks or activities that are valuable *from the external customer's point of view*. This is critical, because almost any step can be justified in someone's eyes. "We do this because the boss wants it" does not mean a task is value-adding to the customer. The three criteria for customer value-adding steps are:
   a. The customer cares and/or would pay us for this activity if they knew we were doing it.
   b. Some change is being made to the service or product. (Moving things around is not value-adding.)
   c. This is the first and only time we're doing it. Fixes, rework, replacements, etc., only correct mistakes made before; they don't *add* value.

All three criteria must be met for a step to be called "value-adding."

2. **Value-Enabling**. There is a class of activities that allow you to do work for the customer more quickly or effectively, meaning you can deliver products or services sooner, at less cost, with greater accuracy, etc. As a simple example, you as the customer don't want to pay to have a company purchase a high-capacity, high-speed computer, but having one may help the company deliver your product or service much more quickly. The steps your business takes to enhance its value-adding capability are called "value-enabling." You need to be careful, though, to make sure you don't automatically label anything that doesn't fit into the "value-adding" as "value-enabling." There are usually very few in this group. (Note that if the high-speed computer crashes and orders have to be reprinted, that is *rework*, not more "value-enabling" work.) Also, satisfying legal, regulatory, and "good business practice" requirements (such as running a credit check before approving a loan) can be considered value-enabling.
3. **Non-Value-Adding**. These are the "rude awakening" aspects of a process, because in most organizations there are lots of non-value-adding steps. The kinds of activities that fit in this category include rework, as well as:
   - Delays
   - Inspections
   - Reviews
   - Setup and preparation
   - Transport (from one location or step in the process to another)
   - Internal report and justifications

The non-value-adding category can seem rather brutal. When you get right down to it, most of what happens in a typical organization doesn't add value in the eyes of the customer. Would your customers pay more because you interviewed 10 candidates instead of five for a customer service agent position? If you spent millions developing a product they don't really need or want? If each supervisor writes a daily report to his or her manager on problems during their shift? More than likely, no in every case.

Process Time Analysis

Dividing process steps into the three value-related categories is one way to see which steps are most important to your customers. Another approach is **time analysis**.

Time analysis can be another shocker: there's often a *lot* of idle time in business processes. It's not the people who are just sitting around, it's the things getting worked on that are idle. They sit in transport bins, file folders, in-baskets, and e-mail attachments doing … nothing. Where cycle time improvement has been a priority, time analysis has been a tremendous benefit to cutting process turnaround to minutes instead of hours, days instead of months. The need for speed—from "just-in-time" delivery to rapid product cycles to time-based competition—has driven some of the most impressive improvements in corporations around the world over the past 15 years. But many processes have been untouched and offer greater potential.

There are two components of process time:

1. **Work time**—the time actually spent *doing* something to the product or service as it flows on its way to the customer.
2. **Wait time**—the time the product or service spends waiting for something to be done. Imagine a bunch of parts, a stack of applications, or truckloads of product all sitting around twiddling their thumbs (if they had them) waiting for someone to come and work on or move them. (This is also called "queue time" or "staging time.")

Before moving on...

At the end of the *Hypothesis Phase of Process Analysis,* you should have:

- ✔ Documentation that shows what portions of the process add value and which do not (from the customer's viewpoint).
- ✔ Documentation on where time is well-spent in the process and where it is wasted (from the customer's viewpoint).

Process Analysis: Verifying Causes

In some respects, verifying causes identified through process analysis is much easier than verifying causes identified through data analysis. Why? Simply because any process step that does not add value for your customers, any step that adds time without adding value, is a *cause* of the defect of "wasted time."

You will have to do more detective work, however, if it's not clear whether confusion or variation in a process step contributes to defects that are noticed by and important to customers. If Bob always places the shipping labels dead center on the box and Alec places them in the lower right corner—so what? It's a "cause" only if it really contributes to late deliveries.

Tips for verifying process causes:

- Use causal logic, just as you would with data analysis (see p. 218).
- If you haven't already done so, create a detailed process map to see if there are important differences in process steps.

"Experimentation" in the form of pilot testing is another good tool for verifying the impact of process defects ... that is, as long as you define "experimentation" as making deliberate changes to a process (on a small scale!) and measuring the impact of that change. If you think that the Shipping Department would be able to process orders more quickly if the labels were always put in the exact same place, try that approach for a week or two and see if processing time drops.

The goal, of course, is to remove anything that does not add value in the customer's eyes, including the time it takes for rework, inspection, etc.

An added bonus for removing many of the non-value-adding steps listed above is that you will probably shorten the overall cycle time of the process, which should make your customers happier, too. Six Sigma teams often remove 70% to 80% of the time taken up by rework and other steps that don't add value from the customers' point of view. However, it will do the team no good to merely speed up a problem process: it must also eliminate the causes of defects that Analysis has revealed. Then it will be time for the Improvement stage.

Before moving on...

At the end of the *Verification Phase of Process Analysis,* you should have:

- ✔ Documentation of process changes that eliminated or reduced the problem.
- ✔ Data or other information that demonstrates a link between process problems (waste, errors, delays) and the targeted output.

Advanced Analytical Tools

In many cases, a Six Sigma team can do just fine with the basic analytical tools described earlier in this chapter. Using Pareto charts to dig deep into the "vital few" contributing factors, using run charts to pinpoint special causes, and having in-depth team discussions about how a process really does operate today (and how it could work better) will go a long way towards reducing defects.

However, there are times when you need more sophisticated tools to test the relationship between X variables and Y outcomes. They can be used when the team's hypotheses about correlation and causes rest on data that indicates change and differences, but leave nagging questions behind: "Are the things we're comparing really all that different from one another?" "Is there a significant difference in the things being measured, or is it just my personal bias?"

These kinds of questions confront Six Sigma teams when they need to

- Test that there is a meaningful difference between sets of data.
- Create a valid hypothesis about the cause of the problem.
- Validate or disprove various hypotheses about causes.
- Prove to someone who insists on numbers, not just graphs, to show the level of correlation and causation.

To answer these questions, statisticians have come up with standards that apply to most Six Sigma projects: they have operationally defined as "statistically significant" anything that has less than a 5% probability of happening by chance alone. To test for this probability, statisticians have devised a number of tests, including ...

- chi-square tests
- *t*-tests
- analysis of variance (ANOVA)
- multivariate analysis

Although these tests use various techniques for different types of data, they all aim to answer the question: is the difference (or cause) I see in the data real (read "statistically significant") or is it just my bias, my perception?

For example, a team decides that the cycle time in a process measured at two different times in a single quarter proves their hypothesis about different methods causing delays in the overall process. In this, the team might analyze the cycle time data using a *t*-test for statistical significance in the two measures. Nowadays all of these tests are run on computer programs like Minitab, which do all of the tedious calculations once you've entered the correct data.

Underlying Causes at Gemini

The *Point of No Returns* Gemini team continued to gather information about several strata of the process, testing several theories of causes as it went along. In one case where the differences between two sets of data were too close to call just by eyeballing the patterns, the team asked its Master Black Belt to help them with a test of statistical significance (ANOVA, or analysis of variance).

The test showed there was no statistically significant difference in the data that the team thought would prove their causal hypothesis. So they had to modify their hypothesis and gather more data.

When their analysis was complete, the team reached agreement that several causes contributed to the complaints about incomplete orders from new customers and small businesses:

- Small businesses and new customers do not use the web site to do their own ordering. Instead, they rely on the Telephone Service Clerks (TSCs) to take their orders. TSCs are encouraged and rewarded by their managers to handle orders quickly, as one way to respond to the sharp increase in customer demand this year. Leaving the Order Verification Desk clerks to verify and correct orders, TSCs seldom use the more up-to-date Web site for current information on product availability, price, or systems compatibility.
- TSCs do not follow a standard script to get customer information about the systems they already own.
- OVD clerks are being overwhelmed by the volume of business and number of call backs and checks they must do on orders based on the Jimmy Gemini database. Pushed by management to do their checks quickly, OVD clerks don't check every order against the web site, but often use the same database as the TSCs—in effect they simply pass the orders along without inspecting all of them carefully.
- The data maintenance support staff updates the Web site information more often than they do the Jimmy Gemini system.

Another set of tools that Six Sigma teams might use to test causal theories are **correlation** and **regression analysis**. These are tests to show the numerical measure of correlation between X variables and Y outputs. If the team has paired data, regression analysis can help measure the degree to which different variables influence the outcomes. For example, a Six Sigma team finds an apparent positive correlation between the speed that phone orders are taken by salespeople and number of defects in the orders taken. By calculating the "correlation coefficient," the team discovers that only about 25% of the defects correlate to the speed with which the orders are taken. This is a powerful clue, but the team will need to keep probing for other causes.

Finally, you might continue the pursuit of causes using a tool called Design of Experiments (DOE). Using some of the tools described above, a team can assess the influence of various causal factors in the process. It does this by combining a number of causal factors at different levels, conducting the experiment, and deciding the level of impact of the factors.

Although DOE can get very sophisticated, anyone who has experimented with cooking recipes by varying temperature, cooking time, and ingredients understands the basic principles.

All of the above can and should be used by Six Sigma teams when the need arises. Unfortunately, these more sophisticated statistical tools are a lot like the foreign languages that you may have studied at school: unless you get a good understanding of them to begin with and then use them every day, your fluency drops in a short time. Most Black Belts will have to consult with a practicing statistician using a computer-based program like Minitab to carry out more advanced tests like those mentioned above.

Getting Ready for Improve

By the end of the Analyze phase of DMAIC, your team should have confirmed the root causes of problems with your processes, products, and services. And now that you've pinpointed the cause, you stand a better chance of implementing changes that will have a lasting effect. Before you move on, however, there are a few last tasks to complete:

1. Document the verified causes.
2. Update your project storyboard.
3. Create a plan for Improve.
4. Prepare for the tollgate review by your Sponsor or Leadership Council.
5. Celebrate.

6σ See the Analyze Checklist on p. 267.

1. Document the Verified Causes

The solutions you put in place in Improve will be based on the causes you pinpointed here in Analyze. Having clear documentation of what causes your team investigated and which of them were verified with data is therefore a key turning point in your project. Knowing which potential causes you investigated is important even if some or most of them proved to not have much impact on the problem, because there may come a time when your team needs to revisit this problem or another team encounters something similar—in which case you can save time by knowing which ideas led up blind alleys.

This documentation need not be elaborate. For instances, some teams just mark or highlight the causes they investigated with an asterisk on a cause-and-effect diagram, and highlight or color code those that were shown to have a big impact. The simpler you can make the documentation, the easier it will be for others not on the team to follow your logic—and for you to convince them that you know what you're doing!

2. Update Your Project Storyboard

The content of the Analyze section of your team's storyboard will vary depending on what tools and approaches you used. As a general rule, include only those charts, graphs, or other data displays that will have the biggest impact on what your team does in Improve—for example, a histogram showing the centering, distribution, and spread of the process and its relationship to customer specifications is standard. Pareto charts drilling down through several strata of data make it easy for team members and others to follow the investigation into root causes. Run charts can show the trends (if any) in the process over time, as can control charts. Cause-and-effect charts show how the team is trying to track down root causes.

Since you now know the root causes of the problem or variation that prompted your project in the first place, you may also want to include a revised and problem statement that links the observed problem to specific causes.

3. Create a Plan for Improve

As with the other DMAIC steps, you can't know for sure how long or what it will take to complete the Improve step, but you should be able to develop ballpark estimates based on your knowledge of the root causes and of how easily your organization adopts changes. Leave time in your plan for:

- Generating creative solution ideas.
- Analyzing the solution options and selecting the best candidate(s).
- Developing implementation plans.
- Trying out solutions on a small scale.
- Learning from the pilot test and implementing the solutions full-scale.

This is also a good time to re-evaluate team members, since you now know what is causing the problem and where likely solutions will be put in place: if you do not have someone on the team who represents the part of the process that has to change, consider inviting someone from that area to join the team.

4. Prepare for Tollgate Review by Your Sponsor or Leadership Council

Before you prepare for the Analyze tollgate review, do a debrief with the team on what happened with your Measure tollgate:

- What improvements did you try out that were different from the first (Define) review? How well did those work? Should you do the same thing this time, or try something different?
- What comments or suggestions did your reviewers (your customers!) have?
- Which of the support materials (slides, overheads, handouts, flipcharts, etc.) worked and which didn't?
- How did you do on time? If it was too long, what could you do this time to make sure you keep it brief? If it was too short, do you need to add more detail? Speak more slowly?

6σ See the Analyze Tollgate Preparation Worksheet on p. 268.

After this general review, start your preparation for the Analyze tollgate review. As before, the key is continuing to link the steps in your project. How did what you learn in Define influence what you did in Measure? How did your Measure work naturally lead to the Analyze work? How is what you learned here going to influence your Improve work? Review the tips given in the Analyze tollgate instructions (pp. 266 and 269).

5. Celebrate

As before, take time to celebrate the *work* and *progress* on your Six Sigma projects. Be sure to point out particular challenges that the team handled well in its Analyze work, for example:

- Innovative data analyses that might set a precedent for other teams in your organization.
- Patience in working through process analyses.
- People maintaining their commitment, carrying through on assignments.

Once these steps are complete, your team is ready for *Improve*.

Chapter 13

Power Tools for "Analyze"

Understanding Problems

THE ANALYZE STAGE OF DMAIC has two components (as described in Chapter 12):

- **Data Analysis**: Using data to find patterns, trends, and other differences that can suggest, support, or reject theories about the causes of defects.
- **Process Analysis**: A detailed look at the existing key processes that supply customer requirements in order to identify cycle time, rework, downtime, and other steps that don't add value for the customer.

The tools in this chapter are clustered in three groups for each of these components, again paralleling the structure of Chapter 12. You'll find both data and process analysis tools for:

1. **Exploring**: Investigating the data or process with an open mind, just to see what you can learn.
2. **Generating theories about causes:** Using your new-found knowledge to identify the most likely causes of defects.

3. **Verifying or eliminating causes:** Using data, experimentation, or further process analysis to verify which of the potential causes significantly contribute to the problem.

Data Exploration

As you probably know by now, the type of data you collected in Measure will shape which tool you can use to explore how your process functions. Be sure to review the background information in Chapter 12 (pp. 204-213) to help you use these tools most effectively. Remember, too, that the key to good data exploration is to ask a lot of questions: "What happens when ____? Are ____ any different from ____? Does ____ have any effect on the outcome?"

Pareto Analysis and Chart

Purpose: To compare the frequency and/or impact of different types or causes of problem. Allows selection of "vital few" improvement priorities.

Application:

- Setting priorities.
- Defining problems/opportunities.
- Determining root cause(s).

Related Tools: A Pareto chart requires a checksheet or datasheet to gather raw data and uses discrete (or attribute) data or continuous data that has been divided into "buckets" or categories.

Instructions (see Figure 13-1):

1. **Determine the process and types of problem or cause to be measured.** (Use brainstorming, customer interviews, and other data gathering as needed.)
2. **Determine appropriate frequency (time-frame) and data collection method.**
3. **Gather and compile data.**
   Hint: The interpretation of your Pareto chart will be more reliable if you collect at least 50 data points total (across all categories).

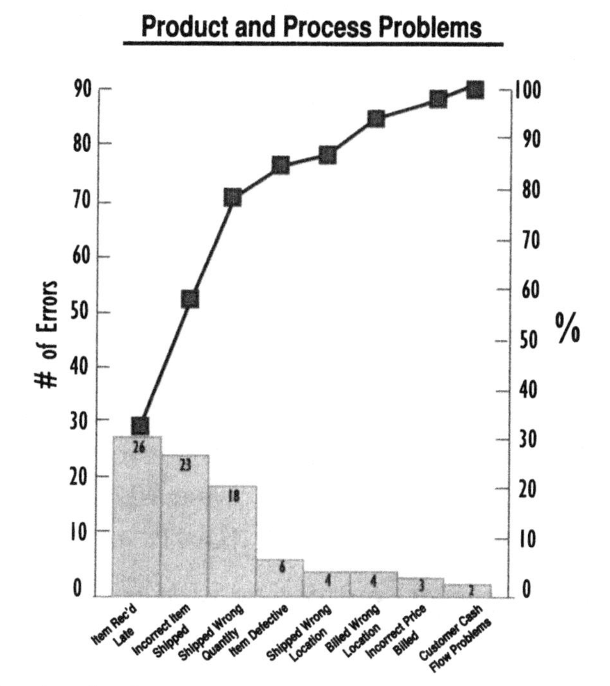

Figure 13-1. Pareto chart

4. **Total each category for the entire period to be analyzed.**
5. **Determine percentages** (optional).
   - Total all categories for entire period. (Sum represents 100%.)
   - Divide the number for each category by the total to determine percentages by category.
6. **Draw the axes of the chart.**
   - Draw a vertical (Y) axis and divide it into appropriate increments. This axis should be at least as tall as the total number of incidences or impact observed across all categories.

     Example: You are counting the occurrence of five types of defects on an Accounts Payable report. After you finish collecting data, you tabulate the results for all five and determine there were 750 errors in all (the total for all five defects). You could divide the Y axis into 50-point increments, ranging from 0 to 750.
   - Draw a horizontal X axis.

7. **Draw the bars in descending order from left to right. Label each bar below the axis.**
   - Draw the first bar, right next to the Y axis, at the height representing the total count (or impact) of the largest category in your data.
   - Leave a small space, then draw a bar representing the second-most frequent category.
   - Continue drawing bars for all categories in descending order of frequency.

   Hint: If all your data falls into, say, 10 or fewer categories, draw a separate bar for each category, even if the counts in some are low. If you have data that could fall into more than ten categories, and there are a number of categories with just a few items, create an "Other" bar at the height representing the total of all categories not represented by individual bars. The "Other" bar always goes at the far right of the chart.

8. ***Optional:*** **Add percentage notation** on vertical axis at left, zero to 100 percent. **Draw a "cumulative line"** by adding the value of each bar from left to right up to 100%.

9. **Use Table 13-1 to help you interpret the chart and take appropriate action.**

Run Chart, Trend Chart, Time Plot

Purpose: To measure and track a key input, process, or output measure over time.

Application:

- Identifying problems/opportunities (trends/patterns/variation).
- Determining potential root cause(s).
- Follow-up and verification of results.

Instructions (See Figure 13-2):

1. **Identify the key process factors** you want to measure and the metric (the way you will measure those factors).

   The metric may be positive (e.g., sales, on-time shipments) or negative (e.g., rework, complaints).

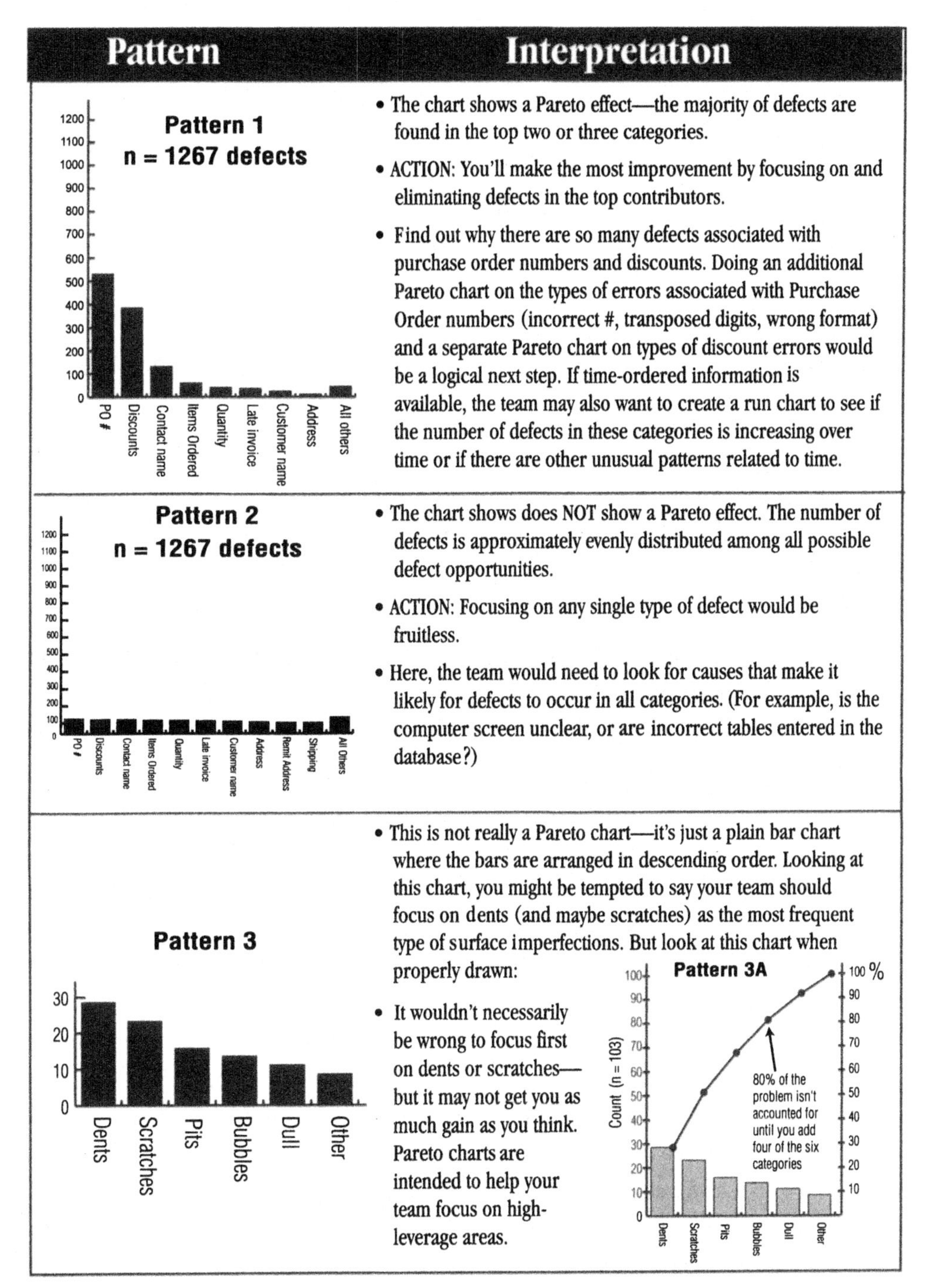

Pattern	Interpretation
Pattern 1 n = 1267 defects (PO #, Discounts, Contact name, Items Ordered, Quantity, Late invoice, Customer name, Address, All others)	• The chart shows a Pareto effect—the majority of defects are found in the top two or three categories. • ACTION: You'll make the most improvement by focusing on and eliminating defects in the top contributors. • Find out why there are so many defects associated with purchase order numbers and discounts. Doing an additional Pareto chart on the types of errors associated with Purchase Order numbers (incorrect #, transposed digits, wrong format) and a separate Pareto chart on types of discount errors would be a logical next step. If time-ordered information is available, the team may also want to create a run chart to see if the number of defects in these categories is increasing over time or if there are other unusual patterns related to time.
Pattern 2 n = 1267 defects (PO #, Discounts, Contact name, Items Ordered, Quantity, Late invoice, Customer name, Address, Remit Address, Shipping, All Others)	• The chart shows does NOT show a Pareto effect. The number of defects is approximately evenly distributed among all possible defect opportunities. • ACTION: Focusing on any single type of defect would be fruitless. • Here, the team would need to look for causes that make it likely for defects to occur in all categories. (For example, is the computer screen unclear, or are incorrect tables entered in the database?)
Pattern 3 (Dents, Scratches, Pits, Bubbles, Dull, Other)	• This is not really a Pareto chart—it's just a plain bar chart where the bars are arranged in descending order. Looking at this chart, you might be tempted to say your team should focus on dents (and maybe scratches) as the most frequent type of surface imperfections. But look at this chart when properly drawn: Pattern 3A (Count (n = 103); Dents, Scratches, Pits, Bubbles, Dull, Other; 80% of the problem isn't accounted for until you add four of the six categories) • It wouldn't necessarily be wrong to focus first on dents or scratches—but it may not get you as much gain as you think. Pareto charts are intended to help your team focus on high-leverage areas.

Table 13-1. Pareto chart interpretation

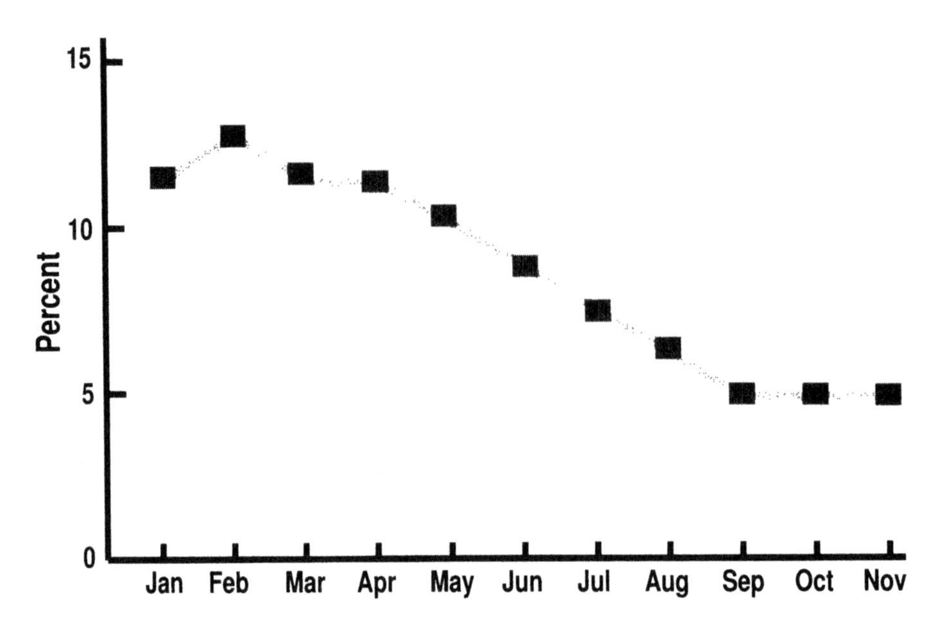

Figure 13-2. Run chart of percent lost cartons/month

2. **Develop data collection method** (see checksheets and datasheets) **and gather data over time.**
   Hint: Interpretation of a run chart will be more reliable if you have at least 20 to 25 data points.

3. **Draw and label the horizontal and vertical axes.**
   Hint: Run charts are easier to interpret visually if they are wider than they are tall. While in theory it shouldn't matter whether the axes are the same length or not, experience has shown that people have a harder time identifying special causes when the data points are crammed together (the X axis is very short) or spread too far apart (the X axis is too long). There is no exact ratio, and knowing what works will largely be a matter of experience. At first, try making the X axis about one-and-a-half times as long as the Y axis is tall. (*Example:* If the Y axis is 2 to 2.5 inches have an X axis that is perhaps 3 to 3.5 or 4 inches long.)
   - The vertical axis should be slightly taller than the range of data values observed. (*Example:* If the data values range from 367 to 906, you could divide the vertical axis into 50- or 100-point increments ranging from, say, 0 to 1200. There is no one right solution.)
   - Mark the horizontal axis off in units of time appropriate for your data.

(*Example:* If you take weekly data starting on Sept. 14, you'd mark the X axis increments as 9/14, 9/21, 9/28, etc.)

4. **Add the data to the chart.**
   - Draw a point above each time interval at the appropriate Y value.
   - Connect points with line to see "run" of performance over time.
5. **Draw a horizontal line at the median value.**
   - The median divides the data exactly in half: half the points fall above the median, half fall below it.
6. **Analyze the charts** using the guidelines below and take appropriate action.

   Hint: Some people confuse histograms and run charts. An easy way to tell which you're using is to ask: If we had new data, would it be added to the right (extending the line)? If so, it's a run chart. If you'd add new data to the count in an appropriate column or bar, it's a histogram or a Pareto chart.

Interpreting Run Charts: Identifying Special and Common Cause Variation

The run chart in Figure 13-3 shows the random variation associated with a process driven by *common cause* variation (see pp. 209-211). Notice that there are lots of ups and down, several points in a row increasing and decreasing, and clusters of a few

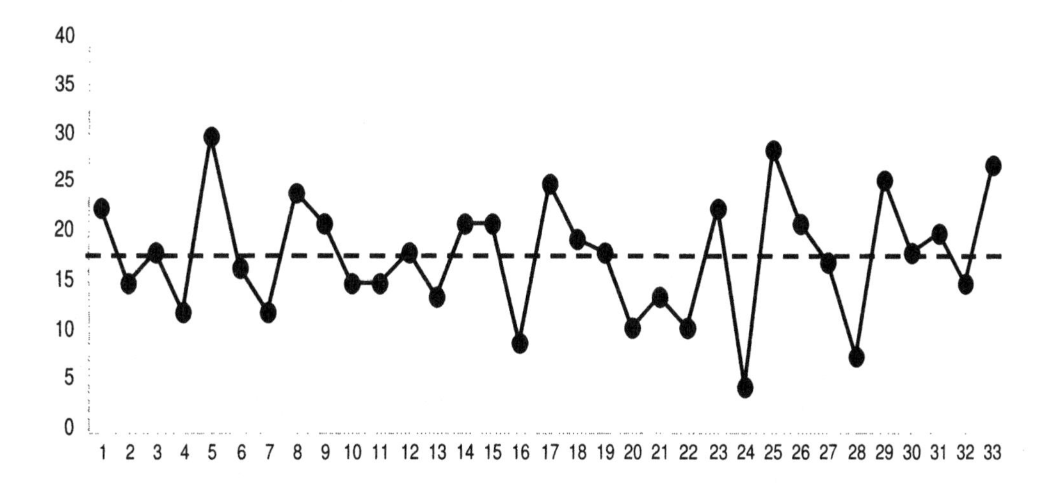

Figure 13-3. Run chart, no special causes

points above and below the median. The challenge for a Six Sigma team is to distinguish this random, common cause variation from special cause variation. The difference is important because the improvement strategies are different:

- If you find a special cause, you need to find out what was *different* about the process when those data points were collected. Something "special" happened, and you need to find out what it was and eliminate it.
- If you find only common cause variation, and still want to improve the process, you need to look at changing factors that are always present in the process.

The charts in Table 13-2 illustrate patterns that signal the presence of special cause variation.

Counting Runs to Identify Special Causes

Run charts got their name because of the practice of counting "runs"—sequences of consecutive points on either side of the median. As Figure 13-3 shows, the random variation in any data set will produce small clusters of points above and below the median. Too many or too few clusters are a signal that something special is happening in the process. Here is how to count runs:

1. **Plot the data in time order.**
2. **Draw a line at the median** (the value that divides the data equally in half).
3. **Count the number of data points *not on the median*.**
4. **Circle clusters of consecutive points** ("runs") above or below the median.
   - Stop the circle when the data line crosses the median.
   - *Ignore* points that are exactly on the median—they do not stop or add to a run.
5. **Count the number of runs and use Tables 13-3 and 13-4 below to determine if there are too many or too few runs.**

   Example: The chart shown in Figure 13-3 has 33 data points total, 29 of which are *not* on the median. (four points are on the median.) There are 17 runs. Locate "29 points" in the left-hand column of Table 13-4 and read across to see that the expected number of runs is between 10 and 20. Since 17 falls between those limits, you can conclude that there is only common cause variation in this process.

Pattern	Interpretation
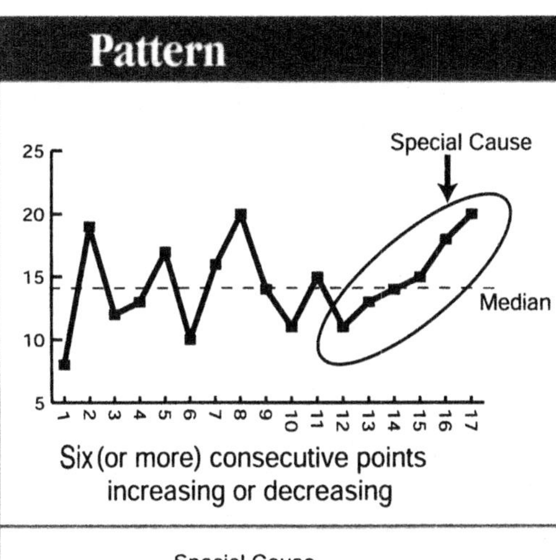 Six (or more) consecutive points increasing or decreasing	• The chart shows a trend. • Something has changed in the process to cause a steady increase/increase. • ACTION: Look for what changed in the process on or shortly before the time the trend began. (Sometimes it takes a while for a process change to show up in the data.) Look for changes in materials, procedures, types of services/products being produced, etc.
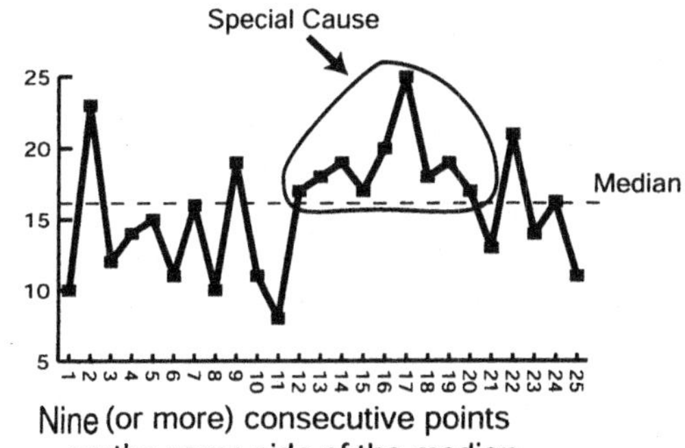 Nine (or more) consecutive points on the same side of the median	• The chart shows a shift. • Something was present in the process during the time those data points were produced that was not present at other times. • ACTION: Look for what was different during the time when the shift appeared. Look for changes in materials, procedures, types of services/products being produced, etc.
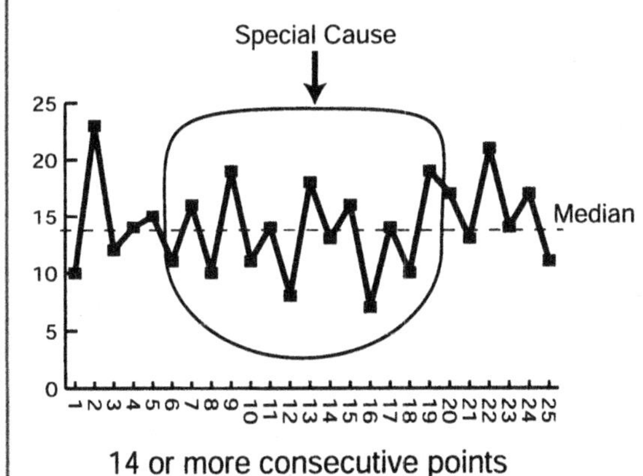 14 or more consecutive points alternating up and down	• The chart shows a sawtooth pattern. • This pattern appears when there is systematic difference in how "units" (materials, product, service deliveries) travel through the process. In manufacturing, for example, it would be likely that alternating data points represent the output of different production lines or machines. The same pattern can appear in services when data come from two different people or groups. • ACTION: Look for what is different between the "up" points and the "down" points.

Table 13-2. Special cause signals on run charts

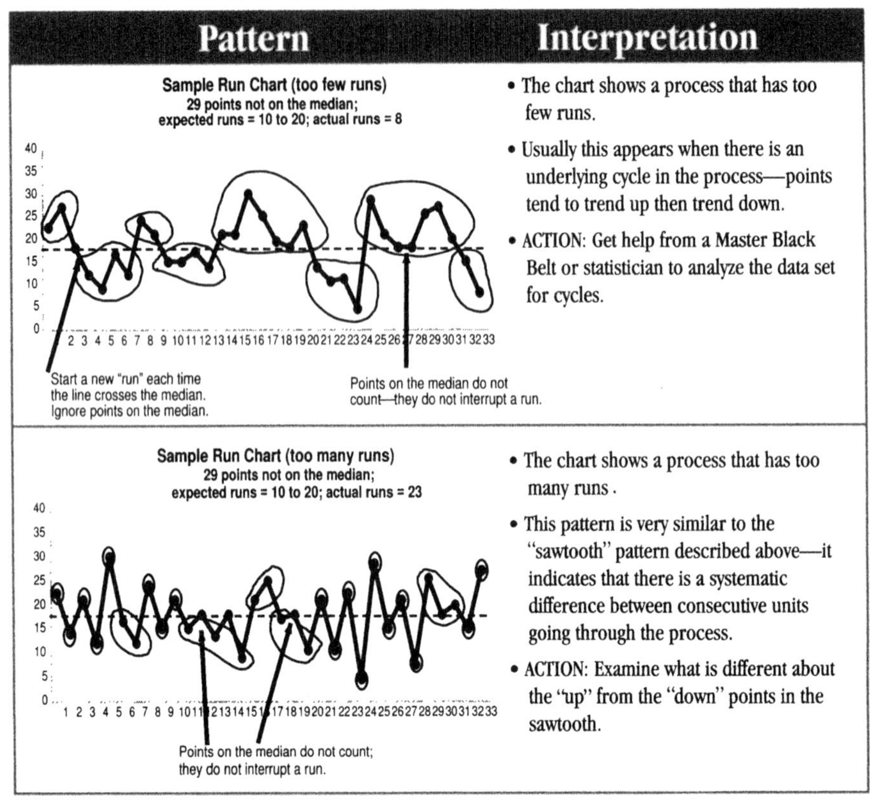

Pattern	Interpretation
Sample Run Chart (too few runs) 29 points not on the median; expected runs = 10 to 20; actual runs = 8 Start a new "run" each time the line crosses the median. Ignore points on the median. Points on the median do not count—they do not interrupt a run.	• The chart shows a process that has too few runs. • Usually this appears when there is an underlying cycle in the process—points tend to trend up then trend down. • ACTION: Get help from a Master Black Belt or statistician to analyze the data set for cycles.
Sample Run Chart (too many runs) 29 points not on the median; expected runs = 10 to 20; actual runs = 23 Points on the median do not count; they do not interrupt a run.	• The chart shows a process that has too many runs. • This pattern is very similar to the "sawtooth" pattern described above—it indicates that there is a systematic difference between consecutive units going through the process. • ACTION: Examine what is different about the "up" from the "down" points in the sawtooth.

Table 13-3. Counting runs on a run chart

Histogram or Frequency Plot

Purpose: To show the range, amount, and patterns of variation in a group of data (aka "population").

Application:

- To see the range and distribution of continuous factors (e.g., weights for each shipment, dollars spent per purchase, size of each hole, reboot time for each computer).
- To see the variation and performance around a customer specification/ requirement (e.g., size, cycle time, temperature, cost).
  Note: continuous factors only.
- To see how many defects occur on each unit in a group of defective items (when there are multiple opportunities for error). These may include discrete characteristics.

Runs Table

Instructions: Count the number of points on your run chart that are not on the median. Locate this number in the table below. Then read across to find the number of the expected range of runs for that number of points. If the count of runs falls within the range given, there is no signal of special cause variation.

Number of Data Points Not on Median	Lower Limit for Number of Runs	Upper Limit for Number of Runs	Number of Data Points Not on Median	Lower Limit for Number of Runs	Upper Limit for Number of Runs
10	3	8	34	12	23
11	3	9	35	13	23
12	3	10	36	13	24
13	4	10	37	13	25
14	4	11	38	14	25
15	4	12	39	14	26
16	5	12	40	15	26
17	5	13	41	16	26
18	6	13	42	16	27
19	6	14	43	17	27
20	6	14	44	17	28
21	7	15	45	17	29
22	7	16	46	17	30
23	8	16	47	18	30
24	8	17	48	18	31
25	9	17	49	19	31
26	9	18	50	19	32
27	9	19	60	24	37
28	10	19	70	28	43
29	10	20	80	33	48
30	11	20	90	37	54
31	11	21	100	42	59
32	11	22	110	46	65
33	11	22	120	51	70

Table 13-4. Runs table for determining whether there are too many or too few runs in a set of time-ordered data

- To see how key "count" characteristics in a group or population are distributed (e.g., customers by number of purchases per year, suppliers by score on our quality audit).

Instructions:

1. **Collect and tabulate the continuous data.**
2. **Determine appropriate increments for the data set and label the X axis.**
   - Determine the lowest and highest observed data values.
   - Subtract the lowest value from the highest value to calculate the range of the observed data.
   - Divide the range by 10 or 15 to determine approximate ranges for clustering the data. Usually the range of data values is too great to represent each individual value on the chart, so the data are typically clustered into groups for purposes of labeling the X axis and plotting the values. For example, you might label the X axis increments at "1 to 5," "6 to 10," "11 to 15," etc. Any data value between 11 and 15 would then be plotted in the "11 to 15" category.
   - Round up or down to get increments that will be easy to interpret.
   - You want the X axis slightly wider than the actual range of data values, so start with a whole number smaller than the lowest observed value. Add the increments to that value and label the X axis appropriately. *Example:* Observed data values range from 137 to 409. That means the range encompassed 272 points. Ten increments along the X axis would lead to increment ranges of about 27 points; 15 increments would lead to increment ranges of 18. Logically, you could therefore use increments ranges of 25, 30, or 20 points. If you choose 20, have "120 to 139" be the first X axis increment, "140 to 159" the second increment, and so on.
3. **Plot the data values above the appropriate increment label on the axis.**
   - Option 1: Use an X, dot, or other symbol for each individual data value. (This style is shown in Table 13-5.)
   - Option 2: Determine the number of data points that fall within each increment, then draw a single bar above the increment representing the count of items. (This pattern is shown above in Figure 13-4.)
4. **Use the patterns in Table 13-5 to help you interpret the chart.**

   Tip: If you are comparing the distributions of two or more "populations" and can't tell visually whether they are different (or you want to quantify the degree of difference), use a more rigorous statistical analysis called **hypothesis testing** (pp. 270-274).

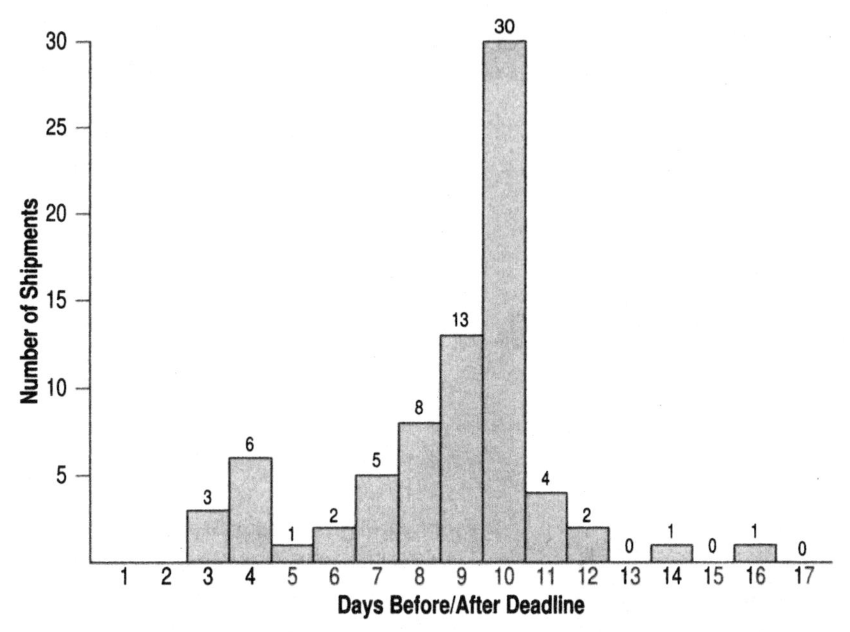

Figure 13-4. Histogram of shipment delivery times

Generating Theories: Cause Analysis

By this point in the team's work, everyone should have...

- A clear understanding of the problem as it it currently defined.
- Detailed knowledge, based on data and observations, about the exact nature of the problem—when and where it appears, where it doesn't, its symptoms, history, and trends.

Throughout the Define, Measure, and Analyze stages, team members have likely been thinking about a lot of potential causes of the defined problem—and been frustrated that they weren't allowed to take any action! At last you've reached the stage where it's time for people to unleash their creativity and ideas. Generally, the first round of "causal thinking" will elicit ideas that have been on people's minds even before your analysis began. The trick is to push your thinking even further, and challenge people to come up with new ideas rooted in the measurement and analysis completed to date.

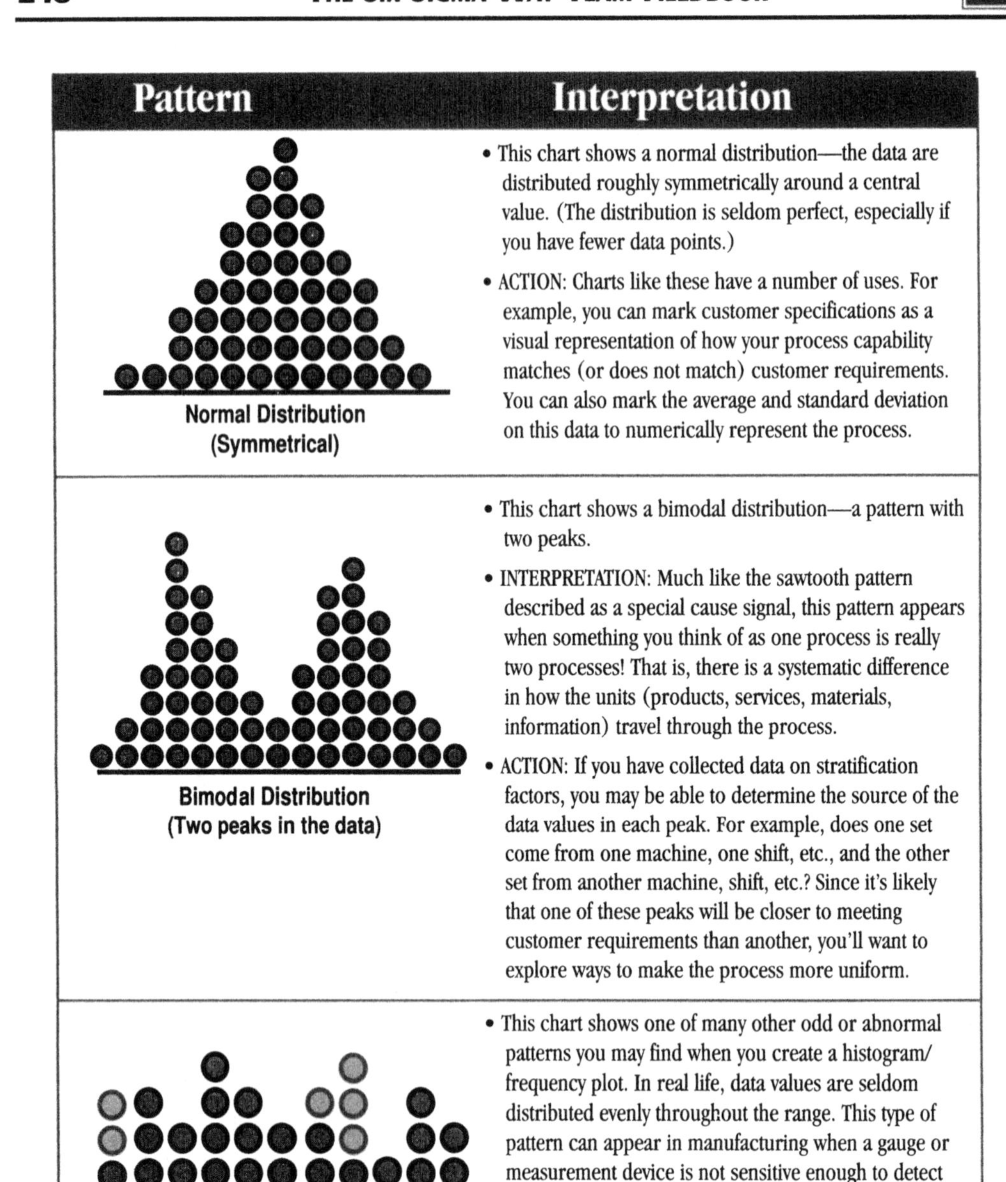

Pattern	Interpretation
Normal Distribution (Symmetrical)	• This chart shows a normal distribution—the data are distributed roughly symmetrically around a central value. (The distribution is seldom perfect, especially if you have fewer data points.) • ACTION: Charts like these have a number of uses. For example, you can mark customer specifications as a visual representation of how your process capability matches (or does not match) customer requirements. You can also mark the average and standard deviation on this data to numerically represent the process.
Bimodal Distribution (Two peaks in the data)	• This chart shows a bimodal distribution—a pattern with two peaks. • INTERPRETATION: Much like the sawtooth pattern described as a special cause signal, this pattern appears when something you think of as one process is really two processes! That is, there is a systematic difference in how the units (products, services, materials, information) travel through the process. • ACTION: If you have collected data on stratification factors, you may be able to determine the source of the data values in each peak. For example, does one set come from one machine, one shift, etc., and the other set from another machine, shift, etc.? Since it's likely that one of these peaks will be closer to meeting customer requirements than another, you'll want to explore ways to make the process more uniform.
Evenly Distributed Data Values (Seldom occurs naturally)	• This chart shows one of many other odd or abnormal patterns you may find when you create a histogram/ frequency plot. In real life, data values are seldom distributed evenly throughout the range. This type of pattern can appear in manufacturing when a gauge or measurement device is not sensitive enough to detect differences between units (like having a ruler that is only marked in whole inches). It can appear in service situations when the people taking measurements are not using the same operational definition. For example, perhaps one data collector starts a stop watch when a customer comes through the bank door, another starts it when the customer steps into the teller line.

Table 13-5. Histogram (frequency plot) patterns and interpretation

Pattern	Interpretation
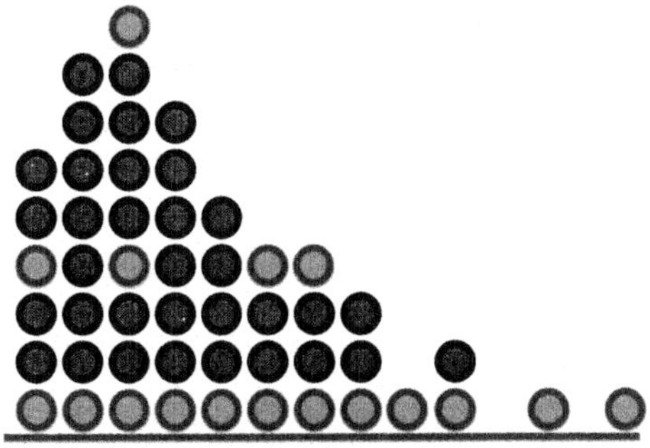 Skewed distribution (Data clusters near one end, tails off in other direction)	• The chart shows a skewed distribution—data points cluster around one end and tail off in the opposite direction. • SERVICE/MFG INTERPRETATION: Charts like these are very common in any type of measure involving time—processing time, cycle time, "days after due date" a product is delivered, etc.—and costs. If the pattern shown to the left related to costs, for example, the interpretation would be that the majority of units can be delivered within a small cost range, but some units cost much more. (If the pattern were reversed, you'd conclude that most orders are costly to deliver, but for some reason a few make it through with significantly less cost.) ACTION: Find out what is different about the units represented by the values in the "tails" of the distribution and either work to eliminate them (if they tail off in an undesirable direction) or copy them (if they are in a desirable direction). • ALTERNATIVE MFG INTERPRETATION: Patterns like these are sometimes (though not often) seen in manufacturing when a measurement device has degraded and simply can't read past a certain value. In other words, what you're seeing is really only part of what should be a "normal" distribution. They are also seen when there is a physical limit to a measurement (you'll never see a "length" or "weight" of less than 0, for instance). ACTION: Check to make sure that any gauges or measurement devices are working properly. • ALTERNATIVE CULTURAL INTERPRETATION: Patterns that have an abrupt cut-off also occur when people have been instructed in no uncertain terms that they are not to fall below or above a certain limit. E.g., you'd see a reverse pattern if managers were told they could not exceed $10,000 in capital expenditure. ACTION: You'll have to tread carefully because you'll be battling cultural forces within your organization. You'll need support from your Sponsor/Champion if you find areas where data values are corrupted because of employees' fear of reprisal.

Table 13-5. Histogram (frequency plot) patterns and interpretation (continued)

Cause-and-Effect Analysis (Fishbone or Ishikawa Diagram)

Purpose: To identify the cause of a problem by applying the experience and expertise of a group in structured brainstorming. Also used to brainstorm possible ways to cause a desired effect to happen.

Application:

- Determining major cause(s).
- Determining potential root causes.
- Determining potential solution options.
- Planning and implementing a process change or solution.

Instructions (see Figure 13-5):

1. **Briefly name the problem or effect** to be analyzed and write it at the "head" of a fishbone diagram.
2. **Determine appropriate cause categories** for the situation.
   - Typical categories are shown in Table 13-6.
3. **Brainstorm potential causes in each category.**
   Tip: When using the fishbone diagram, the team should think not only of

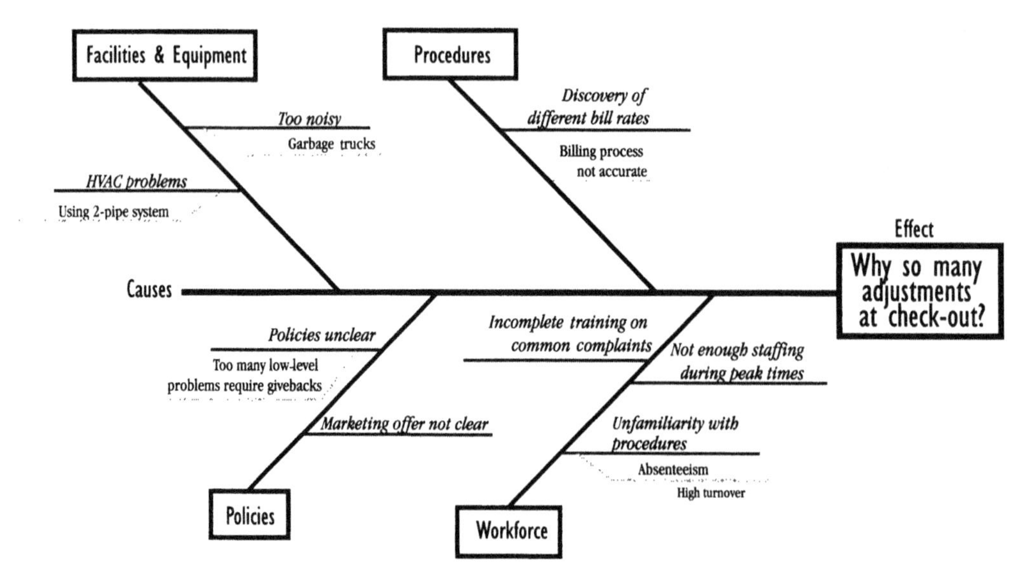

Figure 13-5. Cause-and-Effect Diagram

- *Material:* the raw materials used in a process; in services, usually information or data of some kind.
- *Methods*: processes, procedures, work instructions, the way people do their job.
- *Machines*: equipment and tools of all kinds, computers.
- *Measures*: any methods for measuring the quality of processes and outputs, including inspection.
- *Mother Nature*: really the physical and management environment in which work is done, the marketplace at large, the natural environment.
- *People*: anyone involved in the process, including customers, employees, managers, regulators, and shareholders.

Table 13-6. Typical categories of causes

the different variables and contributing causes that can be added onto any one of these categorical "bones," they should be thinking of the variations that can take place within each of them and how they might affect each other when they vary. A computer system that goes down may affect everything on this list, while poor information from customers will have a direct impact on methods.

4. **Identify the cause-and-effect relationship between factors in each category.**
5. **Construct the fishbone diagram.**
   - The major categories become the "biggest bones" on the fishbone diagram.
   - Arrange each cause and sub-cause on smaller levels of bones. (That is, have the most specific causes written on the smallest bones and feeding into more general causes they contribute to, etc.)
6. **Use data gathering, multivoting, or consensus to narrow down and select most likely or important cause(s) for further investigation.**

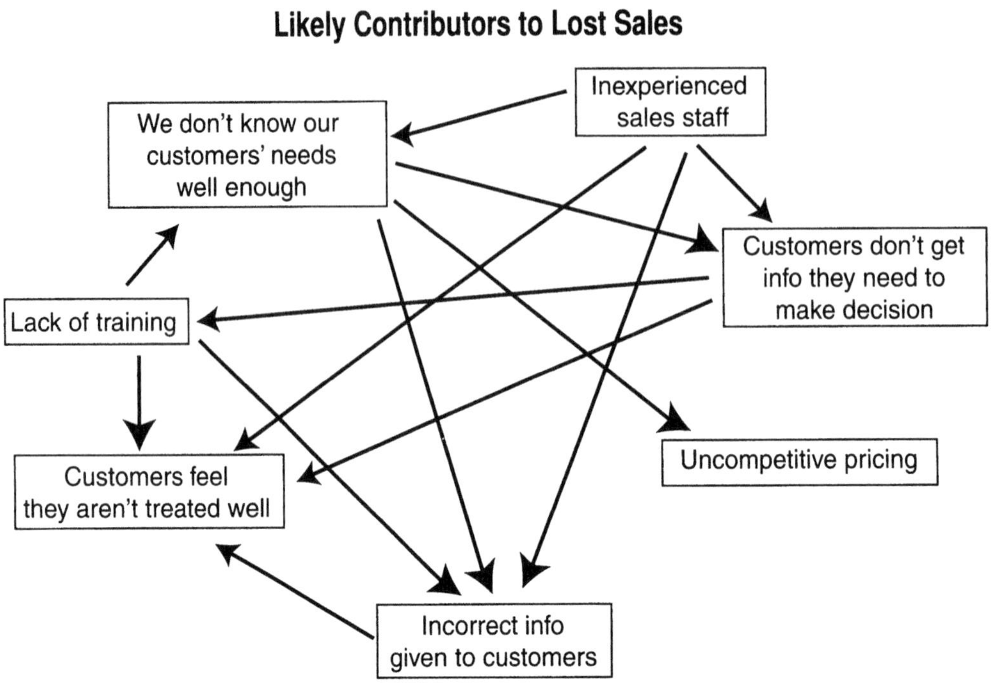

Figure 13-6. Relations diagram

Relations Diagram

Purpose: To help a team identify the drivers (root causes) of a complex problem.

Application:

- To understand complex relationships that can't be analyzed using sequential thinking tools (such as cause-and-effect diagrams).
- To reach consensus within the team on key causes/drivers to investigate further.

Instructions (see Figure 13-6):

1. **Brainstorm** all the issues contributing to the problem. Record these directly on a flipchart or on large self-stick notes.
2. **Sort and prioritize the list** to identify those causes that team members believe are the most likely contributors to the problem.

3. **Write the likely contributors on self-stick notes and arrange those notes in a circle on a large sheet of paper, whiteboard, etc.**
4. **Select any two factors.**
   - Pick any note and compare it to a note next to it. Discuss how the issues captured on the two notes are related, if at all.
   - If you decide that the factor on the first note *contributes to or causes* the impact of the second factor, draw an arrow *from* the *first note to* the *second note.*
   - If you decide that the factor on the first note *is affected by* the issue stated in the second note, draw an arrow *from* the *second note to* the *first note.*
   - If the two factors are not related, do not draw any arrows.
5. **Move to the next note and repeat the process.** Continue working your way around the circle until you have identified the relationships for each self-stick note.
6. **Tabulate the results** by counting the number of arrows leading *into* and *out of* each factor. The issue with the most number of "out" arrows is a **key driver**, and possibly a key cause, of the other issues. This issue deserves more investigation into its effects and impact.

Tips for Relations Diagrams

- A relations diagram is really more of a discussion and summary tool than an analysis tool per se. Even if the diagram doesn't provide any unexpected insights, it will help your team reach consensus on the most likely causes for investigation and improvement. For the team, the diagram may simply summarize extensive conversations.
- A completed relations diagram often looks complicated to people who weren't involved in creating it, but remember that you build the diagram one arrow at a time.

Verifying Causes

A list of factors on a cause-and-effect diagram, a relations diagram, or even a brainstormed list by the team represents *theories,* not fact. To show if and how the problem under study is affected by any of the factors your team identified, you will need to *verify* your theories, preferably through…

- **Data and/or statistics:** Use scatter plots and stratification to confirm cause-and-effect relationships.
- **Process manipulation and/or experimentation:** Make *controlled* changes in the process and observe the results.

The simple tools described in this section will be sufficient for most teams in most situations. However, if you have an especially complex situation, need a high degree of accuracy in your interpretation, or think there might be significant interactions between causes, look at the end of this chapter for some advanced tools.

Scatter Plot or Correlation Diagram

Purpose: To identify/measure possible relationship or "correlation" between two factors or variables.

Application:

- Providing data to confirm a hypothesis that two variables are related.
- Evaluating the strength of a potential relationship.
- Follow-up to cause-and-effect analysis.

Instructions (see Figure 13-7):

1. **If you haven't already done so, construct a data collection sheet and collect paired data.**
   - You must have measured two factors for a single observation or item. Both factors must be either continuous or count-type data.
2. **Draw the axes.** Draw lines representing the horizontal (X axis or independent variable) and vertical (Y axis or dependent variable) lines of the diagram. Label each in appropriate increments.
3. **Plot the data on the diagram.**
   - If data points overlap, circle that point each time it is repeated.
4. **Interpret the data.**
   - Use Table 13-7 to perform a simple visual interpretation.
   - If you need to be more rigorous in defining the relationship between the

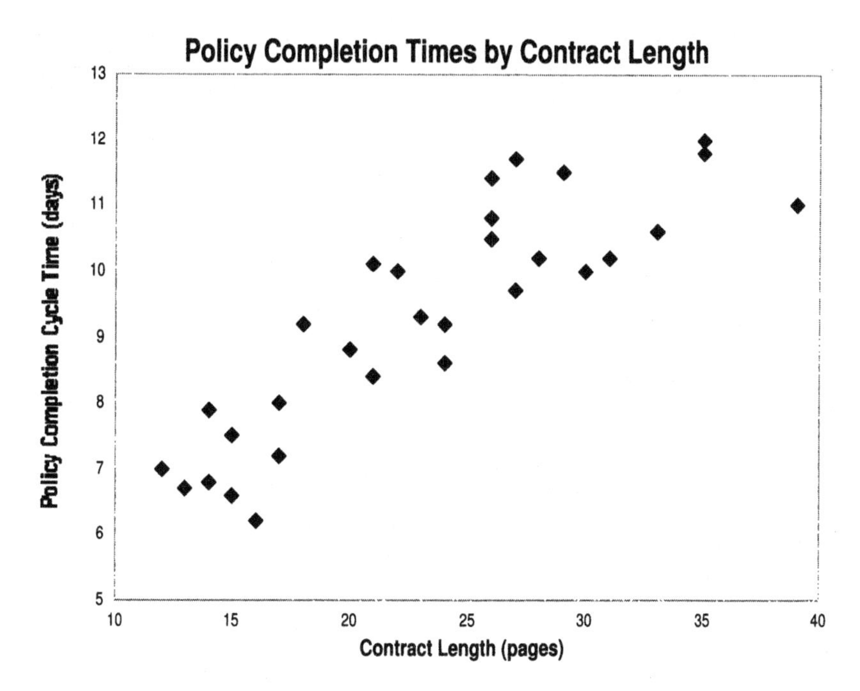

Figure 13-7. Scatter plot

two factors, use **linear regression** (an advanced tool discussed on pp. 274-276). Linear regression calculations are built into most spreadsheet programs, so you may be able to perform such an analysis on your own.

Stratified Charts

Purpose: To look for patterns in the data that link to root causes.

Application:

- Used primarily during data analyses to see if theories about causes or patterns are supported by data.

Instructions (see Figures 13-18 and 13-9): At the simplest level, stratification is simply codifying data points or dividing them into separate sets based on different **attributes**: days of the week, regions, customer or product type, etc. Any time you think there may be a *difference* between subsets of your data, you can stratify it. For example:

- Hypothesis: "Orders are late because new employees aren't as fast as

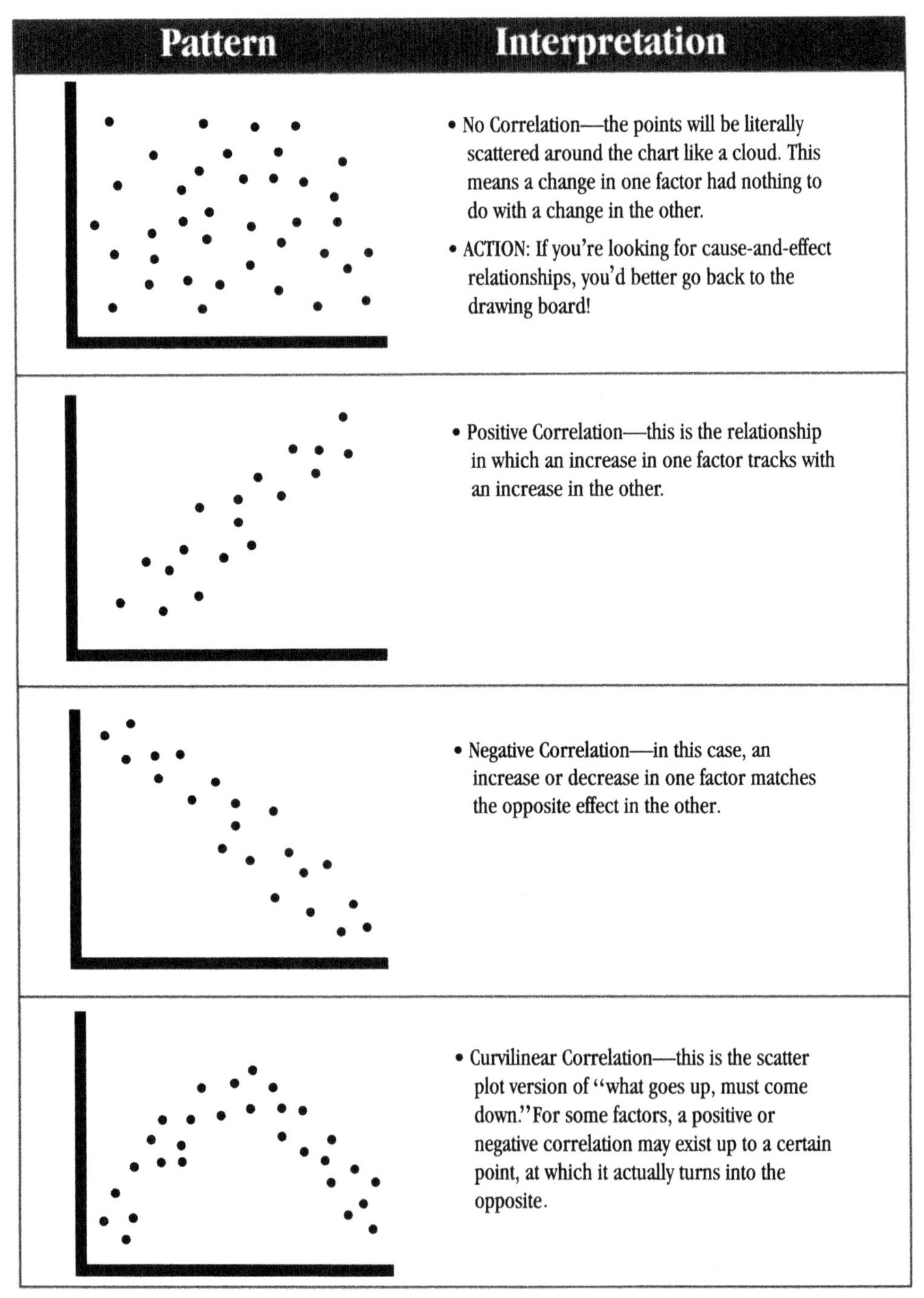

Pattern	Interpretation
	• No Correlation—the points will be literally scattered around the chart like a cloud. This means a change in one factor had nothing to do with a change in the other. • ACTION: If you're looking for cause-and-effect relationships, you'd better go back to the drawing board!
	• Positive Correlation—this is the relationship in which an increase in one factor tracks with an increase in the other.
	• Negative Correlation—in this case, an increase or decrease in one factor matches the opposite effect in the other.
	• Curvilinear Correlation—this is the scatter plot version of "what goes up, must come down." For some factors, a positive or negative correlation may exist up to a certain point, at which it actually turns into the opposite.

Table 13-7. Interpreting scatter plot patterns

experienced employees." Code data on time by "new" vs. "experienced" employee.

- Hypothesis: "Low staffing during lunch hours leads to longer customer wait times." Code data points obtained during the lunch hour and see if they account for the longest wait times. (Alternatively, you could also create separate histograms, one for wait times during the non-lunch-hour periods and another for lunch-hour wait times.)
- Hypothesis: "Raw material from Supplier A produces a better end product than material from Supplier B." Obtain quality-related data and code by supplier.

If you collected the stratification information at the same time as you collected your original data, you can simply go back and code existing charts. If not, you will need to collect new data, then follow the instructions for the type of chart you're using.

There are two ways to create stratified data charts:

- Combine all data onto one chart and code the points by color, shape, etc. (like Figures 13-8 and 13-9).
- Create separate charts for each subset of data (like the example shown in Chapter 12).

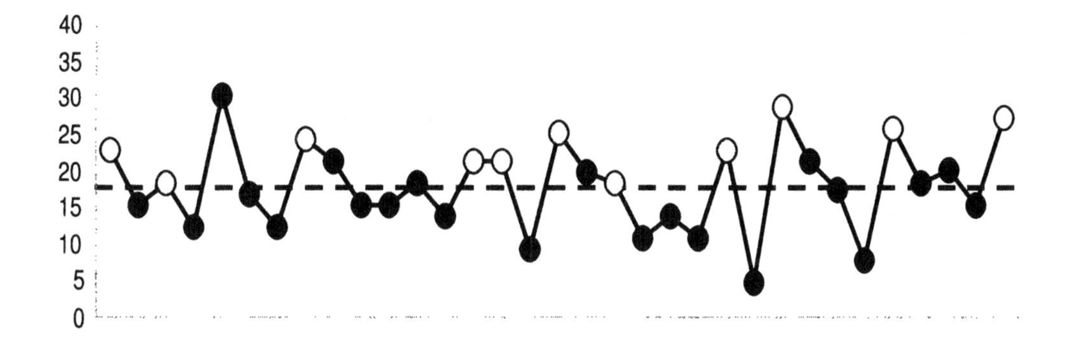

Figure 13-8. Stratified run chart *(gray-filled circle = orders received after 3 pm; black-filled circle = orders received before 3 pm)*

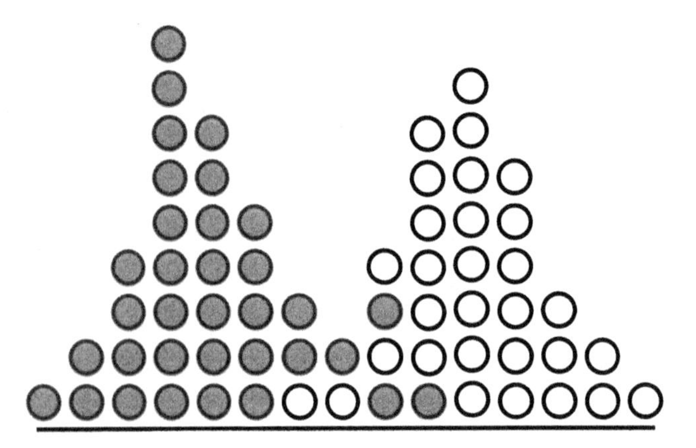

Figure 13-9. Stratified frequency plot (*solid circle = day shift; open circle = evening shift*)

Process Manipulation/Experimentation Worksheet

Purpose: To discover whether a suspected cause contributes to a known problem or defect.

Application:

- Use whenever it is impossible or impractical to use other methods of verifying a suspected cause.

Instructions (see Figure 13-10):

1. **Document specifically what cause-effect relationship is being investigated.**
   Example: "Layout of application is confusing and leads to customer errors."
   Example: "Poor packaging leads to lost sales."
   Example: "Unclear definitions and instructions for special orders lead to excessive wait time."
   Example: "Poor labeling has led to increase in expired chemicals."

2. **Decide what change you will make to test this theory.**
   Example: Change layout of applications.
   Example: Redesign the new packaging.
   Example: Rewrite definitions and instructions.

Worksheet for Using Experimentation to Verify Causes

1. Suspected cause... **... and its effect**

2. Change in product, process, or service that will test this cause

3. How can this be tested on a small scale? (Consider limiting scope, locations, personnel involved, etc.)

4. How will you evaluate success? (Describe data and other observations to be collected)

5. Summarize the plan

Action	Person or Persons responsible	Deadline

Figure 13-10. Process Manipulation/Experimentation Worksheet

Example: Develop a new label format.

3. **Identify how you can test this change *on a small scale.***
   Example: Try the new application in only one branch office.
   Example: Do packaging mock-ups and use in focus groups with customers.
   Example: Test new definitions with experienced staff for one week.
   Example: Use new labels in Chicago area only.
4. **Identify how you will know how well the change worked.**
   - What data will you collect? When, how, and who will collect it?
   - How can you document problems or unexpected issues?
5. **Plan and conduct the experiment.**
   - Develop a plan that details who will do what and when.
   - Identify who will develop the new forms, materials, etc., needed to conduct the test and the deadline for creating it.
   - Train everyone involved in the test.
   - Communicate with other staff, managers, customers, etc., who might be affected by the test.
   - Decide on specific start and stop times.
   - Carry out the plan.
   - Document the results and hold a debrief to learn what worked and what didn't—and whether the suspected cause contributed to the problem.

Note: If you have a complex problem or trouble narrowing down the list of suspected causes, you may want to use a more sophisticated **designed experiment.** See pp. 277-279 later in this chapter.

Process Analysis

The tools for performing process analysis are mostly variations or elaborations on various types of flowcharts. Sometimes these flowcharts will function mostly as a discussion tool—helping your team uncover differences in how people perform the process steps. Other charts can actually be used to help you collect data (on cycle time, for example).

Detailed Process Maps or Flowcharts

Purpose: To document and graphically represent the steps/tasks, sequence, and relationships within a process or system.

Applications:

- Identifying problems/opportunities.
- Defining process scope.
- Defining and documenting a process.
- Analyzing processes for improvements and simplification.

Instructions (see Figure 13-11):

1. **If you have not already identified the SIPOC elements, do so first (see pp. 114-115 and 117).**
2. **Have everyone on the team use sticky notes to identify the action steps that take place in a process step down to the next level of detail.** Post them on the sheet under the title for that step. (A six-step "Process" from a SIPOC map will often become 20 sub-steps.) Usually you will discover that everyone has a slightly different version of the actions taking place. That's OK—all the steps will be verified later by physically tracking the paperwork and/or product through the process.
3. Where there are duplicate steps, **try to combine the different versions into one step**, but don't hide actual variation if it really exists.
4. **Arrange the steps in time sequence.**
5. **Create symbols around each step indicating the type of action taking place.**
   - Circle: Beginning and end points.
   - Rectangle: Action steps or tasks.
   - Diamond: Decision or alternative flow point.
   - Arrows: Direction of flow through process.

   Some decision points will be legitimate—a normal part of the process that help it move ahead. Other decisions will be based on missing information that requires the thing going through the process to be delayed, and a "loop-back" is made to correct the error or find the missing information.

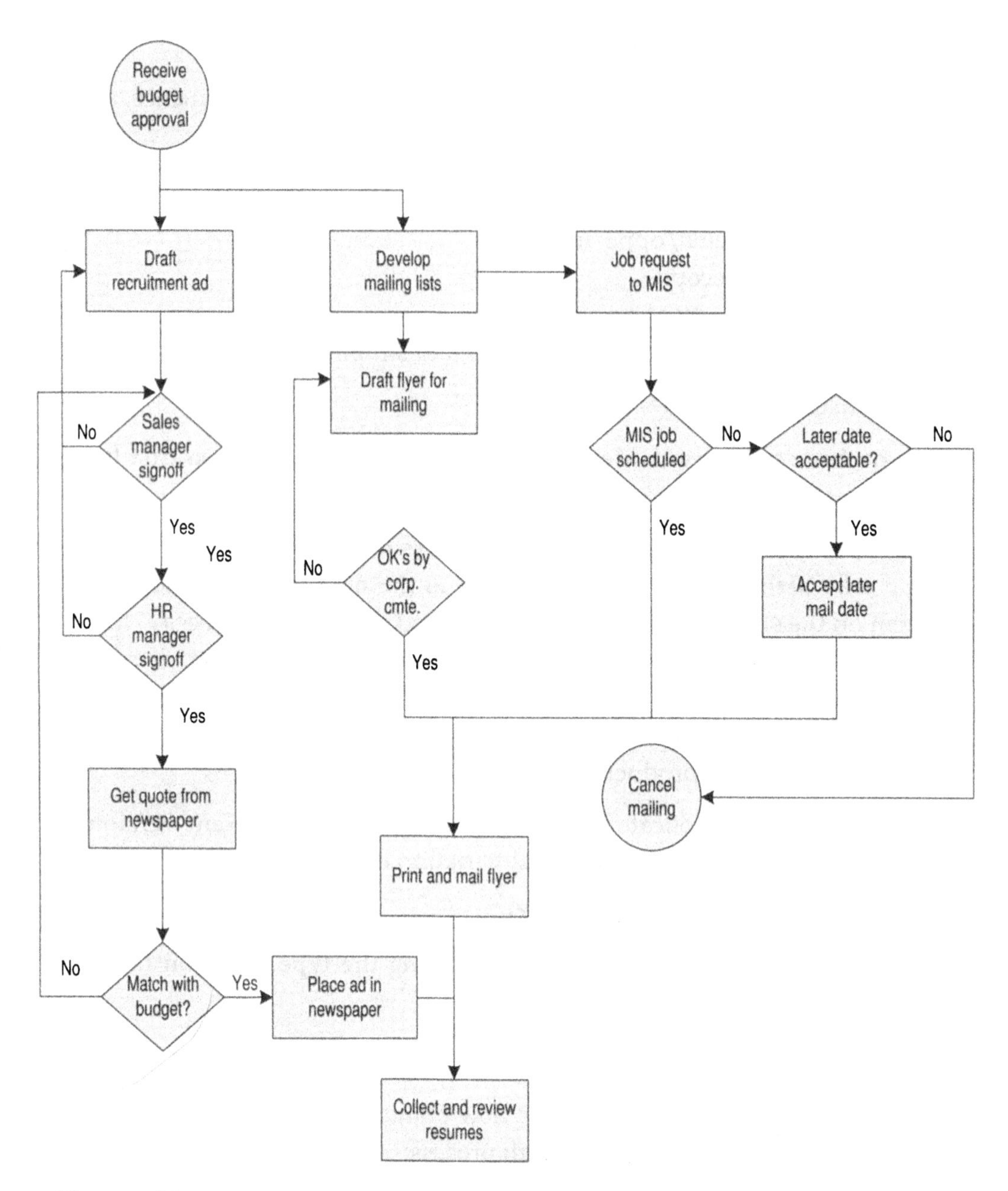

Figure 13-11. Portion of a detailed flowchart for sales staff recruitment

6. **Show the direction of action steps and loop-backs with arrows.**
7. **If needed, gather more data as needed to confirm the process flow.**
8. **Follow up as appropriate:**
   - Validate map with others doing the work but not on the team.
   - Document the process and share the results with everyone working on the process.
   - Examine and analyze the process map/flowchart for improvement opportunities.

It will take some time to complete this more detailed process map. In most processes, you'll probably come up with 15-25 steps. If you come up with more, you're probably getting too far down in the daily details of the process.

Cross-Functional or Deployment Process Map

Purpose: To document and graphically represent the steps/tasks, sequence, and relationships within a process or system—with special emphasis on functional responsibilities and interdepartmental hand-offs.

Application:

- Identifying problems/opportunities in a process with many hand-offs.
- Defining process scope.
- Defining and documenting a process.
- Analyzing processes for improvement and simplification.
- Planning and implementing process change or solution.
- Following up and verifying results.

Instructions (see Figure 13-12):

1. **If you haven't already done so by creating a SIPOC map, identify the process output, customer, and starting point.**
2. **Determine which departments, functions, groups, or individuals (including external suppliers) do—or should—participate in the process.**
3. **Set up rows or columns with a heading for each participant** (group or individual) in the process. The functions are listed down the side of the page, with actions flowing horizontally. In other versions, often called

"deployment flowcharts," the functions are listed horizontally across the top and actions flow down the page.

Note: Include the "Customer" as the top or left-hand function and "Suppliers" at bottom or right.

4. **Map the process flow,** placing each activity in the appropriate row(s) or column(s) according to participating function(s).
5. **Document and follow-up as with standard flowcharts.**

Note: Deployment flowcharts are commonly used in Process Management charts—which are used to control a process—because they help clarify responsibilities. If appropriate, your team might want to start using a deployment flowchart early in your project so you are familiar with it by the time you reach the final stage of DMAIC.

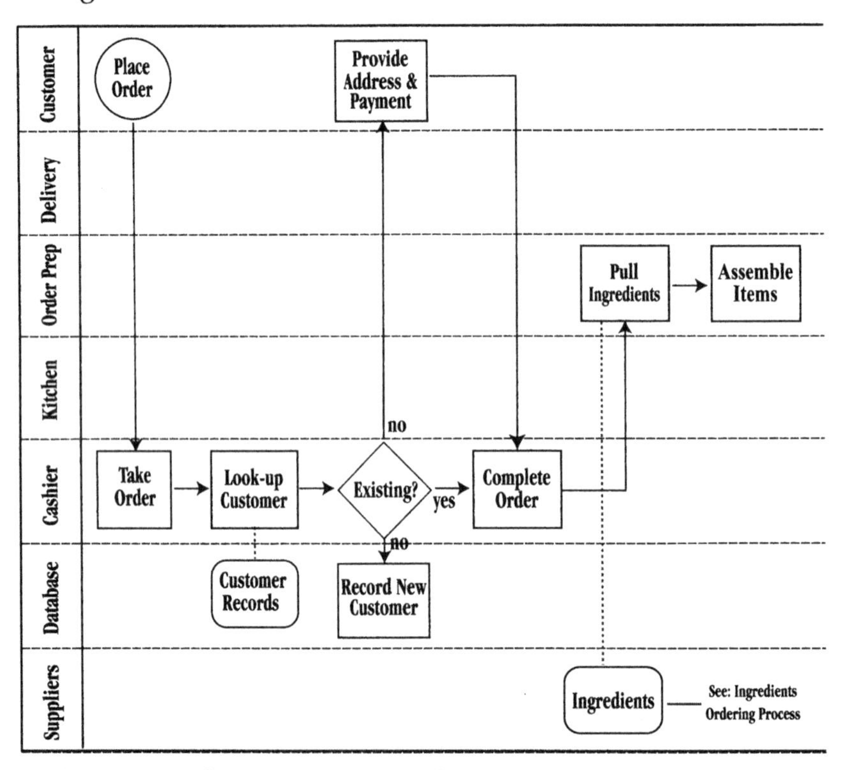

Figure 13-12. Sample Cross-Functional Process Map

Process Value and Time Analysis

Purpose: To identify process steps that add cost and time without adding value for the customer.

Application:

- To identify areas of a process to target for improvement.

Instructions for a Value Analysis (see Figure 13-13):

1. **Identify and map the process to be analyzed.** (See instructions for creating a detailed flowchart, pp. 261-263.)
2. Categorize each step as **value-adding**, **non-value-adding**, or **value-enabling** (see Chapter 12, pp. 225-226 for definitions).
3. **Count the proportion of activities** in each category and review the "balance" of value- to non-value-adding work.
4. **Create a traveler checksheet** (Figure 9-5, p. 147) that lists all the process steps. Use it to gather data on how much time is spent on each step. (Use guidelines in Chapter 9 to determine how much data you should collect.)

Value and Cycle Time Worksheet												
Process Step	1	2	3	4	5	6	7	8	9	10	Total	% Total
VALUE												
Value-Added												
Value-Enabled												
Non-Value-Added												
TIME												
Work Time												
Wait Time												
Total Time												
Time Notation in: ______________				Total Value-Added Work Time								
				Percent Value-Added Work Time								

Figure 13-13. Value and Cycle Time Worksheet

5. **Complete the calculations** to determine what percentage of process time adds value in the eyes of the customer.

6. **Focus on the non-value-adding and/or more time-consuming steps for improvement.**

Completion Checklists

Analyze Checklist

Purpose: To bring a formal end to the Analyze stage of a team's project.

Applications:

- Use during the Analyze work to track progress.
- Use at the end of the Analyze stage to make sure all essential tasks have been completed.

Instructions (see Figure 13-14):

1. **Walk through this checklist item by item at a team meeting.**

2. **Mark a "yes" only if everyone on the team agrees the task has been completed.** If anyone says no, ask him or her to state why they think the task is incomplete.

3. **Reach agreement as a team on each answer before it is marked on the checklist.**

4. **If there is unfinished work, ask for volunteers, assign responsibilities, and set deadlines for completion of those tasks.**

Analyze Tollgate Preparation Worksheet

Purpose: To help a team prepare a presentation for the tollgate review at the end of the Analyze stage.

Applications:

- Use at the end of the Analyze stage to help the team prepare its presentation.

Instructions (see Figure 13-15): Follow the general tollgate guidelines given for

Analyze Checklist

Instructions:

If you can respond "yes" to statement 7 below, and have done many of the tasks described in the other statements, chances are good you are ready to begin developing solutions in the "Improve" phase of DMAIC.

For our project we have ...

1. Examined our process and identified potential bottlenecks, disconnects, and redundancies that could contribute to the problem on which we are focusing. YES NO
2. Conducted a value and cycle time analysis, locating areas where time and resources are devoted to tasks not critical to the customer. YES NO
3. Analyzed data about the process and its performance to help stratify the problem, understand reasons for variation in the process, and identify potential root causes. YES NO
4. Evaluated whether our project should focus on process design or redesign, as opposed to process improvement, and confirmed our decision with the project sponsor. YES NO

For Process Design/Redesign:

5. Ensured that we understand the key workings of the process so we can begin creating a new process to make the organization more efficient and effective. YES NO

For Process Improvement:

6. Developed root cause hypotheses to explain the problem we're solving. YES NO
7. Investigated and verified our root cause hypotheses, so that we are confident we have uncovered one or more "vital few" root causes that create our problem. YES NO

Figure 13-14. Analyze Completion Checklist

ANALYZE Tollgate Preparation Worksheet		
Key messages to cover in the Review (several topics are listed; add to the list as appropriate)	**Best way to present this information** (be creative in finding high-impact visuals–handouts, overheads, flipcharts, storyboard, etc.–to use in the presentation)	**Person or persons responsible for this portion of the presentation**
List promises or commitments made in the previous review.	Identify ways to show progress on the issues raised in the previous review.	
Summarize the logic followed in your analysis.	Identify no more than a handful of charts you can use to illustrate your logic. (Avoid the temptation to include all the charts.)	
List the highlights of your plan for the Control stage. Include estimated timeline and any additional resources needed.		

Figure 13-15. Analyze Tollgate Preparation Worksheet

the Define review (pp. 118-120) to identify priority message to include in your presentation. *In addition:*

1. **Document any commitments or promises** the team made to the Sponsor/Champion, Leadership Council, etc., during the Measure tollgate review.
2. **Compile information about progress** on meeting those commitments.
3. **Develop graphics or other visual aids** to summarize the logic followed in your analysis.
4. **Identify other important messages** you want to convey.
5. **Decide on a sequence for the presentation**, and complete the left column on the worksheet (key messages).
6. **For each message, identify how that information can best be presented** to someone unfamiliar with the details of the project. Be creative! Look for ways to convert messages into data charts, pictures, or other high-impact visuals. Also identify what format that information will take in the presentation (such as handouts, flipcharts, slides or overheads, etc.). Complete the middle column of the worksheet (format for key messages).
7. **Ask for volunteers and/or assign responsibilities** for each section of the presentation. Try to involve the whole team.
8. **Prepare an agenda** for the tollgate review. Identify information that will need to be sent to the reviewers ahead of time.
9. **Do a dry run** of the presentation to make sure it can be completed in the time allotted, and to help team members get more comfortable with their role.

Advanced Analysis Tools

In most cases, the basic power tools described earlier in this chapter will be enough to allow your team to identify significant differences or patterns in your data and verify cause-and-effect relationships that will be at the heart of the solutions you'll identify in the Improve stage of DMAIC.

However, there are times when you need more powerful, rigorous tools. Here are three that are commonly used:

- **Hypothesis testing, a set of calculations for determining the degree to which two or more groups ("populations") are different.** When would this be helpful? Look again at the stratified frequency plot used earlier (Figure 13-9, p. 258).

  In this case, it's easy to tell that the day shift and evening shift must be doing something very different, even without doing any sophisticated statistical calculations. But now imagine that there is more overlap between the two groups—the two peaks are closer together and there are more data values the two sets have in common. It might be difficult then to tell if the two "populations" shown here are truly different. Hypothesis testing can help you make that decision.
- **Correlation** or **linear regression analysis:** An approach for analyzing paired data (like that used for a scatter plot) that helps you **quantify the degree to which two variables are related**.
- **Design of experiments:** A powerful **experimental approach** that allows you to...
  - Test many potential causes at the same time.
  - Identify complex interactions between potential causes.
  - Quantify the degree to which individual causes and their interaction affect the result or outcome you're interested in.

There is a lot of theory and specific procedures associated with each of these tools. You'll find background information on the following pages, but you'll want to work with a Master Black Belt or other statistical expert when it comes time to apply these tools to real-life situations.

Hypothesis Testing: Determining Statistical Significance

If a 10-day heat wave hits your town, people might say "It's global warming." You hit below par on two consecutive golf outings and exult, "My game is really improving!" Sales this month are higher than they were at the same time last year, and everyone is thinking, "This is going to be a strong quarter financially!"

The question, in statistical terms, is whether there is any basis for these conclusions: Is this 10-day heat wave really any different from typical weather seen at this time of year? Is your golf game really getting better? Are the higher sales this month different from the historic sales patterns? Or is all of this just random variation, and therefore has no significance (especially in terms of indicating true change has occurred, and therefore we should change our expectations for the process under study)?

One way to help validate or refute these theories is to perform hypothesis testing, or tests of **statistical significance**. They are some of the most important techniques used by statisticians to look for patterns or test their suspicions about what is really happening in a process, and have a number of applications in Six Sigma projects:

- Confirming that a problem exists or that a process change has led to meaningful improvement.
- Checking the validity of data.
- Determining the type of pattern or "distribution" in a set of continuous data.
- Developing a root cause hypothesis based on patterns and differences.
- Validating or disproving root cause hypotheses.
- Seeing if one result (e.g., process outcome) is really different from typical outcomes.

In short, through hypothesis testing, you're going to be able to reach conclusions about whether the differences you think you see are important for your team. If they are, then you can work to eliminate or preserve that difference, depending on whether it is harmful or helpful to the product, process, or service.

The Heart of Hypothesis Testing: Presuming That Nothing Has Changed

OK, so sales were high for February and March, and people were sure that the new sales training was really paying off. Now it's July and sales are at an all-time low. Staff start circulating rumors about plant closures and layoffs. More likely than not, all of this change is due to random variation (or perhaps annual cycles in sales)—not to the effect of training or the coming Doomsday. Yet it seems to be inherent in human nature to jump to conclusions about causes and effects, and to find patterns even when none exists.

Statisticians guard against the possibility of having "false patterns" lure them into jumping to "false conclusions" by relying on the **null hypothesis:** the proposition that *nothing significant has happened* and any differences are *due to chance alone.* The null hypothesis is the ultimate skeptic, the Devil's Advocate in

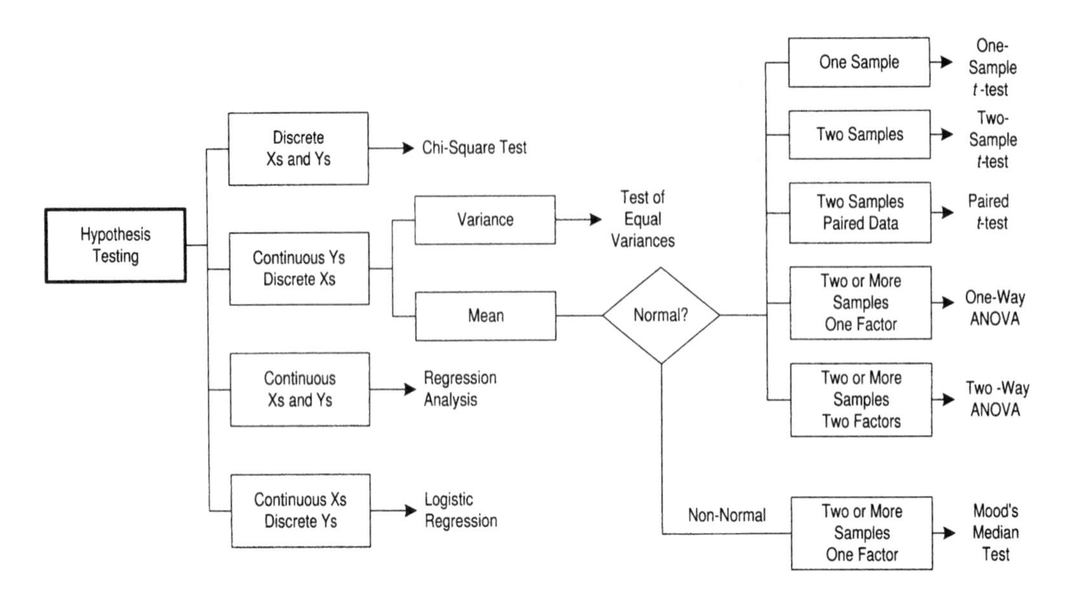

Figure 13-16. Selection tree for advanced hypothesis testing

the face of our pet theories: they won't believe change has happened until we somehow "prove it." In other words, the null hypothesis is often the opposite of what your team hopes is true. *Example:* Your team's hypothesis is that materials from Supplier A are superior to those from Supplier B, and contribute to fewer defects. The "null hypothesis" you will test is that the materials from the two suppliers lead to the same defect rate.

Statisticians shy away from ever claiming to have "proven" anything, but often they can show that *any other explanation is extremely unlikely!* That's the basic approach adopted in hypothesis testing.

Types of Hypothesis Tests

The type of hypothesis test you'll perform will depend on the type of data and type of hypothesis you are testing (see Figure 13-16). Here are the most common tests:

A. **Chi-square (χ^2) test** (rhymes with "pie-square"): A technique used for looking at differences between **discrete** data (and sometimes continuous data that is stratified into groups). For example:
   - *Hypothesis:* The defect rate at Plant A differs from the defect rate at Plant B. (Even if defect rate is expressed as a "continuous" measure, such as

percentage, it is based on counts of defects, which is discrete; here you have the added discrete data of the "attribute" of Plant A or Plant B.)

- *Hypothesis:* There has been a meaningful change in the week-to-week product choices made by customers.
- *Hypothesis:* Staffing levels affect customer satisfaction score.

B. ***t*-test:** Use this method to test for significance when you have two groups of **continuous** data. *t*-tests are a little trickier to use than chi-square tests because the data has to have certain properties, so you'll definitely want to work with a statistician. You can use this test to:

- Compare the cycle time for a key process step at two different weeks in the quarter to see if one is significantly longer or shorter than the other.
- Examine customer income levels from two regions to see if one serves significantly higher- or lower-income customers.
- To test if the seek-time speed in two lots of disk drives is different.

C. **Analysis of variance (ANOVA):** Another test for looking at differences in sets of **continuous** data—but ANOVA can test more than two groups at one time. So if you have four different offices or plants performing the same kind of work, you can simultaneously compare results for all four to see whether they are similar or different. The power of the above *t*-test examples, for example, could be expanded by using ANOVA instead:

- Comparing average cycle time for *each of 13 weeks* in the quarter (not just two).
- Examining customer income levels from *four key regions* (not just two).
- Testing seek-time speed in *five lots* of disks (not just two).

D. **Multivariate analysis:** The first three hypothesis tests are based on comparing a **single factor** across two or more groups (time, income, etc.). Of course, there are likely other important factors that are different between the groups as well, such as the specific types of products or services delivered, differences in customer requirements, different working conditions, and so on. Multivariate analysis allows you to look for differences in **multiple** factors between **multiple** groups at the same time.

Tip: Usually, it pays to focus on a single factor first through ANOVA, then, if something significant appears, look for other related factors you can include in a multivariate analysis.

Prework for Hypothesis Testing

You'll need help from someone experienced in Hypothesis Testing (at least initially) to make sure you set up the data, perform the calculations, and interpret the results correctly. However, your team can layout the groundwork by:

- Clearly defining the outcome you're interested in. (You've probably done this already in the project.)
- Writing down the theory/hypothesis you want to test (called the "alternative hypothesis") along with the null hypothesis.
- Identifying the type(s) of data you either have already or will collect.

Regression and Correlation Analysis

Imagine that you work on a team that has used a scatter plot to verify that the yield of your process is related to the production temperature. Your data is shown in Figure 13-17.

Wouldn't it be great if you could actually *quantify* the strength of this relationship—and perhaps even be able to *predict* (within certain limits) what the yield will be at a given temperature? The power tool you need is **linear regression**, which will help you determine the degree of **correlation** between the cause and the effect you are studying.

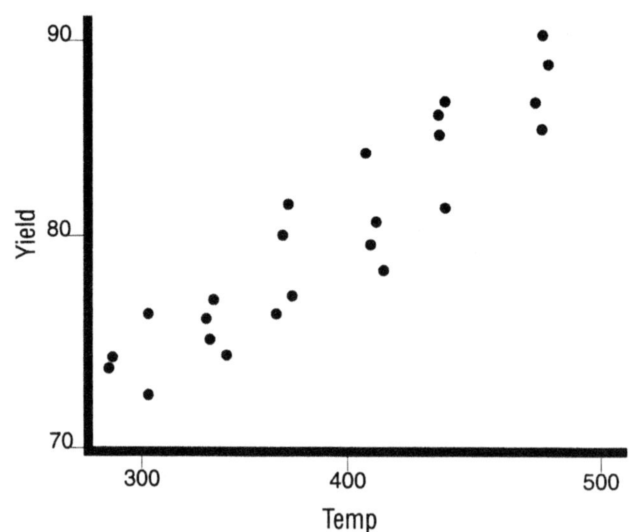

Figure 13-17. Effect of temperature on yield

Linear regression and correlation have multiple uses in Six Sigma projects:

- Testing a root cause hypothesis to see if there is a link between the suspected cause (such as "Temp" in the above chart) and the response or output (such as Yield). Another example might be testing whether the "time between maintenance checks" (the X, or suspected cause) affects the number of copy defects (the Y, or response).

- Measuring and comparing the influence of *multiple* factors (Xs) on one result (Y).
- Predicting the performance of a process, product, or service.

Correlation and regression can be used *only* when you have data for two or more factors matched on *individual* items (like the "paired data" discussed under scatter plots). Note that this use of *individual data points* is one thing that distinguishes regression or correlation analysis from the hypothesis tests described previously in this chapter. In fact, being able to see patterns among individual data points gives linear regression and correlation some advantages over the hypothesis tests: they allow you to see finer, more subtle patterns in smaller samples of data, and to see how a specific change in a process factor or variable will affect individual "units."

Types of Correlation and Regression Analysis

A. **Correlation coefficient (r):** The same data used to construct a scatter plot can be crunched in a single number—the correlation coefficient, denoted as r—that tells you whether and how strongly the factors are correlated. The r correlation coefficients range from -1 (perfect negative correlation) to +1 (perfect positive correlation). Though, you will seldom if ever see r values of exactly -1 or exactly +1, there will always be *some* correlation you can calculate. The question is whether that relationship is strong enough to be the basis of action taken by the team. As a general rule, it pays to investigate any relationships that are stronger than ± .7.

B. **Correlation percentage (r^2):** Many statisticians and practitioners prefer to use the square of the correlation coefficient, r^2, as an indicator of the strength between the cause and the response variables. This correlation percentage, as it called, is interpreted as reflecting the *percentage of variation in the Y (response or dependent) variable that seems to be caused by changes in the X (cause) factor.* Let's say, for example, that you found a positive correlation of .72 between "time between copier maintenance" and "copy defects." Squaring this figure gives you an r^2 value of .52, meaning that roughly 50% of the variation in the number of copy defects is accounted for by the time between maintenance.

C. **Regression:** Various forms of regression analysis concentrate on using existing data to predict future results.
   - The most common type of regression is called **simple** or **linear regression** because it compares *one* potential cause with *one* effect. Figure 13-18, for example, shows the linear regression line for the data shown in Figure 13-17.
   - You can also test multiple causes (Xs) against one or more responses (Ys) through a technique called **multiple regression**.

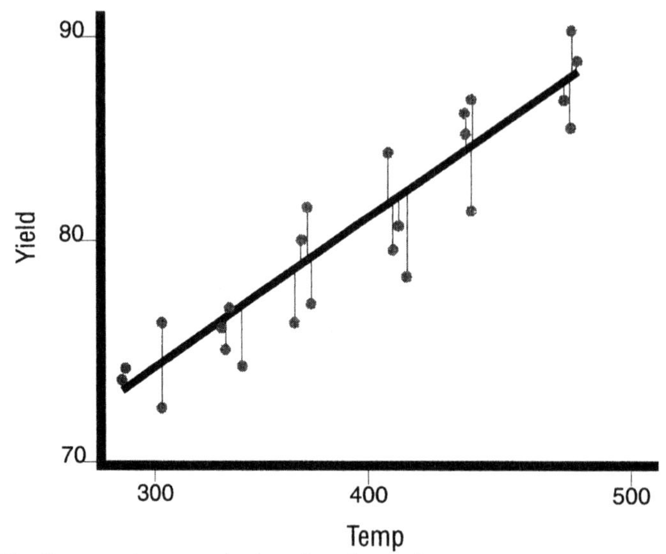

Figure 13-18. Regression analysis of paired data

Prework for Regression or Correlation Analysis

Most spreadsheet or statistical software packages are preprogrammed to calculate the correlation coefficient, correlation percentage, and even linear regression equation for a set of data—so, in theory, you could perform these tests on your own. However, as with hypothesis testing, it would help to work with a Master Black Belt or statistician the first few times to make sure you understand how to perform the calculations, interpret the results, and, most importantly, *take action* in response to the results.

The only preparation you really need to do prior to consulting a statistician is to confirm that you have or can collect **paired data** on individual items (units) going through your process.

Exploring Complex Relationships: Using Design of Experiments to Screen, Verify, or Quantify Causes

Imagine that the defect you're measuring is a "lost sale." The possible reasons for lost sales are numerous: customers don't like your product or service, they object to the price, they didn't like the brochure, the salesperson wasn't skilled in closing the sale, etc. Narrowing down the list of potential culprits for "lost sales" seems daunting enough, let alone confirming which contribute the most.

The same type of situation occurs frequently in manufacturing, where everything from the choice of raw materials to the speed and mechanical settings of the equipment used can affect the type and number of defects that appear.

One statistical tool called **design of experiments (DOE)** is especially well-suited to situations like these. DOE is an entire field of statistics devoted to understanding the often complex relationships in a process and finding the most efficient and effective combination of factors for producing the highest-quality output. It can help you identify not only single factors that influence an outcome, but also *interactions* between factors that help or hinder.

DOE is based on **simultaneously testing multiple factors** that affect a process, service, or product. The first reaction that many people have to that description is that it doesn't sound possible. After all, the "scientific method" drummed into our heads throughout grade school emphasized the need to test one factor at a time (while everything else is held constant). Sound familiar?

DOE avoids the pitfalls normally associated with testing many factors at once by using very specific combinations of factors in very specific patterns. Table 13-8 shows a simple example based on the "lost sales" situation described above.

Based on the factors and levels defined in Table 13-8, the designed experiment would consist of eight trials, where each trial is a particular combination of the factors (see Table 13-9).

Notice that in just these eight trials, the team would get results for all possible combinations of the factors they identified. Analysis of the results (% of lost sales) would help them determine if any one of these factors was more important than the others or if a particular combination had a bad impact on sales.

Simple Designed Experiment	
Outcome of interest:	% or number of lost sales (you want to see this number fall) –or the team could have chosen– % or number of sales (you want to see this number increase)
Factors chosen for study (based on brainstorming and discussion among team members about the most contributors)	• **Sequence** in which customers are contacted • **Experience level** of the salesperson • **Quantity** discounts
Level for the factors Generally DOE tests each factor at two levels, but some designs allow for three or more. (The issue of levels is key in DOE; the factors are not tested at random levels.)	**Sequence** 1: Customers receive a brochure, then a follow-up call 2: Customers receive a call, then a follow-up brochure **Experience level** 1: Less than six months 2: Six months or more **Quantity disounts** 1: 10% discount if purchase totals $100 more more 2: 15% discount if purchase totals $100 or more

Table 13-8. Simple designed experiment

Trial	Sequence	Experience	Discounts
1	Brochure	Under 6 months	10%
2	Brochure	Under 6 months	15%
3	Brochure	6 months +	10%
4	Brochure	6 months +	15%
5	Call	Under 6 months	10%
6	Call	Under 6 months	15%
7	Call	6 months +	10%
8	Call	6 months +	15%

Table 13-9. Simple designed experiment: eight trials

Preparing for a Designed Experiment

You'll need help from someone experienced in design of experiments to properly set up, conduct, and analyze the results. However, your team can lay out the groundwork:

- Clearly define the outcome you're interested in. (You've probably done this already in the project.)
- Identify a manageable number of potential causal factors you want to explore. While sophisticated designed experiments can test 10 or more factors at a time, for your first time out, you'll probably want to keep it much simpler—say two to five or six factors.
- Identify two "levels" for each of the factors.
  - For attribute factors (such as "sequence" in the example above), choosing the levels is simply a matter of deciding what options you want to test. Typically you can categorize the "levels" of attribute factors as "either/or."
  - For continuous data, choosing levels is part science, part art! In the example above, for instance, both "experience level" and "discount" are continuous numbers: you'll find sales people who have worked anywhere from one month to 10 years, and discounts could, in theory, range from 1% to 100% (giving it away free!). Based on their knowledge of the sales process, the Six Sigma team decided that six months was a good dividing point, because it seemed to take people about that long to learn all the ropes. And while their customers would have been delighted to get a 100% discount, the team also knew that fiscally the company could afford 10% to 15% at most.

Here's an important tip: **design of experiments is an extremely powerful and woefully under-used tool available to Six Sigma teams!** If you even just *suspect* that your situation may be suitable for doing a designed experiment, talk to a Master Black Belt or other statistical expert in your organization. You'll be amazed at how efficiently you can learn a lot about your process, product, or service.

Chapter 14
Normal Data and Team Norms
Guiding the Six Sigma Team in the Analyze Stage

By the time your Six Sigma team reaches the Analyze stage, you've probably ironed out many of your personal problems and begun to work routinely by the norms set in its early ground rules. Respect for one another's opinions now comes from working together and the fact that opinions are based on data. The project is becoming clearer as data comes in, and team members are willing to learn new tools to analyze the data.

This doesn't mean that storming can't reappear: it can, but now much of its energy will be focused on problems of analysis and impatience to find the cause of the problem being studied, rather than on the personalities of other team members. Nevertheless, there are ways to prevent the recurrence of storming at the same time the team moves its work along. These include:

1. Draw attention to progress the team's already made.
2. Fix "low-hanging fruit" and build some momentum.
3. Revisit and update team ground rules.
4. Pay more attention to how the team works together.

Let's look at these in more detail.

1. Draw Attention to Team Progress

"When a tree falls in the woods and there's no one there to hear it, there isn't any sound. I know. I was there in the woods when it fell."

There are two kinds of progress the team should be making by the time it gets to Analyze. The first has to do with progress on the Six Sigma project and how to get the word out to others in the organization. The other has to do with the progress the team has made internally, evolving from a group of people into something resembling a team.

Project Progress: Update Your Storyboard

Updating your storyboard is important at every stage of DMAIC, but particularly so here. It is your work in this stage that determines what actions you'll be taking in the next stage—actions that will likely affect many other people outside the team. Informing people of your progress in confirming (or disproving) causes will help them prepare for the upcoming changes.

2. Build Momentum by Fixing "Low-Hanging Fruit"

To paraphrase an old saying, "Nothing motivates a team like success." If a Six Sigma team has to wait six months or longer before its sees some improvements in the process it's working on, there's likely to be a drop in motivation within the team. To prevent this decline, once data begins to throw some light on the process, the team needs to start fixing some of the obvious things in the process that need fixing, the so-called "low-hanging fruit." Where these improvements can be made, teams gain a boost of energy to continue work on the "high-hanging fruit" that will probably be harder to pick off.

The risk, of course, is that in "fixing" obvious problems, the team will be tinkering with the overall process, and add to the variation in it, or otherwise muddy the analytical waters. To avoid making things worse by tinkering, the team should fix things that are obvious and that data show are really causing problems.

If a process map shows a bottleneck or a hand-off so tight it's often missed, the team should make some improvements. The result will be a shortening of cycle time, and perhaps a reduction of defects due to missed hand-offs. Or a new person may not have trained on a particular procedure and is having problems making it work. They need to be trained. The risk of tinkering in such cases is slight. Measuring the impact is important, and overall the team can see some incremental improvement resulting from its efforts.

Major changes in the process should be avoided until essential data has been collected and carefully analyzed. Introducing major new procedures or computer software and the like will have to wait until the problems they might solve had been tracked to their origins.

3. Revisit and Update Team Ground Rules

The ground rules that the team needed in its early days of Defining a problem and project usually need to be revised and updated in the Analyze phase, if not earlier. For example, early rules on "respecting everyone's opinions" will need to be revised as the team comes to rely more and more on the analysis of objective data. How can I respect the opinion of someone on the team who refuses to use data, or analyze it, and relies instead on bluff and bluster to get their way? The old ground rules need to be discussed, amended, and added to as the team matures.

This discussion is all the more important because debate around the meaning of data and how to proceed from it can get heated as the team moves toward implementing solutions for problems. As it begins to pinpoint the cause of problems, the team may have to add "No cow is sacred if it's causing problems" to the list of ground rules. As the team gets closer to making actual changes in the organization, anxiety often begins to build up and otherwise cooperative team members may start drawing lines in the sand. Ground rules will help to prevent those lines from becoming high walls.

And just in case anyone has doubts about the importance of ground rules for the team, these should remain an item on each team meeting agenda. At each meeting, it's worth a few minutes to discuss (not simply read out loud) one of the ground rules. If "Be on Time" is one of the original rules, it's OK in Analyze to discuss **why** it's imperative to be on time for meetings: it saves rework bringing latecomers up to speed, shows respect, gets things done on time, etc. These discussions prevent ground rules from becoming just formalities, and remind people that part of the Six Sigma culture is doing things according to agreed-upon processes. Ground rules are the measurable requirements for the Six Sigma team.

4. Pay More Attention to How the Team Works Together

The ultimate goal of the Six Sigma team is **not** simply to improve a process and increase customer satisfaction, important though these goals may be. Of even greater importance is the new way of doing things in your organization. The Six Sigma way is to gather data, analyze it and then *learn from that analysis.*

The Six Sigma team needs to learn about itself by studying its own activities and processes. How the team operates and how its members work with one another provide opportunities for learning that far too many teams ignore in their desire to solve problems and "get the project over." If this is the mentality of many teams, it's no wonder that things don't really change for the better in most organizations.

How does the team learn about itself?

If the team has been using ground rules and evaluating each of its meetings, it will already have an informal "database" about itself. From time to time, the Black Belt should lead a discussion about this data, pointing out the things that have improved or gotten worse. The Black Belt should also call attention to the following actions when they appear, for they are signs of growing maturity in the team:

- Members of the team other than the Black Belt point out infractions of ground rules.
- Team members volunteer for assignments.
- Team members volunteer to assist other team members in completing their assignments.
- Team members willingly take over the facilitation of a meeting even though the Black Belt is present.
- People avoid the use of "I agree with what you said, but…." Instead, they genuinely try to build on what others have said or done.
- People stick to the point under discussion and avoid tangents.

All of these are signs that the group of people who came to the first meeting with their own interests and agendas are gradually evolving into a team with common purposes and methods.

Thinking about this evolution (or the lack of it!) will help the team learn about itself. The hope is that team members will carry these behaviors back into their "real" jobs and departments, and begin to change behaviors there, too. Where b.s. and bluster continue to rule meetings, we can't expect to introduce tools like run charts and histograms successfully!

These improved team behaviors will be very important as the team moves into the challenging area of developing and implementing solutions for the problems it was chartered to solve. Improvement is the subject of the next chapter.

Troubleshooting and Problem Prevention for Analyze

Failure #1: The Team Bogs Down in the Paralysis of Analysis and Can't Identify Root Causes of Problems and Unwelcome Variation

Why this happens: The tools to analyze the data collected range from simple comparative logic ("Does the data explain why we see this problem in Chicago and not in Los Angeles?") all the way to measures of statistical probability and multivariate analysis. Some teams resist concluding that they have in fact discovered the major sources of variation in their process, and keep looking for "one more proof." The team heaps up reams of statistical proof when it should be moving on to improvement.

How to prevent it: At the end of the day, statistics only offer us probable proof of the causal relationship between key variables and key outputs. These probabilities must be balanced by the common sense and experience of the team. Teams should only use enough statistical analysis to reach reasonable conclusions of cause, and then manage the risks of improvement. To paraphrase an old saying, "If the only tool you have is statistics, everything looks like a correlation coefficient."

Failure #2: Jumping to Conclusions About Causes Before All the Data Is In

Why this happens: Of all the challenges facing Six Sigma team members, reaching this step *without* having already made up their minds about the cause of the problem/defect may be one of the toughest! After all, team members were most likely chosen because they have some relevant experience or knowledge about the process or problem—which means they've likely been living with the "defect" for some time, and naturally have their own theories about what's going on. And in the long run, having well-informed ideas about causes is what's going to lead to a permanent solution.

How to prevent it: While you can't really prevent people from making up their minds early in the project, you *can* prevent their decisions from harming the project by insisting that the team have data to back up its conclusions. Remind the team that they will be asked by their Sponsor and others to show what led them to their conclusions. Make sure that you encourage open-minded thinking during brainstorming discussions, especially when it comes to creating a cause-and-effect diagram, for instance.

The Improve Stage

Chapter 15

Improving the Process

Creating, Selecting, and Implementing Solutions

AFTER ALL THE HARD DETECTIVE WORK that Six Sigma teams do to Define, Measure, Analyze, and discover the causes of problems and variation, do they sometimes drop the ball when they get to Improve the Process? Unfortunately, at times, they do. It's tough to change gears from being a detailed data detective to being a developer of innovative new processes. The skills needed to analyze problems can be very different from those needed to create and implement great new solutions. There's always the danger that your team will simply settle on slight modifications of the process and miss an opportunity for more substantial gains. Or you may fail to pick the best possible choice among the options available, or overlook some potential problem with the new solution that causes it to fail at a crucial moment.

Even when you can develop truly creative possible solutions, it can be tough to go from blue-sky brainstorming to the practical challenge of piloting new solutions, or anticipating and preventing all of the things that might go wrong. In the Improve stage, then, the Six Sigma team must be ready to veer back and forth between outlandish ideas and down-and-dirty details of executing a plan.

To avoid these pitfalls, Six Sigma teams have to keep their improvement options open—not just jump at the first solution idea that comes along—at the very time there's probably considerable pressure on them to solve the problem at last. After having focused for so long on the cause of problems, they need to broaden the scope of their creative thinking.

Fortunately, the Improve stage of DMAIC has its own process steps to help keep the scope broad and the options open as long as possible.

Guidelines for Improve

In Improve, teams need to remember the following guidelines:

- Whatever the team selects as a solution should address the root causes of the problem and the goal the team set for itself in the Project Charter.
- Although the team will brainstorm many possible solutions, one or two will be better than the others; the team must decide which are the best options and determine what it will take to make them work.
- The solutions must not cost so much or be so disruptive that the expenses outweigh the benefits in the long run.
- The chosen solutions must be tested to prove their effectiveness before they are completely implemented.

But even before the team reaches the point of implementing solutions to make improvement, it must struggle with the poverty of imagination that often follows the brilliance of analysis, as the following vignette shows.

Brain Storm, Brain Drizzle, Brain Drought

At Brillatec Marketing, a Six Sigma team just finished a four-month measurement and analysis of why customers who received certain kinds of mailers bought more of the clothing items advertised than other customers. The causes had to do with the use of certain colors in the ads, the weight of the paper stock, and the time of the month customers received the mailers.

When the team met to create solutions, Didi Haines, the Black Belt, announced that it was time to brainstorm some solutions and stood next to the flipchart, ready to capture the storm of ideas she expected.

"What's the big deal?" asked Bill Wannamaker, the team's media guru. "All we need to do is use more aquamarine tints on 20-pound stock, and make sure the customers get them on the last Thursday of the month. It's pretty obvious, isn't it?"

As Didi looked around the team for other ideas, she saw only heads nodding in agreement. It was, as Bill said, pretty obvious. Instead of a brainstorm, the team had a brain drought.

Steps to Workable, Effective Solutions

The goal of the Improve stage is to find and implement solutions that will eliminate the causes of problems, reduce the variation in a process, or prevent a problem from recurring. There are five steps for reaching that goal:

1. Generate creative solution ideas.
2. Cook the raw ideas.
3. Select a solution.
4. Pilot test.
5. Implement full-scale.

Step 1. Generate Creative Solution Ideas: Learning to Be Practically Impossible

Most people pride themselves on being "practical," and that's a good thing most of the time. But when a team is trying to come up with new ways to operate processes, they need to create solutions that seem to be "practically impossible" at first glance. "Pretty obvious" improvements won't do.

How can teams be "practically impossible"? By using techniques to enhance the creativity of their brainstorming. The basic rules of brainstorming were covered earlier in this book (see p. 54); here's how to increase your creativity:

1. **Be clear about what your brainstorm is supposed to produce.** The team needs to be focused on the target.
2. **Set a quota of ideas.** Left alone to brainstorm, most people come up with three or four ideas and quit. If asked for 15 ideas, most people double the "average" output. Go for quantity!
3. **Play off other people's suggestions**. Not easy to do this when you're trying to be brilliant, but even Edison got many good tips from other people. Listen carefully!
4. **List ideas without comment, discussion, or criticism.** Tearing up each idea as it gets written down is not only time-consuming, it's depressing.

Rather than spending 10 minutes talking about the first idea to pop out, get 10 ideas and discuss them afterward. Having people write down 10 ideas on separate sticky notes before slapping them on the wall will discourage premature discussion. Keep the storm moving!

5. **Challenge assumptions and go a little crazy**. Easier said than done, unless you can think like a three-year-old and keeping asking "why" about everything you're working on. Why are tires underneath the car? Why can't they go on top of the car? That would save wear and tear on the treads, especially if they were filled with helium! Get crazy before you get practical!
6. **Brainstorm one day; check back the next day.** The old notion of "let me sleep on it" makes a lot of sense around brainstorming. Ideas just seem to get a little crazier and better if you come back to them the next day.

If at First You Don't Succeed

An experienced Black Belt, Didi didn't give up when Bill Wannamaker put the hammer on the brainstorm session by suggesting some obvious "solutions" to the clothing sales question the team was studying. She wrote Bill's idea on the flipchart, and then spoke to the team.

"Thanks for the idea, Bill. We also need to push the envelope some on this one. Let's try for some wacky solutions, too."

"Wacky how?" asked Terri Crump, another team member.

Didi had to think fast on her feet. "Well, the assumption seems to be that customers have to get something in the mail before they buy the clothes. What if they could find out about the product from a sky-writer?"

"A what?" asked Terri, the mistress of the two-word question.

"Or what if they could order from the catalog by clapping their hands a certain number of times over the phone?"

Didi was on a roll, but the team members were looking at one another like they might have to cut down on her coffee intake.

"Order by clapping your hands?" asked Terri, in her longest question all day.

"Oh, I'm just throwing out ideas," said Didi, a little exasperated.

"Why assume you need to order by phone?" asked Terri. "You could order on the Internet."

"Now you're getting it!" Didi laughed.

Then practicality set in. Bill had a question. "I thought we were just supposed to find out why some customers liked certain kinds of mailer materials and make sure they get the right ones."

Didi switched into her practical gear for a moment. "You're right, Bill, but as long as we're brainstorming, let's open things up a little and see if we can't get outside the box. I'm not criticizing your suggestions, but we've always used mailers to sell the product. Now's a good time to see if we can find some other, better ways to meet our customers' needs."

See Chapter 16 for more on advanced creativity techniques and assumption busting.

Bill folded his arms, and said: "Such as what better ways?"

"Well, let's keep brainstorming and we'll see. Everybody grab 10 sticky notes and put some new ideas down. Then we'll review all of them, including Bill's.

Identifying Process Changes

If your Analyze phase showed that many problems stem from process inefficiencies, there are many options which, depending on the product/service and work being done, can improve that performance. Some principles that apply in many process design situations include:

- **Simplification.** The fewer the steps and the more consistent the path, the better your ability to eliminate defects and control variation. You can have fewer hand-offs, fewer people (too many cooks and all), fewer non-value-adding activities. Simplification can be a reason for *avoiding* automation when it's less complex to do work manually.
- **Straight-line processing.** If tasks can be arranged in sequence, it can avoid communication and coordination issues. The straight-line path is the easiest to track and manage. A big disadvantage of the straight-line path is that it can add time to the overall process by delaying the start of each task until the previous one is finished.
- **Parallel processing.** Doing tasks "in parallel" or concurrently reduces overall process cycle time. For example, in a new product development effort, several components can be designed independently, then integrated into the complete product. The challenge of parallel flows is what you

might call the "right-hand/left-hand" syndrome: changes or decisions are made in one path of the process that other paths don't know about. The result is a problem "downstream" in the process when the paths converge.

- **Alternative paths.** Pre-planned flexibility in how work is done, based on customer needs, product type, technology, etc., is increasingly important in an environment where every product or order is unique. Alternative paths allow you to handle work according to any number of factors. For example, when you go to the hospital, there are different "paths" to be admitted, depending on the urgency of your condition. The risk of alternative paths is having to keep track of and manage various ways to handle an item in the process.
- **Bottleneck management.** In almost any process, there are points where capacity or cycle time causes a slowdown or backup. In bottleneck management, the process flow is "widened" to streamline the entire process. But *beware!* Adding people or equipment may *not* be the best way to widen the bottleneck. Consider also how the product, service, or task/procedure can be changed to eliminate the slowdown. Also, eliminating one bottleneck may create another farther downstream in the process—so bottleneck management should be undertaken with a "whole process" perspective.
- **Front-loaded decision making.** Because decisions can be challenging, there's a natural tendency to defer them until later in the process. That delay may force a lot of work to be based on assumptions that later prove wrong. Pushing decisions upstream in the process can reduce the probability of rushed efforts or rework later.
- **"Standardized" options.** This is a way to simplify decisions yet still offer flexibility by defining a fixed number of options and preparing the process to handle them. The output of this design would be a "semi-custom" product or service. Depending on the number of elements to be selected, there can still be a large number of possible end products. One of the most familiar examples of this approach is in the car business. Manufacturers offer a set of color "packages" and other options that you can choose from—but you can't just get the beige carpet with the blue exterior unless it's part of one of the packages.
- **Single point of contact or multiple contacts.** These are two ends of the

spectrum of customer interface. In the "single contact" option, a customer and/or order is assigned to a person or group which maintains responsibility for the item as it's processed. Another term for this is the "case worker." If you call a customer service number and are told to "always ask for Amy," you're dealing with a single-point-of-contact process. (Unless they have a lot of Amys.) "Multiple contact" processes are usually backed up by strong customer and/or order tracking systems. They allow any person on the system to follow and respond to customer requests and questions. We use a travel service in which we enter an I.D. code at the beginning of the call; then, the agent who takes our call will have our latest travel data on their computer screen when they say "hello." They can then make itinerary changes, answer questions, etc.

Review these options with your team, then hold a brainstorming session using the guidelines given previously, but this time focused solely on "how can we change the process to eliminate lost time, rework or delays, etc.?"

Before moving on...

At the end of *Improve Step 1: Generate Creative Solution Ideas,* you should have:

✔ A wide range of solution ideas (linked to the verified causes from Analyze) that go beyond the team's original thinking.

✔ The ideas should include product, service, and/or process changes as appropriate.

Step 2: Cook the Raw Ideas: Synthesizing Solution Ideas

Now that your team has generated a lot of wild ideas, you need to develop a refined "Solution Statement" that addresses the original challenge defined in the Problem Statement of your Project Charter. To do this, you need to "cook" these raw ideas into some possible and practical solutions.

6σ See the Practicality Scale on p. 310.

A. Refine the brainstormed list.

- Combine related ideas. (If you have a lot of ideas, use an affinity process [pp. 56-57] to identify solution ideas that are closely related.)

- Eliminate non-starters through multivoting (p. 56).
- Discuss ways to build on the elements of impractical ideas to make them practical.

B. **Identify what portions of the problem each of the individual solution ideas will address**, or the degree to which they will address the problem overall. For example…
- Link each solution idea to a particular cause or process problem identified in your Analyze work.
- Make educated guesses as to what impact the solution ideas would have on those causes or problems. For example, "changing the application form" might eliminate defects (delays, rework) in Step 2 of the process, but not improve anything that happens in Step 7.

C. **Use the information from B to generate "complete solution" ideas.**
- Look for solution ideas that are complementary—that is, together they address more of the problem as a whole without overlapping.

D. **Document the "full solution" ideas.**

Another approach that works here is to use a tree diagram and/or an Affinity analysis to help you organize the disparate ideas into logical groups. Instructions for this technique can be found in Chapter 16.

See Tree Diagram for Solution Development on pp. 312-313.

Before moving on...

At the end of *Improve Step 2: Cook the Raw Ideas,* you should have:

✔ Documentation showing how incomplete solution ideas were combined into "full solution" alternatives.

Step 3. Select a Solution

Sometimes deciding which solution to go with is easy: one solution obviously stands out, and the team quickly agrees that it's the one to implement. Other times there are a number of possibilities with no clear front-runner. Then the team will have to use some process to choose the best of the choices.

In either case, the team must be able to explain its choice to others, to those whose work and services will be changed because of the choice made. It's easier to explain a decision if you've used some kind of process rather than simply going with the well-known but hard-to-measure "gut feeling." This is particularly true if the team has to sell an expensive solution to upper management.

You have several options at this point:

A. Perform a "minimum requirement" test.
B. Assess the amount of benefit (impact) for the effort required.
C. Do a formal analysis of pros/cons, costs and benefits.

A. Minimum Requirement Test

If your team has a single solution option that seems head-and-shoulders above all the others, you may not need to perform a detailed analysis of that solution. But there should be a minimum requirement test: "Will this solution eliminate the root causes of the problem? Will it reduce the average variation of the process and product? Will it prevent the problem from reappearing?"

In short, will this solution address the causes of the problems identified in your Team Charter? Of course, the solution has yet to be tested out in reality, but just asking these questions can eliminate non-starters.

B. Impact and Effort Assessment

There's a well-known saying in the rural parts of this country: don't plow the wet end of the field first! For those of us raised in more urban settings, the meaning is that it's a waste of time to put a lot of effort into something that will have little immediate benefit. A simple way to translate this lesson into solution assessment is by using the Impact/Effort Matrix.

To use the Impact/Effort Matrix, your team would need to discuss each potential solution and ask questions about how easily you could implement the effort, and what kind of impact it would have on both the "next step in the process" and your company's customers. In answering these questions, you try to reach con-

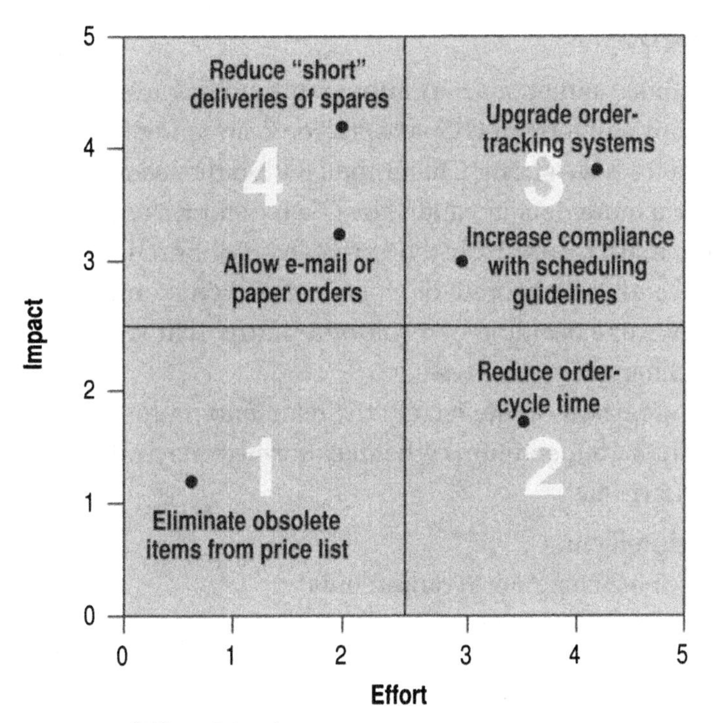

Figure 15-1. Impact/Effort Matrix

sensus within the team, then decide where each option belongs in the matrix (see Figure 15-1).

Naturally, you would want to favor solutions with high impact, especially if they also required little effort (quadrant 4) and avoid those that take a lot of effort with little impact (quadrant 2). But you should still consider options in other quadrants as well. For example:

- It might be that the only way you can truly meet customer needs is to implement a quadrant 3 solution (high effort, high impact)—part of your planning would then be how you might phase it in, or make sure that everything goes smoothly to reduce potential delays or impact.
- At the opposite end, it might be that you'd want to implement a quadrant 1 solution (low effort, low impact), because it is easy to do and will relieve an irritating headache for employees. But don't do a quadrant 1 solution at the expense of something with greater impact.

C. Formal Analysis

In some situations, neither a "minimum requirement test" nor an "impact/ effort" assessment will let you choose between competing solutions—especially if the alternatives have strong Champions within the team! In such cases, you need to perform a more detailed analysis. The tool for this purpose is a **criteria-based decision matrix**, in which the possible solution choices are compared against criteria and then scored on their capability of meeting the criteria. Figure 15-2 shows one example of a solution matrix that shows different alternatives for reducing late deliveries.

The criteria are based on the factors the team and its customers have identified as important to the solution, whatever form it may take. Typically listed among these criteria are:

- Cost to implement.
- Ability to meet customer requirements.
- Impact on other processes.
- Time to implement solution.
- Complexity.
- Expertise required (less is better).

Objective: Reduce later deliveries (root cause: Order LIFO process)

		Alternative A			*Alternative B*			*Alternative C*		
Criterion	**Value**	**Computerize order entry**	**Score**		**Order slip carousel w/ times**	**Score**		**Shorten delivery routes**	**Score**	
Time to implement	7	3 months	3	21	1-2 weeks	10	70	1-2 weeks	10	70
Safety	10	Eye strain, carpal tunnel	6	60	None	10	100	Small risk as drivers learn new routes	7	70
Amount of defect reduction likely	9	Depends on ability of cooks to read orders from screen	6	54	Significant; ensures preparation as orders come in	10	90	Minimal; does not address root cause	2	18
Implementation cost	3	$4,500, plus training	3	9	$75 plus installation	8	24	None	10	30
Operating cost	5	Vendor estimates $150/mo	5	25	$0	10	50	Appox. 4% gas usage increase	5	25
Impact on other parts of the business	2	Will allow faster bookkeeping	10	20	None	3	6	None expected	4	8
			Total	189		**Total**	340		**Total**	221

Figure 15-2. Sample Criteria Matrix (Total = Value x Score)

These factors are laid out in a matrix, and the team scores each option against the criteria, then compares the outcome to see which options have the most potential.

As part of its implementation planning, your team will need to do a detailed risk analysis, but you may want to do a quick risk analysis at this stage just to help you eliminate solutions that carry too much risk. Here are two questions that could help:

- What are the ways in which this solution could fail?
- Does this solution stand a good chance of failing to meet business criteria (such as taking too long to implement or not having enough "return")?

After all the work of comparing solutions against criteria and assessing risks, the team may find that it has reached a decision on which solution to implement without a formal "decision" being made. This is common to "consensus" decisions, but the Black Belt must make certain that everyone on the team is given a chance to express their support or reservations about the decision.

Before moving on...

At the end of *Improve Step 3: Select a Solution*, you should have:

✔ Identified which of your solutions ideas will most effectively and efficiently address the causes identified in Analyze.

✔ Confirmed the feasibility of the selected solution.

Step 4. Pilot Test

Having reached a decision on a solution, the team is ready to implement its choice, but has to remember the "four Ps": **P**lanning, **P**iloting, and **P**roblem **P**revention. (There's a fifth P, but we'll come back to that later.)

A. Planning

After the excitement of creative brainstorming and decision making comes the slightly less exciting task of project management. For many DMAIC projects, a simple Gantt chart or timeline, showing the parallel and overlapping steps to realize the solution will suffice. Gantt charts have been around since the 1920s for a good reason: they're easy to make and easy to understand. You will also want to document process changes by revising any process maps created in earlier steps.

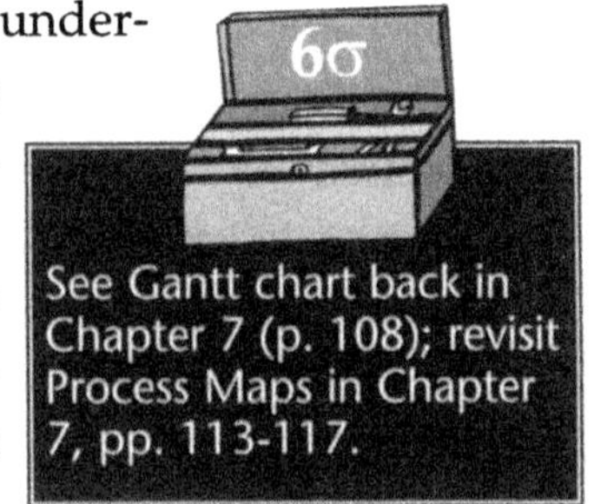

See Gantt chart back in Chapter 7 (p. 108); revisit Process Maps in Chapter 7, pp. 113-117.

As part of your planning, be sure to include how you will know if the solution is successful. For example, what will happen to the measures or indicators you are tracking?

B. Piloting

Failure to try out new things on a small scale and then learn from the problems is the reason many Six Sigma teams come into existence in the first place—because previous "solutions" failed, and something new is needed! As the old saying goes, a plan is just a plan, but a pilot is the real thing. The best plan in the world cannot predict what will really happen when the team executes its solution. Unexpected glitches and unintended opportunities emerge as real people try to do things really differently.

There are various options whenever preparing a pilot. The most sophisticated pilots can be used as "experiments" to compare different approaches and identify the best combination of factors for effective, efficient performance. Broad choices for pilot strategies—which also influence how you eventually implement the process permanently—include:

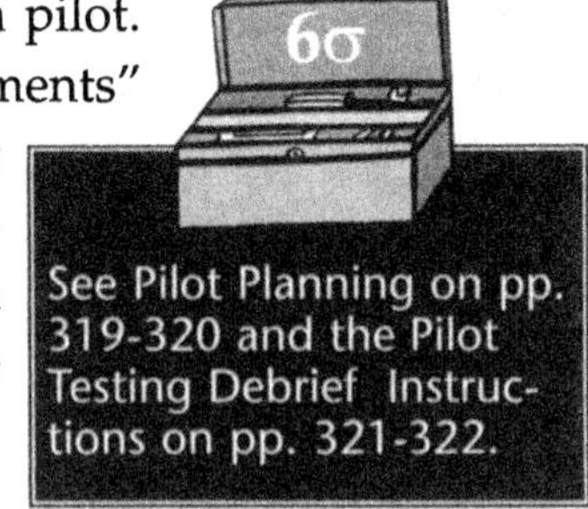

See Pilot Planning on pp. 319-320 and the Pilot Testing Debrief Instructions on pp. 321-322.

1. **Limited time**: Try the new solution out for just a few process cycles. A defined-length pilot offers a couple of advantages:
   - Participants know the test has a defined end point, so they may approach it with more of an open mind.
   - The post-pilot period offers "downtime" for corrections or refinements that may be harder to accomplish if the pilot continues to operate.
   - Comparative measures can be even more revealing. For example, if improvements are seen during the pilot period, but then disappear afterwards, it adds validity to the conclusion that the solution (not some other unknown factor) created the gain.
2. **Selected items or customers:** This approach creates an "alternative path"

in which a certain type or number of real items is sent through the new process. This piloting strategy lends itself well to a "parallel" implementation in which more and more work is moved over to the new process.

3. **Limited scope:** Try out the new idea on only a portion of the people, machines, etc., normally used. (E.g., if there are four labeling machines, pilot the procedures on one; if there are 30 process operators, involve five in the test; if there are 12 locations that do the same work, try it in one or two first.)
4. **Limited resources:** For example, spend no more than 10% of the budget ed allocation for implementation.
5. **Partial solution** *(selected solution components):* Rather than testing the entire new process, different parts of the change can be tried independently. This is the approach that works best as an experimental method. (For more, see information on "design of experiments" in Chapter 13, pp. 277-279.)
6. **Real-life simulation** *(off-line pilot):* Like a laboratory test, in this approach the pilot is really a "dummy" operation that resembles/replicates the real world. The output of this approach may end up not being sold or delivered to customers, but its "quality" can still be evaluated to check the effectiveness of the process. In some companies, a "pilot plant" is used to test new processes and equipment and/or develop products for test (pilot) marketing.
7. **Physical models:** If part of your solution includes rearranging or redesigning the workspace, it helps to build scaled models of the changed layout. Then have everyone who works in that area imagine themselves in that new space and try to anticipate what it would be like. Document their ideas for issues that have to be resolved or for improvements in the layout.
8. **Computer simulation:** Computer simulations are popular in design and engineering situations where product or process variables are relatively predictable and can be programmed into complex models.

The purpose of a pilot is to *learn* what works, what doesn't, and what changes or modifications could help improve the effectiveness of the solution. Be sure to capture these lessons by documenting the data you collect and lessons learned.

C. Problem Prevention

Even before the pilot, preventing problems through the power of negative thinking is a good idea. One way to prevent problems is to ask a series of questions about each of the critical steps in the plan:

- What could go wrong with this step? (Remember Murphy's Law!)
- What could cause the step to go wrong?
- How could we prevent this cause from creating a problem?
- What back-up plan should we have if the problem happens anyway?
- How will we know when to go to our back-up plan?
- What might be some unintended consequences of the solution?

Use the answers to build problem prevention measures into your plans and to develop back-up plans on what to do if particular problems appear.

Knowing when to go to the back-up plan—usually triggered by some missed point on a checklist or a call from an angry customer—is important because back-up plans are almost always more expensive then preventive actions. (Which is cheaper in dollars and stress: root canal surgery or flossing every day?) And yet problem prevention seems to take a backseat to the "fire-fighting" and "diving catches" that characterize so much of managing a modern business.

6σ See Force Field Analysis, pp 317-319.

The Fifth P

Behind and underlying these four Ps is a fifth: **Proactivity.** Rather than waiting for problems to happen or variation to increase, Six Sigma teams should anticipate change and do what they can to proactively direct its course. Everyone knows that, left to themselves, processes tend to degrade, and that customer requirements constantly change, so why pretend that they do not? Planning and prevention are not just "nice to haves" in the Six Sigma organization; they are at the heart of the Six Sigma philosophy.

Conducting the Pilot Test

Depending on the team's original charter, and the progress of the pilot solution, it can take some time to test solutions and measure results. The results of the pilot will need to be charted and compared against earlier data to show that the team's work has indeed made the difference between past and present results. Comparative Pareto charts, run charts, histograms and other charts can display these changes. For the more sophisticated, *t*-tests and other measures of statistical significance may be needed to convince Doubting Thomases that the team, and not mere chance, is responsible for the improvements in the process (see Chapter 13, pp. 270-276).

You will need to conduct a thorough debriefing after the pilot test, and perhaps even conduct a second or third pilot to improve the ideas and resolve unexpected problems. Only when your team is convinced that it has a workable idea should you proceed to full-scale implementation.

Before moving on...

At the end of *Improve Step 4: Pilot Testing,* you should have:

- ✔ Developed and implemented a small-scale test of the selected solution.
- ✔ Documented the lessons from the pilot test.
- ✔ Improved the materials and methods used in the solution.

Step 5. Implement Full-Scale

It's a big mistake to get over-confident after a successful pilot. Compared to full-scale implementation, the pilot is usually a much more controlled situation, with fewer variables to manage and fewer people involved. Other problems are almost sure to arise in the conversion from test to final roll-out of a new process. Some of the critical ingredients—all common sense but worth noting—of a successful launch of an improved process include:

- **Training:** New approaches need to be taught (and learned), old habits broken.
- **Documentation:** References on how to do things, answers to frequently asked questions, process maps, etc., are all important.
- **Troubleshooting:** People need to be clear about

There are no *new* tools used in implementation that you haven't used before. You'll want to use process maps, planning tools (such as Gantt charts and tree diagrams), and the various Measure tools (to help you monitor ongoing performance).You can also adapt the pilot testing checklist and debriefing guides to help you plan and learn from the full-scale implementation.

whose responsibility it is to deal with issues that arise.
- **Performance management:** Watch for needs/opportunities to revise job descriptions, incentives, performance review criteria.
- **Measurement:** Document results.

In practice, the full-scale implementation that signifies the end of Improve tends to blend somewhat with actions that are actually part of Control. In this step, you want to get everything up and running under real-world conditions. Once your team is confident that any problems have been fixed, you can move on to Control, where you develop tools to prevent backsliding, and hand-off ongoing responsibility to the process owner.

Before moving on...

At the end of *Improve Step 5: Full-Scale Implementation,* you should have:

✔ Developed training materials and implementation aids (such as checklists, new process documentation, etc.).

✔ Started using the process under regular working conditions and fixing any obvious problems.

Getting Ready for Control

By the end of the Improve phase of DMAIC, your team should have led the full-scale implementation of a solution that was clearly linked to root causes of the targeted problem. The issue in Control will be how to prevent backsliding to the older methods. Before you move on, there are a few last tasks to complete:

1. Finalize any process documentation.
2. Update your project storyboard.
3. Create a Plan for Control.
4. Prepare for the Tollgate Review by your Sponsor or Leadership Council.
5. Celebrate.

See the Improve Completion Checklist on p. 323.

1. Finalize Process Documentation

In the course of your pilot tests and full-scale implementation, you should have created new process maps and/or other documentation to help people implement the new procedures. As a final check, make sure that any refinements to the procedures have been captured, and useful job aids to help with daily implementation are in place.

2. Update Your Project Storyboard

Under Improve, teams often display the results of their efforts to brainstorm solutions that will reduce or eliminate the causes of their problem. They often include information about experiments and pilot tests of solutions, along with analytical charts showing how they reduced variation and the number of defects produced. Charts designed to prevent problems (based on failure mode and error analysis methods) can be shown here, too. Another sigma calculation and an updated Pareto or histogram are good ways to show improvements. A cost benefit analysis showing what the payoff of the solutions might be rounds out the Improve section.

3. Create a Plan for Control

Planning for Control is really planning for the long-term maintenance and improvement of the process. The key aspect here is that you will probably want to involve more people than just those on your team—because it's not just team members who will be responsible for making sure everything is kept in place! Your plan should include:

- Time to identify likely ways in which the new procedures may fall apart.
- Time to develop measures to prevent that backsliding.
- Ways to involve everyone who works with the process.
- Time to develop ways to monitor the process performance (including the quality of the product/service output).
- Methods for making sure that the hand-offs are complete and responsibilities clear.

4. Prepare for Your Tollgate Review by Your Sponsor or Leadership Council

Before you prepare for the Improve Tollgate Review, do a debrief with the team on what happened with your Analyze tollgate:

- What improvements did you try out that were different from the Measure review? How well did those work? Should you do the same thing this time, or try something different?
- What comments or suggestions did your reviewers (your customers!) have?
- Which of the support materials (slides, overheads, handouts, flipcharts, etc.) worked and which didn't?
- How did you do on time? If it was too long, what could you do this time to make sure you keep it brief? If it was too short, do you need to add in more detail? Speak more slowly?

6σ

See the Analyze Tollgate Preparation Worksheet on p. 268.

After this general review, start your preparation for the Improve Tollgate Review. The most important components of this review will be making sure that the solutions you put in place were clearly linked to the causes identified earlier, and you have data to show that they created the desired improvements. Review the tips given in the Analyze tollgate instructions (pp. 266 and 269).

5. Celebrate

By now, your team should have actual results to celebrate! Since a portion of your Improve work most likely involved people not on the team, consider holding an event (meeting, lunch, etc.) where all those involved are invited:

- Highlight data that shows improvements have been made.
- Thank *everyone* involved in the pilot test and full-scale implementation.
- Acknowledge team members for their continued commitment to the project.
- Invite everyone to help the team maintain the improvements.

Once these steps are complete, your team is ready for *Control.*

Chapter 16

Power Tools for "Improve"

Getting Better and Better

PERHAPS THE BIGGEST CHALLENGE in Improve is to keep from feeling a bit like Jekyll and Hyde as you bounce back and forth between creativity and practicality. The tools described in this chapter will help ease that transition by guiding your team both through creative brainstorming and detailed planning. The chapter covers the same steps described in Chapter 15:

1. Generate creative solution ideas.
2. Cook the raw ideas.
3. Select a solution.
4. Pilot test and implement full-scale.

Some of the tools you'll need to complete your Improve work have been covered earlier in this book and are not repeated here. You may want to revisit:

- Gantt charts, pp. 107-108
- Process mapping, pp. 113-117
- Tree diagrams, pp. 312-313

However, several other tools already covered—like brainstorming—have a different twist when used in this context, so are described again.

Step 1: Generating Creative Solution Ideas

What good will it do if your team puts in all that work to Define, Measure, and Analyze a problem only to end up implementing the same ideas that have been tried before? The answer is obvious! To avoid falling into that trap, you, as the Team Leader, really have to push your teammates to be creative, to think beyond what is already accepted. At this point, don't worry about keeping things "practical" and "realistic." The time to make the ideas workable will come next; here, you want to break free from the same-old, same-old methods that led the problem in the first place!

Advanced Creativity Techniques

Purpose: To help a team think out-of-the-box.

Application:

- Particularly useful when developing solutions.
- Can be applied in any meeting where creative thinking is a goal.
- Several of these methods are alternatives to meeting-based brainstorming sessions, so would be particularly useful for teams that have a hard time scheduling face-to-face meetings.

Instructions: A number of creativity methods are described on the next few pages. Each is accompanied by its own instructions.

Channeling

Uses: To expand the variety of ideas; to ensure all options are considered

Overview: "Channels" are essentially categories of ideas. By identifying channels, a team can do several rounds of idea generation, each focused on or building on the potential types of solutions.

Instructions:

1. **Identify three to seven "channels."** You can do this several ways:
   - List/brainstorm categories up front.

- Do initial brainstorming, then an affinity analysis (pp. 56-57) to create channels.
- Use a cause-and-effect diagram with cause categories as channels (ideas = causes to achieve the desired effect).

2. **Conduct channel-based brainstorming.** Options include:
   - Do a standard brainstorming. When ideas slow down, change channels!
   - Conduct idea generation one channel at a time.
3. **Try for both range** (ideas in all channels) and **depth** (a good quantity of ideas in each channel).

Anti-Solution

Uses: To open people's minds and see things differently.

Overview: Brainstorming the opposite of what you want to accomplish (improvement's evil twin).

Instructions:

1. **Define your brainstorming objective.**
   - *Example:* "How best to speed invoice preparation."
2. **Create a new objective, opposite of the "real" one.**
   - *Example:* "How best to slow invoice preparation to a crawl."
3. **Brainstorm based on the "anti" objective**. Have fun and be wild.
4. **Examine each "anti" idea and see what positive idea it suggests.**
5. **Record the positive ideas and add to them as possible.**

Analogy

Uses: To approach an issue from a perspective where they can be more creative.

Overview: Identifying a similar (analogous) situation and generating ideas there first.

Instructions:

1. **Define your brainstorming objective.**
   - *Example:* "How best to speed invoice preparation."

2. **Identify a similar but different situation that would be a more "free" brainstorming opportunity.** For example:
   - Boost sales <—> Increase yield in an orchard.
   - Reduce application defects <—> Get school kids to follow directions.
   - Offer more options to customers <—> Expand the menu at a restaurant.
3. **Conduct brainstorming** on the *analogy* situation, using preferred method.
4. **Examine ideas and create a "parallel" idea** for the *real* product, service, or process.

Chain Letter (done instead of a meeting)

Uses: To get a large quantity of possible solutions outside of a formal meeting.

Overview: In the chain letter, team members generate and pass ideas around via memo or e-mail.

Instructions:

1. **Define your brainstorming objective.**
   - *Example:* "How best to speed invoice preparation."
2. **Establish a medium and distribution method.**
   - Paper or electronic, fax or internal mail or e-mail.
   - What's the order of routing?
   - Set a time frame for response ("Forward in one day").
3. **First round, each person writes one or two ideas,** passes to next person on distribution list.
4. **Next person builds on and/or adds to the ideas.**

Billboard (done instead of a meeting)

Uses: To gather ideas from a broader range of people, in a non-meeting formal.

Overview: The billboard is a public brainstorming tool—manual or electronic.

Instructions:

1. **Define your brainstorming objective.**
2. **Post a message in a public place asking for ideas.**
   - Include the objective. (Some parameters may be helpful.)

- Place flipchart page on wall of break room, hallway, cafeteria, etc. (Place a marker on a string next to the billboard for convenience.)
- Post on intranet or send e-mail. (Beware of e-mail overload!)
- Set a time frame.

3. **Gather ideas at end of time frame and narrow/select.**
4. **Remember to thank people for contributing.**

Assumption Busting

Purpose: A questioning process that helps identify—and eliminate—preconceptions that inhibit viable solutions.

Applications: Primarily used when a team is developing solutions.

Instructions (see Table 16-1): Assumption busting is not so much a step-by-step method as it is a technique you can use at any point in a discussion of solutions. Here are two ways to practice assumption busting:

1. **Listen for assumption-based phrases, then turn them around (Table 16-1).**

Assumptions	Turn it around...
We can't.	We could.
They'll never buy it.	We can convince them.
We've always done it this way.	Maybe it's time to change how we do this.
I've never seen it before.	I can imagine it.
It won't work.	It can work.
Just thinking about it makes me break out in a cold sweat.	Just thinking about it gives me a sense of inner peace.

Table 16-1. Turning around assumptions

2. **When you feel an idea is good, but it makes you uncomfortable:**
   a. Identify what premise, rule, or experience creates your discomfort.
   b. Ask: "Is this premise valid?" "Could I/we be wrong?"
   c. Consider ways to test the premise: ask customers, talk to team players, get expert advice, see how others do it, etc.
   d. Identify actions you can take to make the new assumptions work.
   e. Take *time* to get used to a new way of seeing things.

Step 2: Cooking Raw Ideas—Synthesizing Full-Solution Ideas

During the creative work of brainstorming solution ideas, it's important to *not* pay too much attention to what was practical or realistic. But now your team needs to switch gears completely and turn even the most wild or craziest ideas to workable solutions.

Similarly, during a brainstorm, people aren't concerned with whether their ideas address all of the root causes identified by your team. So the other challenge here is to bring together all the bits and pieces of a solution into coherent, fully fledged options.

The two tools describe here each address one of those challenges. In addition, you may also want to use an affinity process (Chapter 5, pp. 56-57) and multivoting (Chapter 5, p. 56) to help you synthesize and prioritize the ideas.

The Practicality Scale

Purpose: To help a team get the most from its creativity.

Application:

- Use after any brainstorming session to see if there are elements of even the wildest ideas that can be put to use.
- Particularly useful in developing solution ideas.

Instructions (see Figures 16-1 and 16-2):

1. **Draw a scale** on a flipchart like the one shown in Figure 16-1.

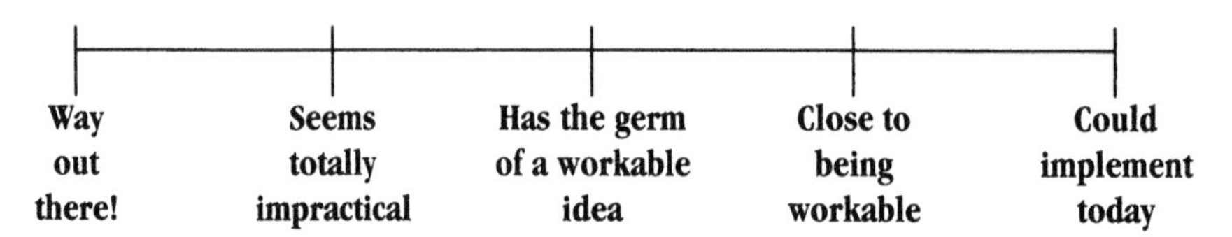

Figure 16-1. The practicality scale

2. **Copy the essence of your brainstormed ideas onto self-stick notes.**
3. **Place the notes onto the appropriate portion of the scale.**

4. **Focus on the ideas on the "won't work" part of the scale (Figure 16-2) and ask, "Is there anything here we can make workable?"**

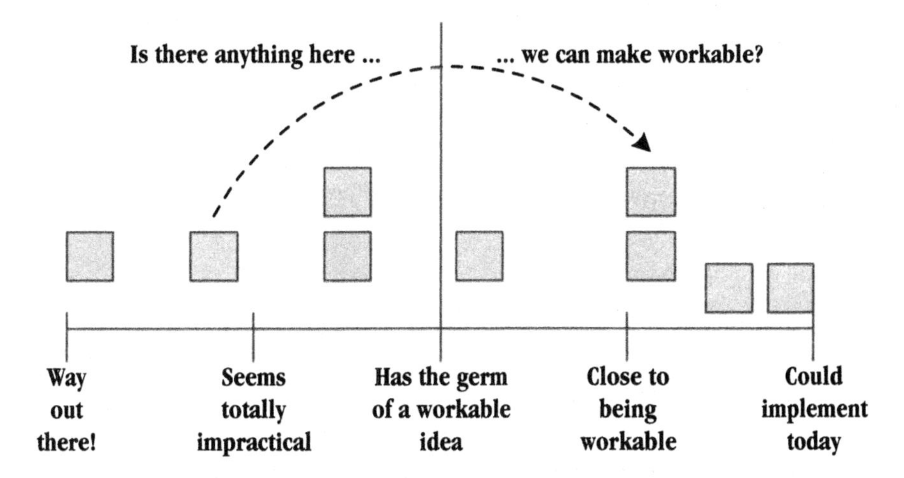

Figure 16-2. Making "impractical" ideas useful

- Read the solution idea out loud to the team.
- Ask, "What is this idea trying to accomplish?" Solicit suggestions about the need or function that idea meets.
- Ask "How *could* we meet that need using today's technology and capabilities?"

  Example: One member of a team struggling with numerous process inefficiencies suggested the team "build a new factory from scratch!" Obviously, that wasn't going to happen. When asked, "What was that idea trying to accomplish?" the answer was "Give us an opportunity to redesign the total layout of the plant so machinery, equipment, tools, materials, etc., could all be arranged more logically."

  When challenged to think of ways the team *could* develop a better layout without building a whole new factory, they came up with a number of smaller-scale ideas that might work: redoing the layout of the tool room (an area over which they did have control), moving material storage areas so as to not interfere with the main process flow, etc.

5. **Capture the refined ideas and add them to the mix of options your team will consider.**

Tree Diagram for Solution Development

Purpose: To organize the components of a goal or target into related sets of tasks.

Applications:

- Identifying strategies and tactics needed to achieve a specific goal.
- Identifying indicators and measures that relate to an identified goal or target.

Instructions:

1. **Write one of your problem-solving goals in the left-hand box** of a blank form that looks like Figure 16-3. Each goal should relate to a root cause verified through your Analyze work.
   Tip: Create separate tree diagrams for each problem-solving goal.
2. **Brainstorm solution ideas** that will contribute to achieving that goal.
3. **Sort the ideas into hierarchical groups**—the most specific, limited actions, broader tactics, and general strategies.
4. **Enter the general strategies as the first level** of branches on the tree.
5. **Enter the tactics needed to accomplish the strategy as the next level** of branches. (Write these as horizontal branches off the first-level boxes.)

For the purposes of synthesizing a "complete solution" idea, your team can stop here. If you're reasonably sure that the ideas you're fleshing out will be part of the final solution, it's worthwhile to push for two more steps:

6. **Repeat step 5** until you reach a level where each step can be acted on—that is, the action step will be clear to team members.
7. **Assign responsibilities** for steps, actions, or improvement targets.

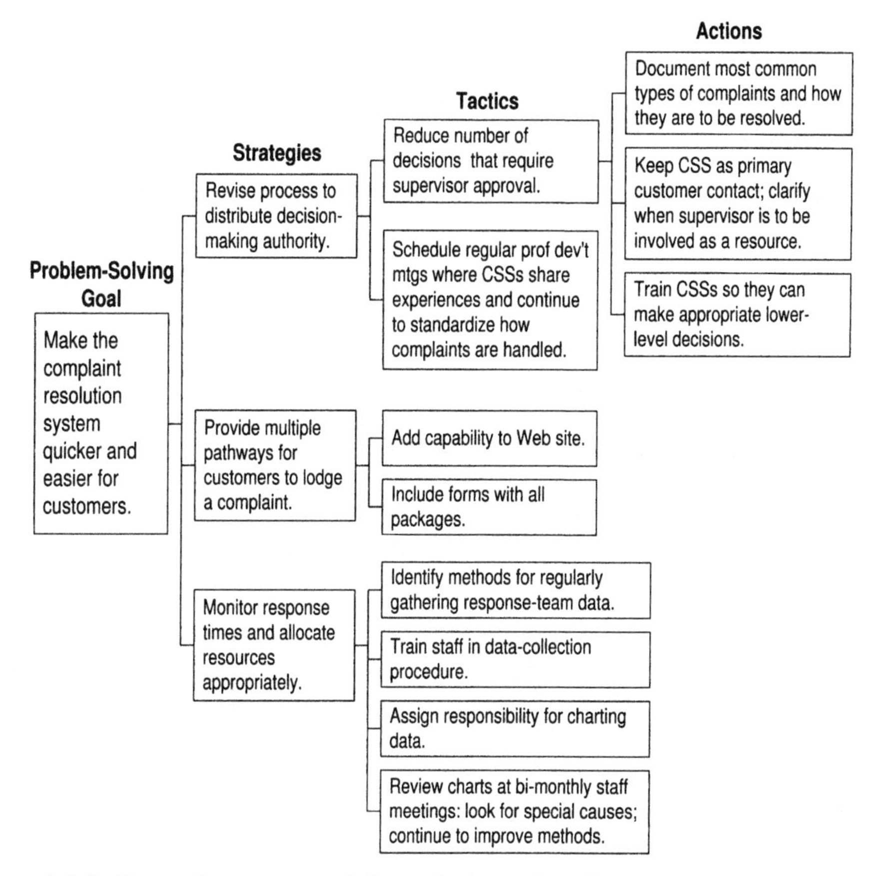

Figure 16-3. Tree diagram used for solution planning

Step 3: Analyzing and Selecting Solutions

There are several tools available for analyzing and selecting solutions alternatives, but all boil down to the same approach: comparing your options against key criteria. Typically, this information is captured in a table form, like the impact/effort matrix or criteria matrix (also called a decision matrix or solution matrix) described here.

> 6σ
>
> If you have several potential pieces of a solution and have a hard time deciding on the best combination, consider using Design of Experiments (pp. 277-279), which will allow you to efficiently test multiple combinations.

The biggest challenge is to make sure that the criteria and their importance are objective—to keep the work of developing criteria completely separate from evaluating the options! By this point, most team members will have their own solution favorites, and may unintentionally favor criteria that fit that favorite solution.

Impact/Effort Matrix

Purpose: To help a team make informed decisions about which potential solutions to implement.

Application:

- Use any time your team is faced with competing actions or solutions.

Instructions (see Figure 16-4):

1. **Compile a list of solution ideas suggested by team members.**
2. **For each alternative, discuss both what impact it will have** and **how much effort it will take to implement.** Here are some starter questions…

   Impact
   - Would our customers notice an immediate benefit?
   - Would making this change bring noticeable relief to those working downstream in the process?
   - Does this solution address the biggest sources of trouble we identified in Analyze?

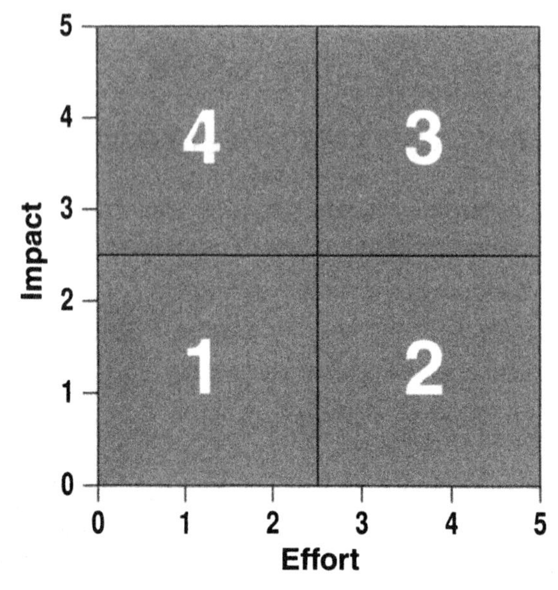

Figure 16-4. The Impact/Effort Matrix

Effort

- How close are we to being able to implement this solution?
- Would a lot of training be required to effectively implement this solution?
- Do we have the resources/materials needed to implement this solution?
- How many people would need to change how they do their jobs?
- Are we technologically capable of implementing this solution?

3. **Arrange the solutions on the matrix (Figure 16-4).**
4. **Decide which of the solution(s) to implement.** (See Chapter 15, pp. 296-297 for guidelines.)

Criteria or Decision Matrix

Purpose: To make effective choices and decisions among alternatives, based on key success criteria.

Objective:						
	Criteria					
Alternatives						Total

Figure 16-5a. One format for a Criteria/Solution Matrix

Application:

- Setting priorities.
- Choosing best process design.
- Selecting best solution.

Objective:

		Alternative A			Alternative B			Alternative C		
Criteria	Value		Score			Score			Score	
			Total			Total			Total	

Figure 16-5b. Alternative format for Criteria/Solution Matrix

Instructions:

1. **Identify objective of selection.**
2. **Determine appropriate criteria.** Be thorough in listing keys to success.
3. **Discuss and weight relative importance of the criteria.**
4. **Construct a matrix:**
   Option 1: List solutions down the left and criteria across the top, as in Figure 16.5a.
   Option 2: List criteria down the left and solutions in columns on the right, as in Figure 16.5b. (For filled-in example, see p. 296.)
5. **Evaluate each option for each of the criteria.** Mark the key information directly on the matrix.
6. **Analyze choices by weighting them within each criterion.** Mark the final rank with a number on the matrix.
   Tip: You need to have enough range in the weights to allow you to see clear distinctions between options. For example, rate what you think is the best option for each criterion as a 10 and the worst as a 1, then fit everything else in between.

7. **Determine the final choice(s)** by multiplying the criteria value horizontally with the alternative weights, then totaling these cross-products vertically for total scores for each choice.

8. **Review "concerns" associated with best choice(s), as needed.** Determine whether or not to complete a risk assessment or failure modes and effects analysis before continuing on with your decision process.

Step 4: Pilot Testing and Implementing

The only differences between a pilot test and implementation are scale and potential complexity. Otherwise, they are the same: for both, you need to decide what changes will be made, prepare for those changes, develop a plan, etc.—which means you'll end up using basically the same tools for both!

The first tools described in this section deal with contingency or proactive planning: trying to anticipate what might go wrong or where you might encounter resistance, and developing strategies you can call on should any of the potential problems appear. The last tools deal specifically with planning and learning from your pilot test, but they can just as easily be used during your full-scale implementation.

In addition to the new tools described here, you may also want to use:

- Detailed and/or deployment flowcharts (pp. 261-264) for documenting new procedures.
- Gantt charts (pp. 107-108) and tree diagrams (pp. 312-313) for planning.
- Measurement planning (Chapter 10) and data analysis tools (Chapter 12) to identify what data you'll collect and how it will be analyzed to determine whether the changes have the desired effect.

Force Field Analysis

Purpose: To help a team proactively build support and neutralize blocking factors to an impending change.

Application:

- Determining solution(s).
- Identifying and removing obstacles.
- Planning and implementing a process change or solution.

Objective: Convert to digital color separation process for No. 3 and No. 4 presses

Driving Forces	Restraining Forces	
Faster make-ready.	Equipment requires more space.	*Constraint*
Can handle more jobs.	Digital perceived as lacking "warmth."	*Illusion*
Better color accuracy.	Fear of taking on unprofitable "specialty jobs."	*Illusion*
More customizing features and effects.	Contract with press manufacturer prohibits "3rd party" add-on components.	*Block*
Two competitors already have digital.	Challenge to train make-ready staff.	*Illusion*
Faster make-ready.	Beyond current year capital budget.	*Constraint*
Less waste during print runs.	Risk of "downtime" and maintenance problems.	*Illusion*

Figure 16-6. Example of Force Field Analysis

Instructions (see Figure 16-6):

1. **Clarify the change or solution to be analyzed.**
2. **Divide a flipchart page or whiteboard into two halves** down the middle.
   - Label one side as "Driving Forces: help/support the improvement."
   - Label the other side as "Restraining Forces: hinder the improvement."
3. **Brainstorm forces for and against the change** and write them on the appropriate side of the flipchart.
   - A "force" can be anything: market trends, unmet customer needs, a manager or executive with strong opinions (and lots of authority!), new equipment purchases, strategic business objectives, advances in technology, improved quality, and on, and on. Encourage creativity by the team.
4. **Start with the driving forces and identify ways to...**
   - **Reinforce that support** (such as by communicating with key supporters).

- **Make the support more visible in your plans** (such as tying measures to a priority customer requirement).
- **Link your solution to that support** (for example, the solution might make it easier to reach a key business objective that's important to the CEO).

5. **Categorize the restraining forces as follows:**
   - **Blocks**: Things for which there is no apparent action to reduce or eliminate (such as regulations, policies, laws).
   - **Constraints**: Things that limit the amount of progress or success you would hope to achieve (such as resources, budget, time).
   - **Illusions**: Untested assumptions or expected resistance ("That's not the way we've done it here before!").
6. **Identify steps to "neutralize" restraining forces:**
   - **Blocks:** Look for ways around or to change policies or regulations.
   - **Constraints:** Seek alternative approaches that conserve resources, or possible sources of additional time, money, etc.
   - **Illusions:** Try to educate yourselves or others to better understand the situation and overcome concerns.
7. **Determine which neutralizing actions are necessary/feasible and build them into your plans:**
   - Focus on items in your sphere of "influence and control" or within the control sphere of your Champion.
   - Identify people who can be approached to help your team implement items or remove obstacles.

Pilot Planning Checklist

Purpose: To make sure the pilot test is a successful learning experience (see Figure 16-7).

Application: Any time a team wants to try out process changes before taking them full-scale.

Instructions: Use the checklist to help your team develop and implement its plans for conducting a pilot test.

Pilot Planning Checklist

The timeline and plans are set.

- ❑ We have set deadlines and responsibilities for all the preparatory work needed.
- ❑ We have deadlines for starting and stopping the pilot.
- ❑ The plans include how to respond to unanticipated problems.
- ❑ The plans account for constraints (budget, resources, time) placed on the pilot.
- ❑ We have scheduled a debriefing session and everyone knows how to prepare for that session.

We have defined the new procedures to be tested.

- ❑ New procedures are documented with visual aids (flowcharts) as well as written instructions.
- ❑ An other needed materials, instructions, etc., are prepared.

All stakeholders have been prepared.

- ❑ Everyone involved in the pilot test understands his or her role and responsibilities.
- ❑ We have reviewed the new documentation and explained the new procedures to those involved with the pilot.
- ❑ Anyone affected by the change has been informed of when the test will start and stop and what it involves.

We know how we will measure success or failure and capture lessons learned.

- ❑ We are prepared to monitor both methods and results.
- ❑ We have a data collection plan in place that will allow us to monitor key indicators.
- ❑ We have methods/tools for documenting what works, what doesn't, and who we respond to unanticipated problems.

Figure 16-7. Checklist for conducting a Pilot Test

Pilot Testing Debrief

Purpose: To help a team cover all relevant information in its debriefing after a pilot test.

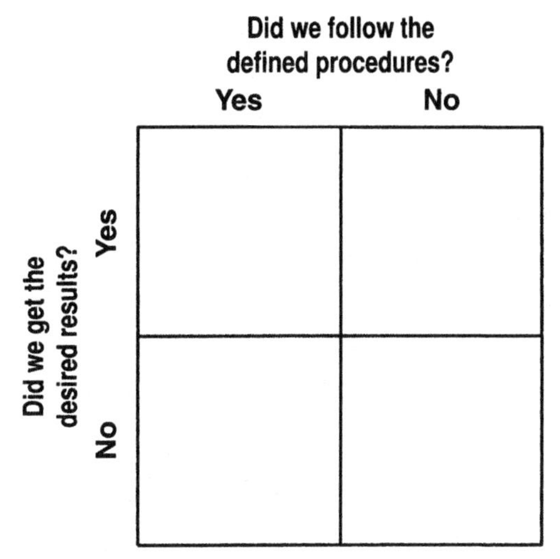

Figure 16-8. Pilot Testing Debrief Procedure

Application: Use any time the team tries new methods, implements solutions.

Instructions (see Figure 16-8):

1. **Prework:** Make sure your plans for a pilot text, experiment, etc. include ways for people to capture their thoughts about what has worked and what hasn't.

2. **Hold a meeting with everyone involved in the pilot test.**

3. **Cover two main points in the meeting:**
   - Did you follow the methods you defined?
   - Did you get the results you wanted?

4. **Plan appropriate follow-up actions** depending on the four possible answers:
   a. **Plans followed, results obtained**: That means the methods you described work effectively. Proceed to full-scale implementation.
   b. **Plans followed, results *not* obtained**: The methods are inadequate to address the problem you're trying to solve. Return to your solution

ideas if necessary and try to find ways to either improve the effectiveness of the methods you tried and/or identify additional ideas to test. Conduct another pilot with the revised methods.

c. **Plans *not* followed, results obtained**: You lucked out! Somebody did something that worked in spite of what you had planned to do. You need to find out what people *really* did, document those procedures, then try them again to confirm that they bring the desired improvement. Finalize the documentation; then proceed to full-scale implementation.

d. **Plans *not* followed, results *not* obtained**: Find out why people did not follow the plans. Were the methods unclear? Where they too difficult to implement as described? Did they just not realize the importance of using the new methods? Develop a revised plan that includes ways to ensure the plans are followed. Conduct another pilot; then debrief again.

Completion Checklists

Improve Checklist

Purpose: To bring a formal end to the Improve stage of a team's project.

Applications:

- Use during the Improve work to track progress.
- Use at the end of the Improve stage to make sure all essential tasks have been completed.

Instructions (see Figure 16-9):

1. Walk through this checklist item by item at a team meeting.
2. Mark a "yes" only if everyone on the team agrees the task has been completed. If anyone says no, ask him or her to state why they think the task is incomplete.
3. Reach agreement as a team on each answer before it is marked on the checklist.
4. If there is unfinished work, ask for volunteers, assign responsibilities, and set deadlines for completion of those tasks.

Improve Checklist

Instructions:

If you can respond "yes" to each statement below, you've achieved success with your improvment, and are ready to plan to "Control" your process/solution.

For our project we have ...

1. Created a list of innovative ideas for potential solutions. YES NO
2. Used the narrowing and screening techniques to further develop and qualify potential solutions. YES NO
3. Created a "Solution Statement" for at least two possible proposed improvements. YES NO
4. Made a final choice of our solution based on success criteria. YES NO
5. Verified our solution with our sponsor and received buy-in and the go-ahead. YES NO
6. Developed a plan for piloting and testing the solution, including a pilot strategy, action plan, results assessment, schedule, etc. YES NO
7. Evaluated pilot results and confirmed that we can achieve the results defined in our Goal Statement. YES NO
8. Identified and implemented refinements to the solution based on lessons from the pilot. YES NO
9. Created and put in place a plan to expand the solution—with refinements—to a full implementation. YES NO
10. Considered potential problems and unintended consequences of the solution and developed preventive and contingent actions to address them. YES NO

Figure 16-9. Improve Completion Checklist

Improve Tollgate Preparation Worksheet

Purpose: To help a team prepare a presentation for the tollgate review at the end of the Improve stage.

Applications: Use at the end of the Improve stage to help the team prepare its presentation.

Instructions (see Figure 16-10): Follow the general tollgate guidelines given for the Define review (pp. 118-120) to identify priority messages to include in your presentation. *In addition:*

1. **Document any commitments or promises** the team made to the Sponsor/Champion, Leadership Council, etc., during the Analyze Tollgate Review.
2. **Compile information** to document progress on meeting those commitments.
3. **Develop graphics or other visual aids** that summarize the solution ideas you considered, how the final choice was selected, and how it links to the causes identified in Analyze.
4. **Summarize the challenges** you faced in implementation and lessons learned in the pilot.
5. **Most important:** Finalize the data charts or other information that shows the implemented solution has had a positive effect on the key measures associated with the problem.
6. **Complete the rest of the worksheet and prepare an agenda** as you have for other tollgates.

IMPROVE Tollgate Preparation Worksheet		
Key messages to cover in the Review (several topics are listed; add to the list as appropriate)	**Best way to present this information** (be creative in finding high-impact visuals–handouts, overheads, flipcharts, storyboard, etc.–to use in the presentation)	**Person or persons responsible for this portion of the presentation**
List promises or commitments made in the previous review.	Identify ways to show progress on the issues raised in the previous review.	
Be prepared to defined the solution(s) selected for implementation. Show how it relates to the causes verified in Measure. What were the key criteria and who did you decide which option was best?	Consider showing your solution matrix, but keep it as simple as possible.	
Summarize the challenges faced with implementation, what you learned in the pilot tests, and how that affected full-scale implementation.		
Be prepared to show documentation that the implemented changes led to improvement. What is the new sigma level?	Prepare "before" and "after" versions of data charts. Recalculate sigma and compare it to baseline.	
List the highlights of your plan for the Control stage. Include estimated timeline and any additional resources needed.		

Figure 16-10. Improve Tollgate Preparation Worksheet

Advanced Improve Tools

FMEA (Failure Modes and Effects Analysis)

Purpose: To anticipate problems so you can take steps to counteract them, and reduce or eliminate the risks.

Application: Identifying ways in which a change in a process, product, or service may cause unintended problems—so appropriate countermeasures can be developed.

Instructions (See Figures 16-11 and 16-12):

1. **Brainstorm all the potential ways the anticipated change in a process, product, or service could fail** (= failure modes).
   Example: In changing an application form, it may be that customers won't understand it, that sales staff won't be able to explain the changes to customers, that it may not capture some essential information, that the new forms won't be available at the promised time or in sufficient quantity, etc.

2. **List these failure modes down the left side of a chart** like that shown in Figure 16-11; describe in a few words what the *impact* of that failure would be.

3. **Identify the potential cause of these failures** (column 3 on the worksheet).

4. **Develop a rating scale** (usually from 1 to 10) for each type of risk associated with that failure:
   - **Severity (S)**: what is the effect or impact?
   - **Occurrences (O):** How often you think type of failure could occur?
   - **Detection (D):** What controls or measures are in place that would increase your chances of detecting this failure?

   Brainstorm examples of failures for each of these risks, then assign them numbers as shown in Figure 16-12.

5. **Rate each of the above factors** according to the scales you have created.
   - Enter the numbers into the S — O — D columns.

6. **Multiply the three numbers together to get the Risk Priority Number (RPN).**

FMEA Worksheet

Project or Person ____________________ **Date** __________

Process, Product, Service ____________________

Feature or Step	Failure Mode	Cause	Effect/Impact (severity data)	Frequency (occurrence data)	Ratings			RPN SxOxD	Response		
					S	O	D		Action	Who	When

Figure 16-11. Worksheet for conducting an FMEA

7. **Starting with the failures that have the highest RPNs**, discuss for each failure ways to eliminate its causes, reduce the chance of it occurring, increase the chance of detection, and/or reduce its impact should it occur.

8. **Assign responsibilities** for carrying through on the actions you just identified in Step 7.

9. ***Optional:*** You can add other columns to this chart to track what actually occurs when the change is implemented. Check failures that happened, make notes on how often they appeared, what actions were taken, and what impact those actions had.

Severity Scale	
Rating	**Criteria: A failure could...**
10	Injure a customer or employee.
9	Be illegal.
8	Render product or service unfit for use.
7	Cause extreme customer dissatisfaction.
6	Cause a major performance loss.
5	Cause a loss of performance that is likely to result in a complaint.
4	Cause a minor performance loss.
3	Cause a minor nuisance but one that can be overcome with no performance loss.
2	Be unnoticed and have only minor effect on performance.
1	Be unnoticed and not affect performance.

Detection Scale	
Rating	**Criteria: Detectability of defect**
10	Defect caused by failure is not detectable.
9	Occasional units are checked for defect.
8	Units are systematically sampled and inspected.
7	All units are manually inspected.
6	Manual inspection with mistake-proofing modifications.
5	Process is monitored and manually inspected.
4	SPC is used with an immediate reaction to out-of-control conditions.
3	SPC as above with 100% inspection surrounding out-of-control conditions.
2	All units are automatically inspected.
1	Defect is obvious and can be kept from affecting the customer.

Occurrence Scale		
Rating	**Time Period**	**Probability**
10	More than once per day	>30%
9	Once every 3-4 days	≤30%
8	Once per week	≤5%
7	Once per month	≤1%
6	Once every 3 months	≤.03%
5	Once every 6 months	≤1 per 10,000
4	Once per year	≤6 per 100,000
3	Once every 1-3 years	≤6 per million
2	Once every 3-6 years	≤1 per 10 million
1	Once every 6-100 years	≤2 per billion

Figure 16-12. Example scales for FMEA

Design of Experiments (DOE)

The basics of DOE were described back in the Analyze tools chapter (pp. 277-279), where it helps teams verify cause-and-effect relationships. Another primary use for designed experiments, however, is in finding **optimal operating conditions** to achieve a desired output. If your analysis exposed a number of process factors that have to be controlled to minimize defects and improve quality, considering using DOE to identify what combinations of those factors is best. Review the description of DOE in Chapter 13, then consult with your Master Black Belt or other advisor for help.

Chapter 17

"At Last We're a Team!"
Guiding the Six Sigma Team in the Improve Stage

It's usually not until a team reaches the Improve stage of DMAIC that it begins to show the characteristics of a "performing" team:

- Team members are fluent in the language of DMAIC. People who couldn't tell a histogram from a telegram now discuss data and graphs as they would the weather.
- Team members work together comfortably and supporting each other. Internal competition (if it existed) is replaced by a sense of learning together.
- People are planning and *doing* things, not just measuring and analyzing.
- Excitement and a sense of responsibility are building as the team takes action to implement its solution.
- Solidarity and group loyalty may emerge as the team collectively faces healthy opposition to its work and its solution.
- Team members have come to respect one another's approach to their mutual work, and may even kid one another about their personal quirks: Old "Show-Me-the-Data" Bob is about to strike again!

The Improve (performing) stage is one of great energy in DMAIC. After twisting and turning its way through Define-Measure-Analyze, the team is excited about the chance to actually **do** something! The energy and excitement are good because the team needs lots of both as it works even harder to plan, pilot, and troubleshoot its solution.

But the team needs to be cautious about this energy, lest it take action for the sake of action only. The energy needs to be channeled into the work of planned implementation, not simply "doing stuff." The Black Belt will have to take the lead in reminding the team of this, while avoiding the appearance of holding them back.

In order to channel the growing energy found in implementing a solution, the team will find the following actions useful:

- **Review ground rules:** By now the original ground rules assembled by the team will need revision and refinement. Directions about working together need attention now as the team goes into action. All members of the team should be taking responsibility for enforcing ground rules, making the job of the Black Belt more that of a facilitator than an "enforcer."
- **Pay attention to detail:** Although there's plenty of creativity in the Improve stage, the team needs to measure improvements carefully and critically. In this stage there's a risk that all of the team members will "get on board" with their solution and lose objectivity. The team needs to retain its objective, critical edge in order to sell its product to others in the organization.
- **Revisit team membership:** Paradoxically, at the time the team is growing more mature and tight, it may have to add one or two new people to its nucleus in order to better handle the planning and work needed to accomplish a successful implementation. Whether to add new members—and choosing whom to invite—should be a team effort, not simply the Black Belt's decision. If you do add members, make every effort to share your team history and show how they can further the team's efforts. In this way the "newbies" can come to feel some ownership for the solution. In rare cases, it might also be time for some current members to leave the team. Such action should be taken *only* if the team member agrees that it's a good idea; *forcing* someone to leave at this time can have untold side effects in poor morale.

- **Stick to your schedule:** Introducing new processes and procedures can eat up lots of time, and team projects tend to start stalling in this stage unless the team proactively takes preventive action against the causes of slowdowns. The Black Belt will have to pay special attention to assignments and their completion.
- **Keep the Champion involved and committed:** In a sense, a key reason for having Six Sigma teams is so that Champions—usually managers—can learn a new, data-based way to manage. To achieve that overarching goal, the team needs to keep the Champion involved and up to date with its work, especially as changes are made in a process. At the very least, the Black Belt should talk with the Champion every week, perhaps daily, as the project moves toward completion. These talks should not simply be summaries: the Champion needs to know about the details of how and why the team is taking the actions it is.

Speaking of Champions ... getting good improvements *implemented* is not just a matter of luck or even planning; it's also a matter of influencing decision makers in the organization. For that, you need to understand who your stakeholders are and how to influence them.

Building Stakeholder Support (The Politics of Successful Change)

> *Have you ever noticed that when you want to try out something new in an organization you have to go out looking for supporters, while the opponents are already out looking for you?*
>
> —The Wise Old Sigma Team Leader

Although Six Sigma and DMAIC are based on data-gathering and the scientific method, teams should never forget that introducing change—even needed, well-designed change—is never easy. So, when it's time to implement improvements, the team is going to have to get into a marketing mode. Selling improvements to key stakeholders is just as important as using good data and useful statistical tests. Anyone who doesn't understand this is liable to get lost now and then in the corporate swamp.

Obviously the team Champion is both a key stakeholder and a safari guide to the team in that jungle. He or she will probably use some old-fashioned clout along the team's way to help spring some data or team members from the clasp of other managers who are also busy. The Champion is also a good consultant when it comes to figuring out a political strategy to get the new solution accepted when the time comes.

One thing the team can do is use a Stakeholder Matrix (Figure 17-1, p. 336) to plan its marketing strategy. The matrix is simple. You list the names of key stakeholders who can influence or are influenced by the change down one side, then mark each for whether they support or oppose the change or are indifferent about it. The team identifies who these people are, and then identifies where each falls on a scale running from "strongly opposes change" (Blocker) through "indifferent to change" (Neutral) to "strongly supports change" (Supporter).

Then the team determines what it can do to maintain supporters, change the minds of opponents, or at least "neutralize" their opposition. The result is a mini-marketing campaign in which team members pitch tailor-made approaches to the stakeholders. This campaign needs to take into account such things as the reasons for stakeholders' support or opposition and their decision-making style. (Are they bean counters? People people? Policy wonks?) Conducted just before and during the early stages of the pilot, this marketing campaign should help avoid the unpleasant surprises that often overtake the politically naïve.

Scoping out Stakeholders at Southern Healthcare

Southern Healthcare is a chain of 12 hospitals and nursing homes. A Six Sigma team at Southern's headquarters has just spent four months reducing the average time to process insurance claims from three weeks to two days, while reducing the error rate per 100,000 from 34,000 to 8,000, a sigma level of 2.9—a long way from 6σ, but a great leap forward anyway.

Nevertheless, the team's Black Belt, Carrie Paggioli, knows that not all the stakeholders will welcome the improvements being piloted because they involve some significant changes in the way the claims are processed, and because of the blurring of functional responsibilities that accompany the changes.

So Carrie is meeting with the team to analyze the positions of the various stakeholders as the team looks down the road toward full-scale implementation.

One of the team members has drawn a simple matrix on the flipchart with spaces for the names of the stakeholders and boxes next to them to indicate their guess at where the stakeholder stood on their solution:

(--) strongly opposed, (-) somewhat opposed, (0) indifferent, (+) somewhat supportive, (++) strongly supportive.

"Let's look at each stakeholder and see if we can agree on where they stand," says Milly Aldiss, the team member acting as meeting leader today.

Surprisingly, it doesn't take long for the team to agree on where each stakeholder stands.

Carrie starts the analysis.

"Tim's support is a no-brainer. He's been behind us all the way on these improvements, and says he'll help us work on Phyllis. Why exactly do you think she's so opposed?"

Terri Handel, a supervisor in the claims area, answers. "She's said from day one that IT can't manage the changes we want in how the claims are handled in the computer system."

"Doesn't she realize by now that we're actually making her job easier?" asks Will Oppenmaier, another team member.

"I think she's afraid that front-line people are going to have too much discretion, and they'll make more serious mistakes," answers Carrie.

"More mistakes?" replies Will. "Things were so complicated before they couldn't make any more mistakes if they tried. She's just a micro-manager."

"Maybe so," nods Carrie, "but we need her support to expand from our pilot. What do you think we can do?"

From her position next to the matrix, Milly points to Ted Gonzalvez's name as she spoke.

"I know Phyllis listens to Ted on a lot of things. Maybe he could talk to her and try to win her over."

"You're right," says Will, "but before we ask him to run interference we'd better make sure he's a solid supporter of our changes. Why do some of you think he's close to neutral on this one?"

And so it goes for nearly 30 minutes until the team has agreed on a plan of action for each member. They would get further advice from their Champion, and then a team member would lobby each person, presenting a variety of information. They agree to ask Ted for his assistance and find out what further information he thinks Phyllis needs to resolve some of her opposition. They agree that to increase Ted's own support, they need to show him specifically how the changes they are trying out will make him and his staff look good because they have to spend less time reworking claims with wrong and missing information. Ted is known to have some ambitions to move up in the Southern administration.

 Before they left the meeting room, Carrie takes the time to erase the matrix from the white board.

Is it manipulative to "psych out" supporters and opponents of change? Not really. If people are opposed because they don't have information, the team needs to supply it. If others are opposed because they cannot see how they or their customers would benefit, this also needs to be explained. And if the team cannot show a benefit to stakeholders and customers, it should *expect* to run into opposition!

Improvement Stakeholder Analysis Worksheet

Purpose: To help you understand and effectively deal with the people who are interested in and/or affected by your project.

Applications:

- Early in a project to identify people who will need to kept informed of the team's work and/or who have information valuable to the team.
- Later as a team plans for implementation, to know what individuals or groups are affected by process changes and therefore should (minimally) be kept informed of and trained in the new methods and (preferably) be involved in designing those changes.

Instructions (see Figures 17-1 and 17-2):

1. **Identify any individuals and groups who are affected by or who can affect the team's work** and list them on a worksheet, as in Figure 17-1. These are your stakeholders.

2. **Discuss the type of support/opposition exhibited by each stakeholder.** Identify whether—at the current moment—they would be most likely to **block** the team's work, **be neutral** to the team's work, or actively **support** the team's work. Mark the appropriate column next to each stakeholder.

3. **Discuss which stakeholders are most critical to the team's success** and develop strategies for dealing with each key stakeholder.
   - For Supporters, the strategy will be one of keeping them informed of

Stakeholder Analysis Worksheet				
Stakeholder Name	Blocker	Neutral	Supporter	Strategy

Figure 17-1. Stakeholder Analysis Worksheet

your project, so they can intelligently champion your cause with others in the organization. Your strategy might also involve using the resources that a given Supporter represents.

- For Neutrals and Blockers, you'll need to decide if it's worth the effort to try to convert them into Supporters. You might decide that you don't have to pay particular attention to a Neutral manager since they are not actively opposing what you do.
- If you would like to turn a Neutral or Blocker into a Supporter, develop strategies that will help demonstrate the benefits of your project to the organization and, if possible, to them individually.
  Hint: One way to get people to support your team is involving them in the team's work. For example, ask a Neutral manager for his/her advice on an issue, or invite him/her to attend a team meeting so team members can do so.
- Ask your team Coach, Sponsor, and/or Leadership Council member for help dealing with Blockers. It is their job to help your team overcome organizational barriers to success, and they will likely have more clout to make things happen.

4. **Document your strategies** for dealing with each stakeholder and assign responsibilities to individual team members for carrying through on those strategies.

Stakeholder Analysis Worksheet						
Stakeholder Name	Blocker (--)	(-)	Neutral (0)	(+)	Supporter (++)	Strategy
Tim Wilson (Champion)					✔	Get his advice on dealing with Phyllis and Ernie. Make sure he's happy with support level of communication.
Phyllis Claymore	✔					After speaking with Ted, ask him to speak to Phyllis.
Ernie Jones	✔					Give him an opportunity to voice concerns. Realistically, won't get his support until we have result.
Ted Gonzalvez				✔		Ask Ted what reservations keep him from being a strong supporter.

Figure 17-2. Sample entries on a Stakeholder Analysis Worksheet

If a Tree Falls in the Woods...

Now more than ever, the Six Sigma team has to get the word out about its proposed process improvements. Chapter 5 reviewed the use of a storyboard to keep the organization abreast of the entire project.

In the Improve stage, the team should consider the use of a storyboard devoted entirely to the improvement being piloted—what's being done, who is involved, what measures are being made, requests for suggestions, and the like.

Even though the process is not ready for the final implementation full-scale, every effort should be made to explain to other people in the organization via e-mails, newsletters, and meetings what's being done.

Six Sigma is all about a new way to manage the business at large and the daily routines, putting them on a more scientific base. If word about this work in progress is not widespread in the organization, it will take much longer to make those hoped-for changes.

Improving the Team's Own Processes

If it hasn't done it before, it's during this stage that the team really needs to pay attention to its own meeting process and how it works together. Meeting analysis should have been going on from the very first meetings: "What went well at today's meeting? What can be improved?" Now it's worthwhile to have one member of the team observe the rest of the group at a meeting and give feedback on how team members show the following skills:

- Asking questions without challenging personalities.
- Making suggestions without bossing others.
- Sharing information without being a know-it-all.
- Using humor to defuse stress without being the "class clown."
- Bringing out the best in others.
- Thanking others for assistance.
- Summarizing progress.
- Building on others' opinions.
- Addressing disagreements without arguing.
- Listening openly to differing opinions.

A team observer can use these skills in a checklist, and near the end of the meeting give feedback about who used them, how they were used (or not) to move the team's work ahead.

Too often teams assume that these skills "just happen," when they are actually skills that team members can recognize, learn, and use frequently to improve both the efficiency and the effectiveness of their own meetings and, by gradual extension, all of the meetings they attend elsewhere in the organization.

We can close this chapter with another quote from the Wise Old Sigma Trainer:

If a meeting takes place in the woods, and there's no one there to analyze it, will the next meeting be any better?

Troubleshooting and Problem Prevention for Improve

In the Improve stage, the Six Sigma team is more likely to fall short of an optimal solution than it is to fail outright. It's as if the team had expended too much energy to Define, Measure, and Analyze the problem before it.

Failure #1: Not Getting Very Creative with Solutions

Why this happens: Sometimes the team is under pressure to finally come up with a solution after all of the measurement and analysis. Customers are waiting, and defects are still piling up. So the team resorts to a less-than-brilliant fix. Or the team simply may not know a lot about brainstorming, and goes with the best of a not very good set of top 10 improvements. Or the team may feel that it doesn't have permission from its Champion to really challenge the assumptions underlying the process being improved.

How to prevent it: The team should expect to spend as much time and energy in the Improve stage as it has in Define, Measure, and Analyze *combined*. Why not? It will take all the ingenuity it can muster to create an improved process that won't relapse into old problems. The Champion must defend the team in Improve against organizational pressure to "Just do it!" long enough for the team to come up with a robust and long-lasting solution. And Champions need to be assumption-busters themselves and set the example for the team. Finally, the team needs to experiment with a variety of brainstorming methods and keep reminding itself that there is no such thing as "The One Best Way." There is always "A Better Way."

Failure #2: Failure to Pilot the Chosen Solution Small-Scale Before Doing a Full Implementation

Why this happens: Similar to # 1 above: pressure to deliver a solution from outside the team and impatience from within the team to show the organization what it can do.

How to prevent it: The Champion and Black Belt need to remind the team that only a pilot can actually shake out the problems still hidden in the solution. Keeping the problems small keeps them manageable.

Failure # 3: Failure to Win Support and Defuse Opposition to the Solution

Why this happens: Lack of corporate political savvy is probably at the heart of this failure. Because DMAIC is such a rational, analytic, data-driven approach, teams sometimes fall into the trap of thinking their solution must be accepted because it is, well, the *right* thing to do. In fact, even the right things to do need to be supported by the right stakeholders to be implemented.

How to prevent it: Team needs to identify key stakeholders in the solution, lobby them, and show how the solution is in their best personal and professional interest, or at least won't matter to them. Of course, the Project Champion should be laying the groundwork for acceptance from day one of the project.

Improve Do's and Don'ts

Do

- **Take the time to pilot your solution.** Everything the team does is theoretical until it actually tries something new in the real world. Theory and reality seldom match up perfectly.
- **Measure results carefully and objectively.** There's a tendency to give the pet solution the team has developed "the benefit of the doubt" when it doesn't perform as planned. The same measures that defined the dimensions of the problem must show results in a clear light, or expose the failure of the solution to solve the problem after all.
- **Celebrate!** Too many teams "celebrate" success with a pizza party. That's OK, but part of the "celebration" should include a careful review of how the team applied (or failed to apply) the tools of DMAIC. After all, using DMAIC is all about learning how we do things. Remember to remember!

Don't

- **Settle for routine improvements.** Too many teams "solve" problems by putting in place slightly improved versions of what was in place before they attacked the problem. The fact that the old process was a "solution" to an earlier problem should be a warning that simply approximating what existed before may lead to another problem in the future. Push the envelope in Improve and, if the solution is not capable of delivering what customers require, it's time to consider the application of process design and redesign.

The Control Stage

Chapter 18

Control and Process Management

Hold the Gains

DOING SOMETHING NEW AND DIFFERENT for a little while is usually not too hard. With a little extra effort, people will bend their habits to fit the solution and new practices put in place by your DMAIC team. But what often is hard is to continue using a new approach or solution. Old habits, as they say, die hard.

The purpose of Control is simple: once the improvement's been made and results documented, continue to measure the performance of the process routinely, adjusting its operation when the data clearly indicates you should do so or when the customer's requirements change. In this chapter we'll look at how this simple concept can be implemented.

Lest there be any confusion, by "Control" we mean maintaining a process whose operation is stable, predictable, and meets customer requirements. Getting your process to this state is what DMAIC is all about. Without Control efforts, the improved process may very well revert to its previous state, undermining the gains you *thought* you'd achieved and making your work for naught.

Control has four parts:

1. Discipline.
2. Documenting the improvement.
3. Keeping score: establishing ongoing process measures.
4. Going the next step: building a process management plan.

Part 1. Discipline

Maintaining a stable and predictable process requires discipline at both the personal and organizational level. Let's start with the latter. Unless the organization clearly rewards disciplined maintenance of improved processes and discourages "taking your eye off the ball," individual employees will decide on their own if they want to measure and monitor process operations religiously. So discipline starts with the very processes whereby companies select, train, track, and especially evaluate and reward their own employees.

If disciplined and proactive processes are not in place, control of improved processes will be hit-and-miss, left to the individual employee and subject to luck, not skill. Occasional "rally the troops" meetings in which management exhorts employees to "be more accountable!" "pay more attention to detail!" and "shape up!" are often wastes of time. They provide a poor substitute for rewarding for people to regularly measure and analyze process operations. Instead, your organization should clearly identify a **process owner**, the person who is responsible for that process—not in the sense of doing all the work him- or herself, but rather in the sense of making sure that the process is continually monitored, studied, and improved.

Discipline at the personal level is difficult unless you and every other person who works on the process understand the *reasons* and *benefits* for monitoring, control, and improvement. Without that understanding, gathering data and creating charts will be seen as busywork—and will become the tasks dropped off the "to do" list at the first sign of pressure. That's why the organization should also make sure that employees are trained in the use of process management tools.

Part 2. Documenting the Improvement

Once the pilot phase of Improve is nearing completion and the improvement appears to be effective and sustainable, you need to make sure the improved process is thoroughly documented.

If you've ever left instructions for a house-sitter while you're on vacation, you know how easy it is to give bad or incomplete directions. Did you think to let the sitter know how to tap the thermostat in just the right place to turn the air conditioning on (in case of a rare December heat wave)? Did you remember to mention that the hot and cold water faucets in the shower are backwards, or that the key to the front door is under the doormat by the side door?

Without good directions, house-sitters and inheritors of improved processes are in for a series of painful surprises. Documentation will help prevent those surprises, if you Keep It Simple, Please. Here are a few hints:

- Write clearly and use pictures—flowcharts, photographs, charts, and videotapes—whenever possible. Avoid jargon and technical terminology, unless you can correctly assume that the person looking after the improvement is as fluent as you are with "insider" language.
- When you're unpacking the new computer, which one do you read, the two-inch thick manual or the 10-page "Highlights for Operating Start-up"? (Or maybe you don't read either one!) Keep it brief!
- Anticipate problems and warning signs that cropped up during the pilot phase. These are the document equivalent of those glass panels that say "Break in Case of Fire and Remove Axe."
- If someone has to take the company shuttle over to the third floor of building 16 or spend an hour on the computer to dig out required procedures, they probably won't bother, and undesired variation will creep back into the process.
- If the process is too complicated, people will stop updating what they're doing, until what the documents say and what people actually do become very different things. Some companies handle this problem by having a department specially devoted to "documentation control," but we recommend you keep control of documentation in the hands of people who actually manage the process and know what needs to updated or dropped.

6σ

See the Process Documentation Checklist on p. 355.

One last hint: your documentation is more likely to be used if those who will be using the documents help create them. Show rough drafts to people who work with the process every day and ask them for their input.

Documenting the Gas Extrusion Team Upgrade

Reducing the air bubbles in high-pressure hose manufacturing had been the project focus of the gas extrusion DMAIC team at Steinway Rubber Hose Company for the last six months. The team finally discovered why there was air mixing with the liquid rubber in the complicated machine that shaped and extruded a continuous stream of hose from its tungsten nozzle.

The team engineered and built a small device into the extruder that prevented any air from mixing with the liquid rubber. The new device needed some attention from the extruder operators, so the team wanted to make sure that the operators understood exactly what it was, why it was there, and how they could maintain its operation.

The team first put together a "home video" team explaining what they had done working on the project, plus a short section on how the device operated and kept air from mixing with and weakening the high-pressure hose. This video would be shown to all employees already on site, and to all new employees who worked the extruder.

On the extruder itself there was a large plastic-laminated card that had a series of colored photos showing how to operate the extruder, and drawing special attention to two digital screens on the extruder, one that warned when air was mixing with the rubber and another to indicate pressure inside the new device. The card included instructions on what to do if the screen showed a problem on the device itself.

There was also a computer terminal near the machine that gave information on the machine's operation and possible problems and cures, as well as a space for people to enter ideas for improvements. The same computer could display control charts showing how the extruder was operating over time. All the extruder technicians practiced with the computer during training.

Over the next eight months of operation, there were only minor adjustments to the new device, and air bubbles stayed well below customer and safety requirements.

Part 3. Keeping Score: Establishing Ongoing Process Measures

Earlier in DMAIC, the team measured to define the problem, to count defects, to analyze for the cause of variation, and to gauge the success of improvements. In Control, the team must identify the key measures that will enable them and their successors to maintain and manage the process over time. How will the everyday people keep score? There are three obvious starting points where your team should look for measures:

a. **Examine a SIPOC map of the improved process.** As usual, customer requirements are the starting point. Obviously the Six Sigma team must **measure process outcomes** for conformance to customer requirements, defects, and process variation.
b. **Decide which upstream process measures are linked to the improvements**, measures that will predict problems downstream in the outputs. For example, if "on time" is a key customer requirement and process measurement shows that two of the key process steps are taking longer and longer, the team needs to find and eliminate the cause of the trend before it causes the output to be late—a defect.
c. **Look at critical input measures** that help to predict the quality of process steps and key outputs.

It may take a while in Control to nail down the chain of measures that best characterize the process. Once established, this chain becomes the basis of a scorecard monitored on a regular basis (each day or month, for example) by whoever "owns" the team's improved process.

Charting the Data

Knowing what you want to measure is half the battle when it comes to maintaining a high-performing process. The other half is defining when and how those measures will be gathered and what to do in response to the data.

To get started, have your team review the new SIPOC of the improved process and discuss:

- How will we get feedback from a representative cross-section of customers of this process? (Feedback should not be limited simply to happy customers or chronic complainers.)
- Where exactly will measures be taken?
- How difficult will it be to collect data?
- How will measurement data be displayed?

The decision on how to display data depends on what you want to show and who wants to know. Charts already discussed in this book—run charts, histograms, Pareto charts—can be used to summarize and analyze measurements.

There is one other power tool that is extremely useful at this stage: the **control chart**, a type of souped-up run chart that provides additional ways to detect abnormal variation.

Control Charts

Claims at Safest Insurance

Mel is the manager of an insurance claims office and serves on a Six Sigma team that recently made improvements aimed at speeding up the number of claims processed per hour. He volunteered to establish a method for the team and process owners to monitor claims-per-hour.

To get started, he gathered data for just over a month (using good sampling methods, of course!) and plotted the data in time order on a run chart similar to the ones seen in Chapter 13. Figure 18.1 shows his first chart.

Based on the tests Mel had learned earlier to detect special causes on a run chart, he didn't notice any patterns that would require action. But then an Master Black Belt showed Mel how to determine the expected amount of variation in the claims process, and to plot the resulting "control limits" on the chart (see Figure 18.2).

When looking at this new control chart, Mel saw that early in his data collection effort, one data point was beyond the upper control limit—

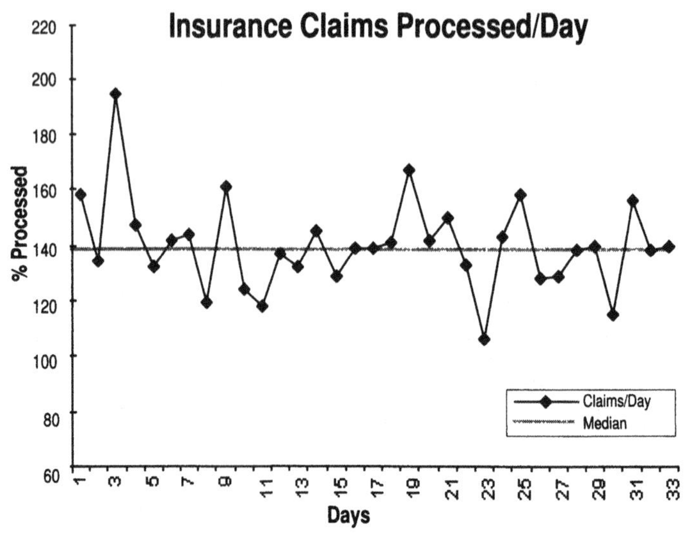

Figure 18-1. Run chart of insurance claims

that means it could not be explained by common cause variation. With this extra information, he knew the claims process wasn't quite as "in control" as everyone hoped—there was still unpredictable variation appearing.

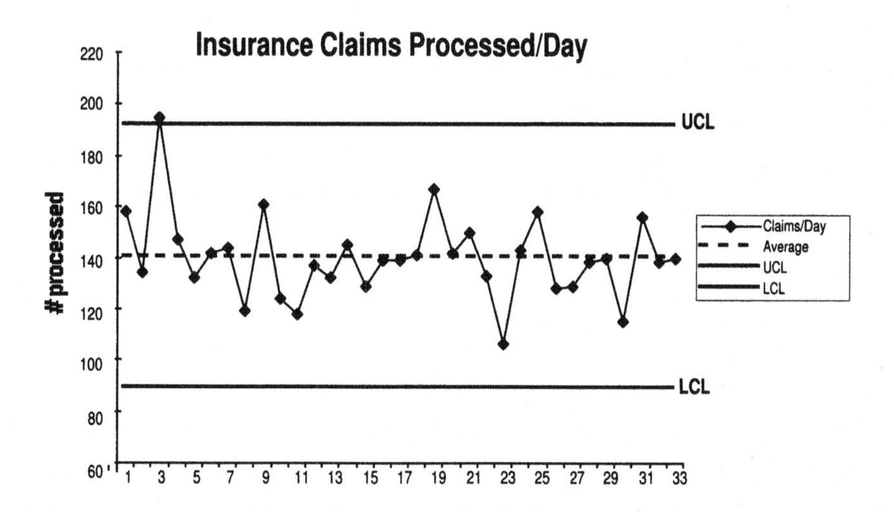

Figure 18-2. Control chart of insurance claims

As Mel discovered, control charts are even more powerful than run charts in detecting special cause variation. The control limits the Master Black Belt helped him calculate—approximately three standard deviations on either side of the average—provide the basis for judging more quickly when certain types of special causes have appeared. You can apply all the same rules that you apply to run charts (see Chapter 13, Table 13-2; remember to skip the points on the median) ...

- Nine points in a row on one side of the average line.
- Six points in a row increasing or decreasing.
- Fourteen points in a row that alternate up and down.

... Plus several more that are based on dividing the control chart into "zones" of standard deviations (see Figure 18-3). If the data is normally distributed (as it should be in most "in control" processes), a bit more than 34% of the data points will fall in each Zone C (within one standard deviation of the average), about 14% in each Zone B (between one and two standard deviations), and a little over 2%

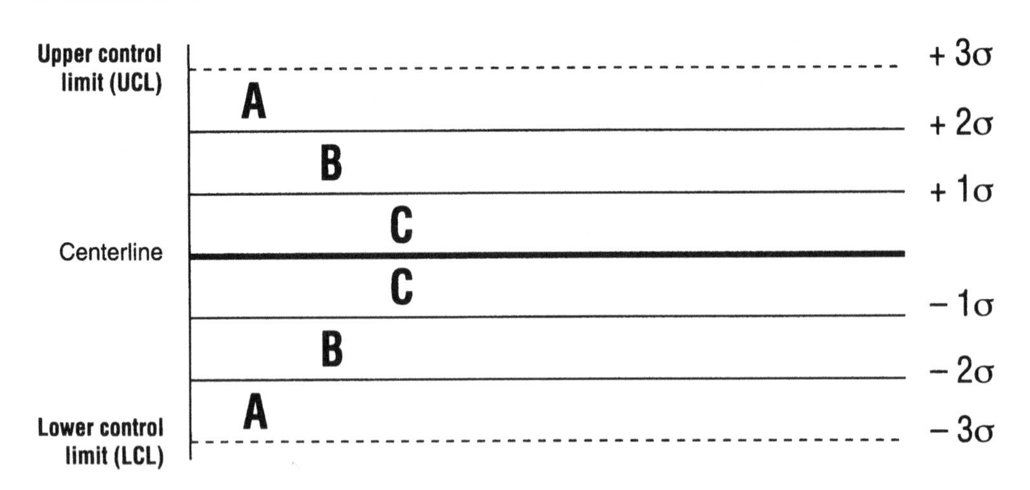

Figure 18-3. Control chart zones

in each Zone A (between two and three standard deviations). When a process is out of control, points can fall outside the control limits (the most familiar of the control chart tests), bunch together in Zones B or A, bounce back and forth across several zones, or even "hug the centerline" (that is, cluster in Zone C). The specific rules and their interpretation are given in Chapter 19.

Finding any of these patterns or signals in your data means something unusual is affecting the process and should be investigated. Once the cause is discovered and removed, the process should return to normal variation within the control limits.

Customer Requirements and Control Limits

Beware! Many people think that if process data stays within the control limits, the process is producing good outputs for its customers. Wrong! Control limits are calculated *from the process data.* Customer requirements or specifications come from the customer and have nothing to do with statistics. In other words, control limits simply show the natural variation of the process as it now exists, which could be very different from what the customers require.

For example, an auto repair shop could use a control chart on its oil change process and find that it is in control (no signs of abnormal influences), but takes an average of 49 minutes to do an oil change when customers have been promised their cars back in 30 minutes!

The goal of the Six Sigma team, then, is to maintain a process that is "in control" and predictable, *and* meeting or exceeding customer requirements.

Putting Control Charts into Action

See Control chart descriptions in Chapter 19, pp. 354-361.

Choosing the right control chart is not difficult; most guidebooks to Statistical Process Control (SPC) can guide the Six Sigma team to the right one. The trick with control charts is to use them to trigger improvements if something unusual is happening to the process. Too many businesses, sadly, construct control charts and then file them away, without taking action when needed.

One other warning: control charts are only as good as the data going into them. From time to time, whoever is managing the process needs to test the validity, reliability, and repeatability of the data collection methods being used to collect data. (These issues were discussed in Chapters 5 and 9.)

Part 4. Going the Next Step: Building a Process Management Plan

Having monitoring and measures in place is just a prelude to actually *using* process performance data in an ongoing management effort. A key element in that management activity is a **process management plan** (a portion of which is shown in Figure 18-4). Even a good process may have its problems. Having a warning and response plan to use if problems occur is part of the overall Six Sigma "Be Prepared for Problems" proactive stance. The process management plan covers the following:

- **Current process map.** The manager of the improved process needs to see at a glance the flow of activities and decisions in the process. A concise process map provides this visual aid.
- **Action alarms.** Develop a process response plan that clearly marks the points in the process where measures can take the pulse of inputs, process operation, and outputs. With thresholds indicating when the quality of any of these is degrading, the process response plan lets the process manager know when to take action. For example, the failure to complete three customer orders on time could signal the need to go to a contingency plan for deliveries.
- **Emergency fixes.** Once an action alarm goes off, it's important to have emergency fixes or back-up plans already spelled out so that employees

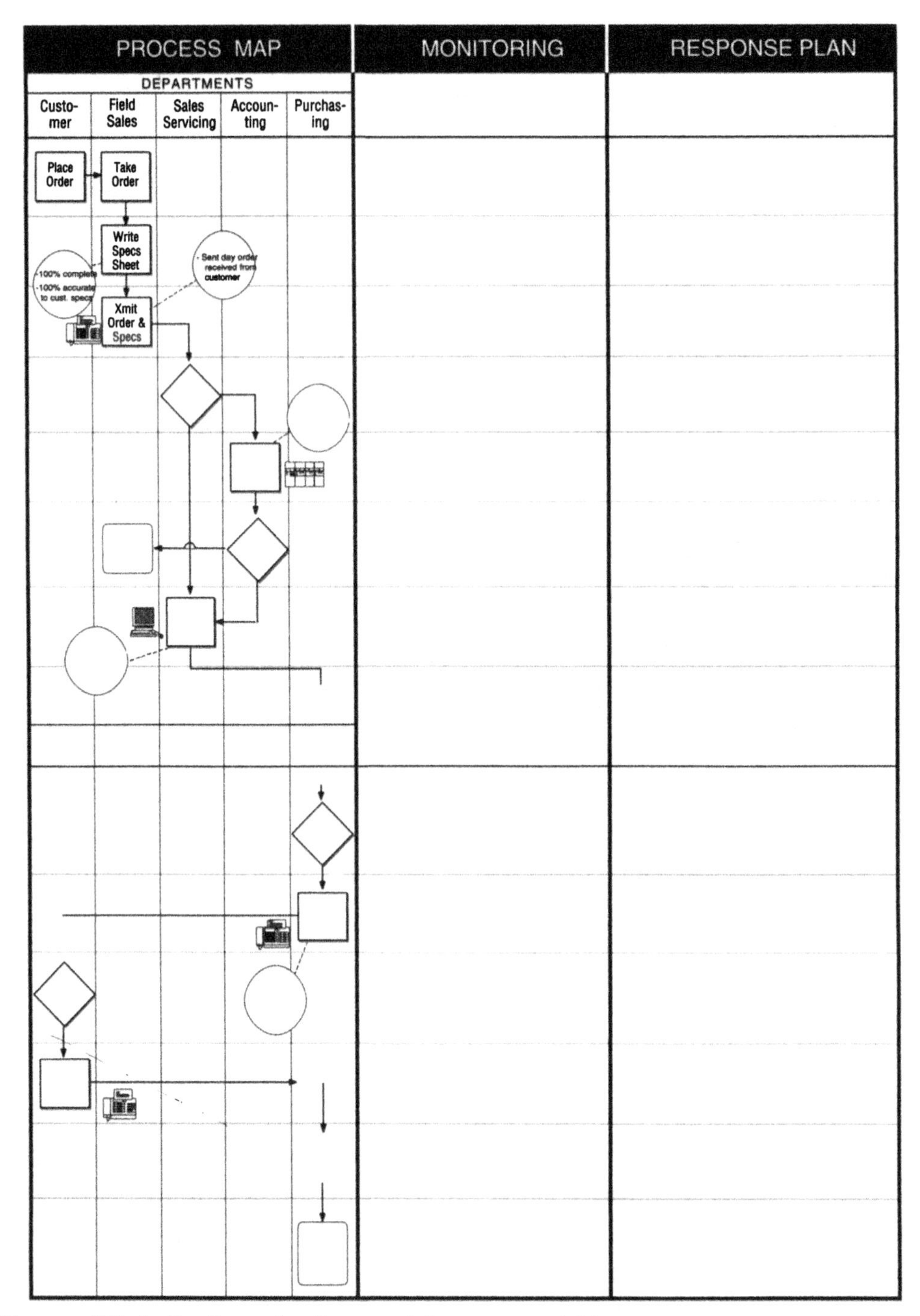

Figure 18-4. Portion of a Process Management Plan

don't have to improvise them. Contrary to the old saying, very few people work well under pressure.

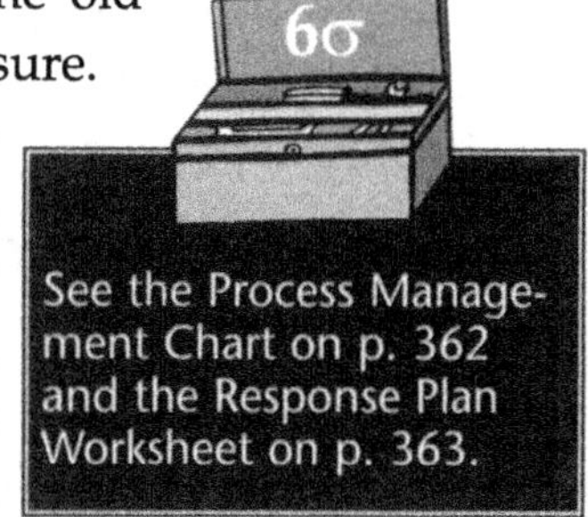

See the Process Management Chart on p. 362 and the Response Plan Worksheet on p. 363.

- **Plans for continuous improvement.** By tracking the occurrence of problems in the process, the process response plan also gives a basis for deciding on the need to have a Six Sigma team overhaul weak parts of the process. Once enough process response plans are in place, and information from them is collected and analyzed, managers will have an extensive pool of projects to be attacked by future Six Sigma teams.

Another tool that aids in process management is called a **process dashboard.** You may be familiar with dashboards used at executive levels of management, where they serve to summarize a number of indicators of organizational health. The principle is the same here: you identify a handful of indicators that relate to the *most important operational aspects* of your process. Often, they include an output measure (such as customer satisfaction), an indicator related to speed or efficiency (such as cycle time), and some measure of process quality (perhaps defect levels).

See the Process Dashboards on pp. 364-365

Ending the Project

As tired as team members get of the project, sometimes it's hard to bring everything to a close. After all, by this time most teams have a high level of camaraderie and are likely working very well as a unit. Still, even good things must come to an end. In this case, that means:

1. Completing your storyboard.
2. Preparing for the final tollgate review.
3. Celebrating the end!

Completing Your Storyboard

The final stage, Control, usually contains an updated process map and some of the control or other charts that will be used to make sure the improved process doesn't degrade to its original state once the team has finished the project. The latest sigma calculation can appear here. The name of the "owner" or manager of the improved process is also displayed here.

Preparing for the Tollgate Review

Often times, an organization may invite people to the final review who have not attended previous reviews and therefore may not know much about the project. You'll need to find out if this is the case in your organization so you know whether to include a overall project summary in your presentation.

In addition to describing the result of your Control work, make time in the presentation for two other key items:

- **Lessons learned**: about the project; the process, product, or service; the company's customers; working on a Six Sigma team, etc.
- **Opportunities not addressed in the project:** For Six Sigma teams who find that ...
  - they can't resolve all the causes that contribute to the problem identified in their charter (some causes may have been beyond the authority of the team) or
  - they learned things during the project that exposed additional opportunities for improvement.

Be sure to summarize these opportunities for management.

The Final Celebration

The last team celebration can be a private affair within the team, or your team can use it as a way to acknowledge the support they received from people who were not official team members. (The latter strategy will help build goodwill and support for Six Sigma projects within the organization!) Either way, don't let the project drag on unnecessarily.

Chapter 19

Power Tools for "Control"

Keeping Things on Track

As noted in Chapter 18, there are four elements of Control:

1. Discipline.
2. Documenting the Process.
3. Ongoing Process Measures.
4. Completion Checklists.

There's no *tool* that can help you with discipline—that depends largely on how committed your organization is to continuously doing a better job of meeting customer needs. In contrast, there are very specific tools for the other three elements of Control. All of them build on concepts covered earlier in this book, such as...

- Using process maps (flowcharts) to document procedures.
- Using data to monitor performance.
- Being *proactive* in planning for potential problems.
- Assigning specific responsibilities for performing the work.

While these principles have been present in quality improvement efforts for decades, what makes the difference here is that they are captured in practical tools your team and coworkers can use on the job to make sure the ideas are acted on.

Tip: Some of these methods will seem cumbersome at first because this level of devotion to process efficiency and effectiveness isn't present in many organizations. Just be aware that you might encounter some resistance or reluctance to use the tools described here—in fact, you yourself may not be convinced that this detailed attention to process performance is worth it! But keep at it, and you'll see that the results are worthwhile; your organization will keep getting better and better at maintaining the gains achieved by projects like yours, and may even see continuous improvements.

Documenting the Process

Process Documentation Checklist

Purpose: To increase the chances that process documentation will be both useful and used.

Application:

- Use in Control to ensure that improvements are maintained.
- Use whenever a process lacks documentation, or the process changes.

Instructions:
Review the checklist shown in Figure 19-1 as you create and use your process documentation.

Ongoing Process Measures

Control Charts

Purpose: To monitor a process for quick detection of abnormal variation (see Figure 19-2 for an example).

Application:

- Use in the Control phase of DMAIC to establish an ongoing method for monitoring process performance.

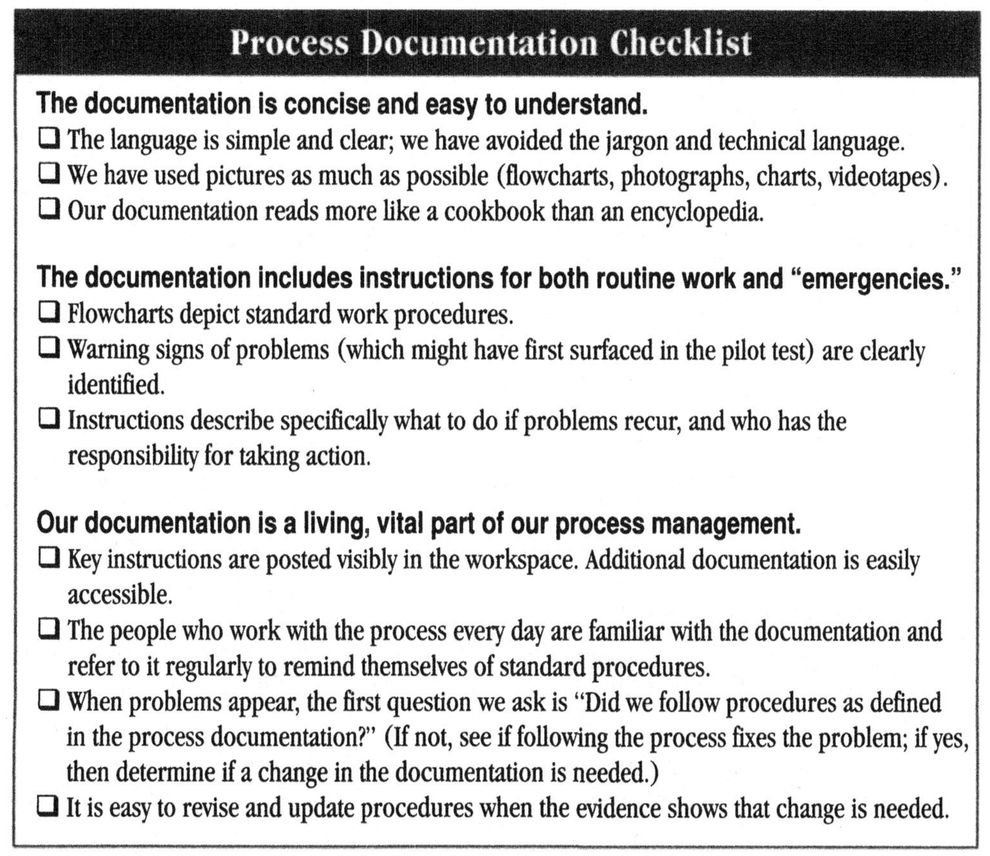

Process Documentation Checklist

The documentation is concise and easy to understand.

- ❑ The language is simple and clear; we have avoided the jargon and technical language.
- ❑ We have used pictures as much as possible (flowcharts, photographs, charts, videotapes).
- ❑ Our documentation reads more like a cookbook than an encyclopedia.

The documentation includes instructions for both routine work and "emergencies."

- ❑ Flowcharts depict standard work procedures.
- ❑ Warning signs of problems (which might have first surfaced in the pilot test) are clearly identified.
- ❑ Instructions describe specifically what to do if problems recur, and who has the responsibility for taking action.

Our documentation is a living, vital part of our process management.

- ❑ Key instructions are posted visibly in the workspace. Additional documentation is easily accessible.
- ❑ The people who work with the process every day are familiar with the documentation and refer to it regularly to remind themselves of standard procedures.
- ❑ When problems appear, the first question we ask is "Did we follow procedures as defined in the process documentation?" (If not, see if following the process fixes the problem; if yes, then determine if a change in the documentation is needed.)
- ❑ It is easy to revise and update procedures when the evidence shows that change is needed.

Figure 19-1. Process Documentation Checklist

Control Chart Basics

- The term "control chart" refers to any of a half-dozen or more types of charts that display the following features (see Figure 19-3):

  a. **Data points plotted in time order.** These data points can be either continuous or discrete (attribute) data. Each data point can represent either an "individual" measurement (such as plotting the completion time for *each* contract) or a calculation based on multiple data points (such as plotting the *average* completion time per week or the *percentage* of defective products per batch).

  b. **A centerline** (CL), which is usually the *average*. (Run charts, in contrast, often have the *median* plotted, not the average.)

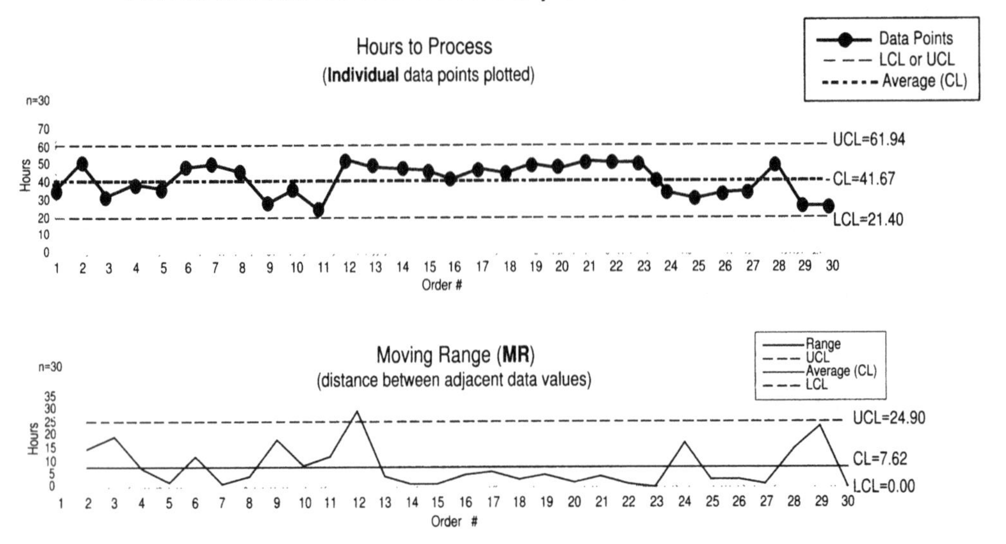

Figure 19-2. Sample Control Chart
This is an I-MR chart, charting of "individual" data (single measurements on individual items) and moving ranges (the differences between adjacent data points).

c. **Control limits** (UCL and LCL) that represent the amount of expected variation in the data. By statistical convention, they are plotted at approximately three standard deviations from the average. The control limits reflect the **process capability:** how well the process will perform, day in and day out, given the way it is currently structured and implemented (providing, of course, that there are no special causes of variation present).

- There are different ways to calculate the control limits depending on the type of data you have.
- The purpose of using control charts is similar to the reasons for using run charts: there are patterns you can look for that will tell you when the process is "out of control" (that is, exhibiting special cause variation).
- In many cases, you'll be using a statistical program (such as Minitab) to create a control chart for you. However, if you haven't used control charts before, consider doing a simple one by hand to help you become familiar with its features and underlying statistical principles. You should also keep hand-drawn charts by workstations where it is helpful to plot each data point as it is generated. That way, employees can just pencil in the

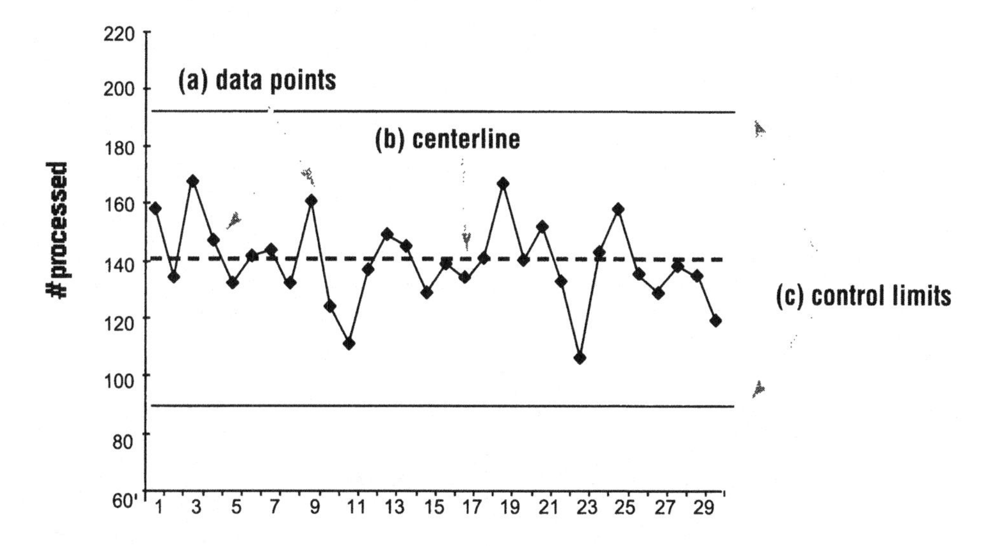

Figure 19-3. Features of a Control Chart

next point, then get back to their other work.

- What will be important is for you to understand how to collect the data used for the control chart.

Instructions: Providing detailed instructions for creating all the various types of control charts is beyond the scope of this book. Here is a quick overview of the basic procedure.

1. **Determine what kind of data is most relevant to your project,** then review Figure 19-4 and Table 19-1 to determine what type of control chart is appropriate for that data.
2. **Collect the data.**
   - Use information from Chapters 9 and 10 to develop useful data collection sheets. This is necessary even if the control chart will be created by a computer program.
3. **Plot the data in time order just as you would for a run chart.**
   - If using a computer program, make sure you review how the spreadsheet will interpret the data so you enter it correctly.
   - If working by hand, aim for a chart that is approximately 2 to 3 inches tall and as wide as necessary to accommodate all your data. You can

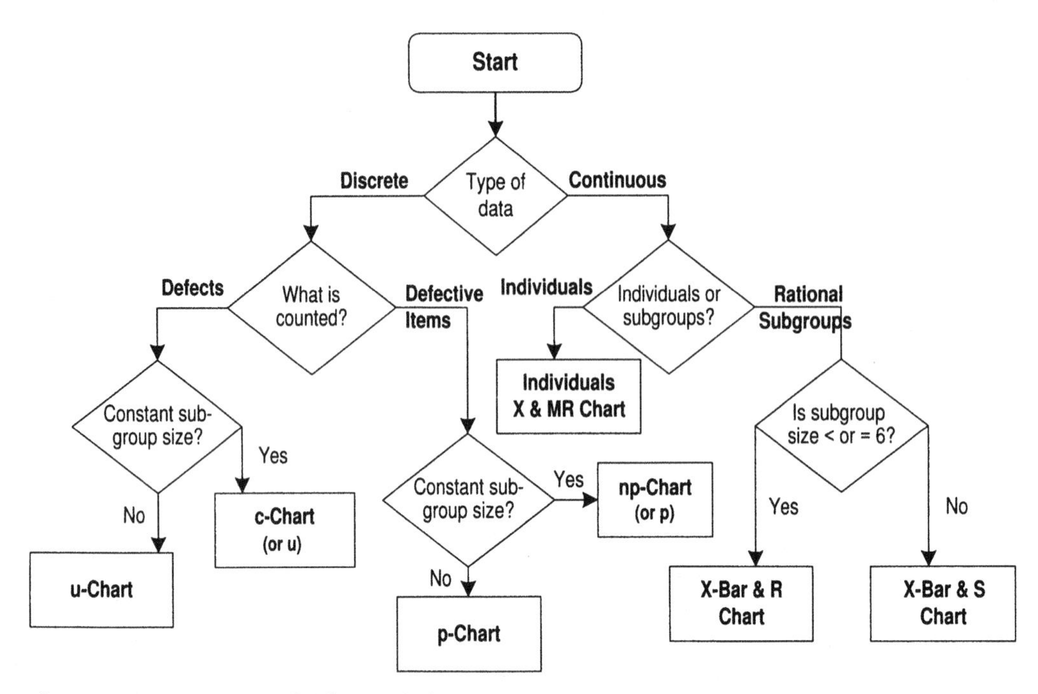

Figure 19-4. Control Charts Selection Tree

use either regular graphing paper or special SPC (statistical process control) worksheets.

4. **Calculate and plot the centerline and control limits on the chart.**
   - The calculations and plotting will be done automatically if you are using a statistical software program.
   - If you are doing this by hand, ask a Black Belt or Master Black Belt for help in determining the appropriate calculations for your type of data.

5. **Apply the special cause tests to monitor process performance and determine if and when special causes appear** (see Tables 19-2a and 19-2b).

Common Types of Control Charts		
Type of Chart	**Appropriate Data**	**Examples of Use**
individuals chart (I chart) often charted along with the moving range (MR) chart	Continuous data collected on individual items. Optional: calculate and plot the moving ranges (differences between adjacent points).	Individuals charts are some of the easiest to create, and can be used in a wide variety of situations.
$\overline{X}$**, R chart** and $\overline{X}$**, S chart** ("X-bar, R" or "X-bar, S" charts) where $\overline{X}$ = the average of subgroups of data, and R = ranges within the subgroups, and S = standard deviations of the subgroups.	Continuous data collected on a subgroup of items. Determine and plot the averages for those subgroups ($\overline{X}$). Also plot either the range of each subgroup or standard deviation.	Common in manufacturing. The use of subgroups to determine X-bar values evens out short-term variation so long-term variation (special causes) will stand out more.
p-chart and **np-chart** (where p = proportion defective; n = sample size)	Collect a sample of data including the number of defectives and sample sizes. Determine the proportion defective.	*p-chart:* Percent defective, percent of sales lost, etc. (sample sizes vary). *np-chart:* Tracking number of mistakes per 100 applications (sample sizes are constant).
c-chart (where c = counts of relatively infrequent events for a standard area of opportunity)	Counts of defects or errors when that number is fairly small compared to the total number of services or products produced.	Number of late shipments per week. Number of dents per car. Number of safety incidents per month.
u-chart (where u = counts when the area of opportunity varies)	As with a c-chart, you are counting occurrences, but the opportunity for the defect to occur varies each time you count.	Counts of purchase order errors per day (where the number of POs varies from day to day).

Table 19-1. Overview of common types of control charts

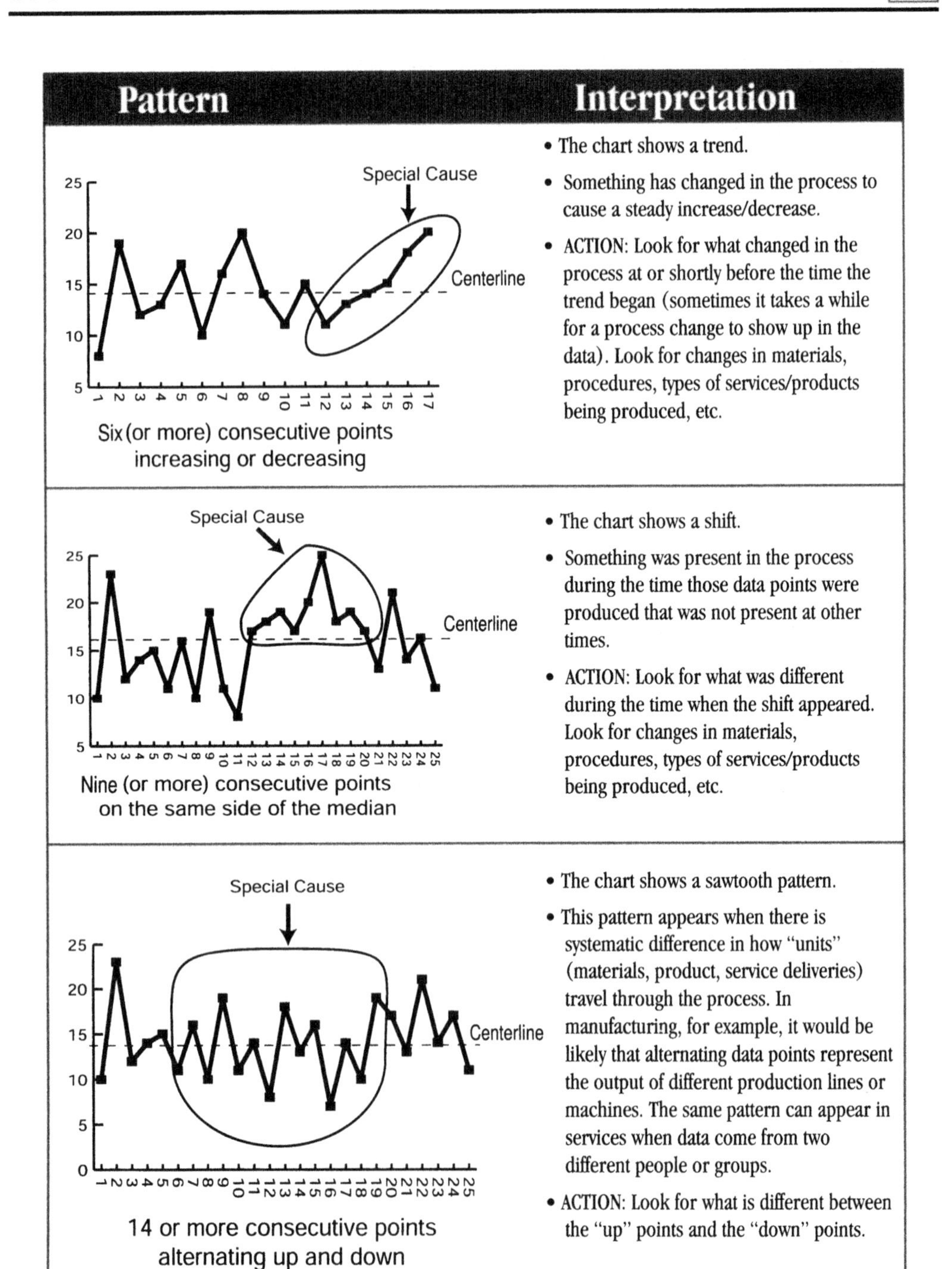

Pattern	Interpretation
Six (or more) consecutive points increasing or decreasing	• The chart shows a trend. • Something has changed in the process to cause a steady increase/decrease. • ACTION: Look for what changed in the process at or shortly before the time the trend began (sometimes it takes a while for a process change to show up in the data). Look for changes in materials, procedures, types of services/products being produced, etc.
Nine (or more) consecutive points on the same side of the median	• The chart shows a shift. • Something was present in the process during the time those data points were produced that was not present at other times. • ACTION: Look for what was different during the time when the shift appeared. Look for changes in materials, procedures, types of services/products being produced, etc.
14 or more consecutive points alternating up and down	• The chart shows a sawtooth pattern. • This pattern appears when there is systematic difference in how "units" (materials, product, service deliveries) travel through the process. In manufacturing, for example, it would be likely that alternating data points represent the output of different production lines or machines. The same pattern can appear in services when data come from two different people or groups. • ACTION: Look for what is different between the "up" points and the "down" points.

Table 19-2a. The tests for special causes used for run charts can also be used on control charts. (This table is similar to that seen in Analyze Chapter 13.)

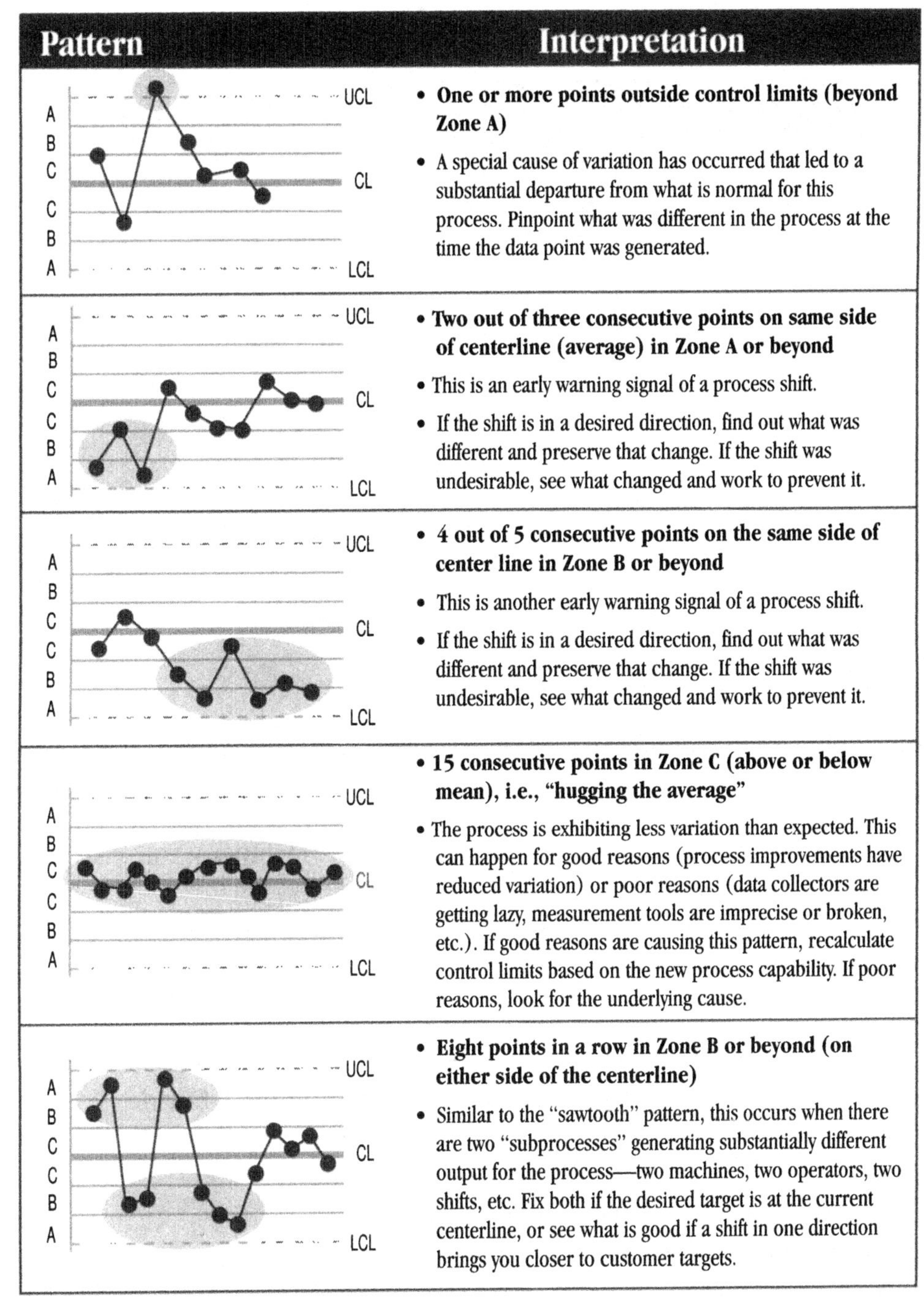

Pattern	Interpretation
	• **One or more points outside control limits (beyond Zone A)** • A special cause of variation has occurred that led to a substantial departure from what is normal for this process. Pinpoint what was different in the process at the time the data point was generated.
	• **Two out of three consecutive points on same side of centerline (average) in Zone A or beyond** • This is an early warning signal of a process shift. • If the shift is in a desired direction, find out what was different and preserve that change. If the shift was undesirable, see what changed and work to prevent it.
	• **4 out of 5 consecutive points on the same side of center line in Zone B or beyond** • This is another early warning signal of a process shift. • If the shift is in a desired direction, find out what was different and preserve that change. If the shift was undesirable, see what changed and work to prevent it.
	• **15 consecutive points in Zone C (above or below mean), i.e., "hugging the average"** • The process is exhibiting less variation than expected. This can happen for good reasons (process improvements have reduced variation) or poor reasons (data collectors are getting lazy, measurement tools are imprecise or broken, etc.). If good reasons are causing this pattern, recalculate control limits based on the new process capability. If poor reasons, look for the underlying cause.
	• **Eight points in a row in Zone B or beyond (on either side of the centerline)** • Similar to the "sawtooth" pattern, this occurs when there are two "subprocesses" generating substantially different output for the process—two machines, two operators, two shifts, etc. Fix both if the desired target is at the current centerline, or see what is good if a shift in one direction brings you closer to customer targets.

Table 19-2b. Additional special cause tests for control charts (See Chapter 18 for an explanation of "zones.")

Process Management Chart

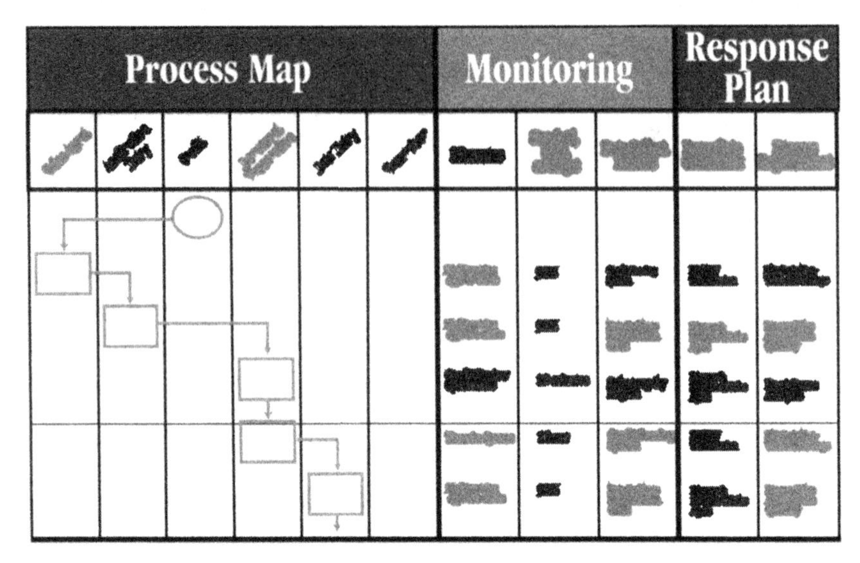

Figure 19-5. Format for a Process Management Chart

Purpose: Maintain a smoothly operating process.

Application: Use to monitor any key process in your organization.

Instructions (see Figure 19-5):

1. **Create a process map for the process to be controlled.** (You may have already done this as part of your Improve work.)
   - Use either a detailed flowchart or a deployment (cross-functional) flowchart (see pp. 262-264). A cross-functional chart will help you track what person or group is responsible for various process steps.
2. **For each critical step in the process, describe a monitoring plan.** How could someone tell whether or not it was done correctly? Decide ...
   - What data or observations should be tracked.
   - Who will collect the data/information and how.
   - How it will be charted and who will chart it.
   - How the information/data will be used.
3. **Develop a response plan** for each monitoring action. Identify how people should respond if they discover problems.

- What should people do if they find a problem?
- Use the Response Plan Worksheet (Figure 19-6). If the descriptions of the responses are complex, keep the complete response plan as a separate document and just cross-reference appropriate sections on the Process Management Chart.

Response Plan Worksheet

Response Plan Worksheet			
Response Plan for: ______________________			
Damage Control	Process Adjustment	Assessing Effectiveness	Continuous Improvement

Figure 19-6. Format for a Response Plan Worksheet

Purpose: To minimize the harm from unanticipated problems by providing for immediate responses.

Application: Use as part of a complete process management process.

Instructions (see Figure 19-6):

1. **Clarify what problem or set of problems is covered by this response plan.**
2. **Focusing on that specific problem, identify...**
   - **Damage control measures:** What should people do *first* when they notice a problem in order to prevent it from affecting customers?
   - **Process adjustments:** What can be done to adjust the process to counteract the problem?

- **Assess effectiveness:** How you will be able to tell if the process adjustments are effective. What will appear or disappear in the data charts, for example?
- **Continuous improvement:** How what you learn about this problem can be captured and incorporated into permanent process changes so it won't appear again.

3. **Assign responsibilities for acting on each of these response measures.**

Process Dashboards

Purpose: Monitor the most important indicators of quality, cost, and effectiveness associated with a key process, product, or service.

Application:

- Linking together the key measures that allow for monitoring of all aspects of a process (inputs, process steps, and outputs).

Instructions (see Figure 19-7):

1. **Identify the most important monitoring points in your process.**
   Hint: Use the SIPOC map developed in the Define stage. Identify what you could measure on inputs, process steps, and outputs.
2. **For each monitoring point, identify appropriate measures** and develop data collection plans. Decide how the data will be charted.
3. **Identify acceptable performance levels or improvement targets.**
   - Think about the gauges on a car dashboard. If your car is operating well, there are acceptable limits for temperature, oil pressure, engine speed, gas consumption, etc.
   - If your purpose is to drive further improvement, identify a measurable improvement target (e.g., a certain level of defects, desired cycle time, etc.).
4. **Develop methods for creating and maintaining the dashboard.**
   - Imagine you have the data in hand. Who will update the chart? How often?
   - Where will the dashboard be posted?

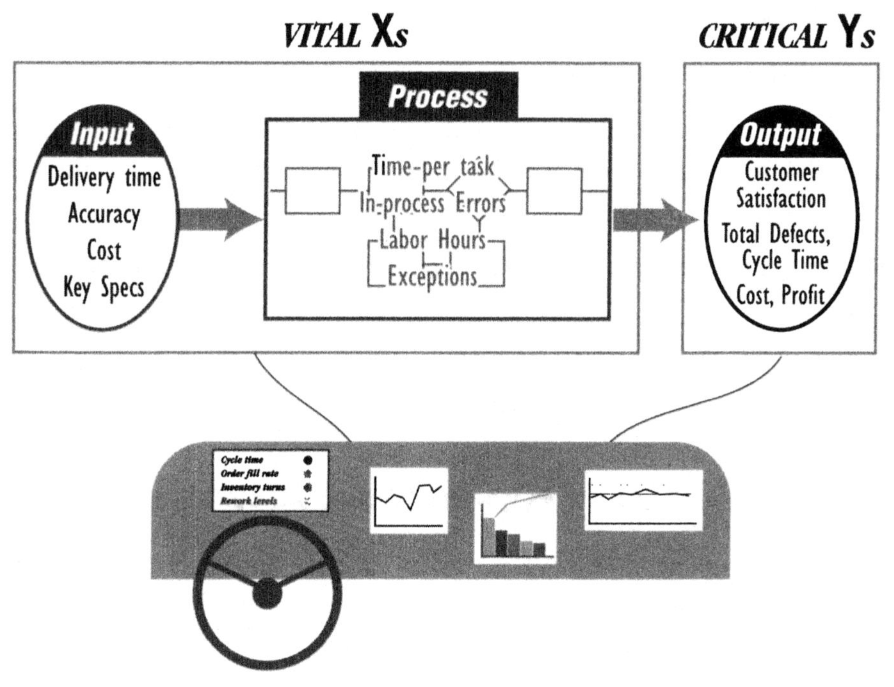

Figure 19-7. Process Dashboard

5. **Discuss and confirm how the dashboard will help you manage the process.**
   - How will the dashboard be used (e.g., reviewed at a monthly group meeting)?
   - What will happen as a result of having the dashboard?

Completion Checklists

Control Checklist

Purpose: To bring a formal end to the Control stage of a team's project.

Applications:

- Use during the Control work to track progress.
- Use at the end of the Control stage to make sure all essential tasks have been completed.

Instructions (see Figure 19-8):

1. **Walk through this checklist item by item at a team meeting.**
2. **Mark a "yes" only if everyone on the team agrees the task has been completed.** If anyone says no, ask him or her to state why they think the task is incomplete.
3. **Reach agreement as a team on each answer before it is marked on the checklist.**
4. **If there is unfinished work, ask for volunteers, assign responsibilities, and set deadlines for completion of those tasks.**

Control Tollgate Preparation Worksheet

Purpose: To help a team prepare a presentation for the Tollgate Review at the end of the Control stage.

Applications:

- Use at the end of the Control stage to help the team prepare its presentation.

Instructions (see Figure 19-9): Follow the general tollgate guidelines given for the Define review (pp. 118-120) to identify the priority message to include in your presentation. *In addition:*

1. **Document any commitments or promises** the team made to the Sponsor/Champion, Ladership Council, etc., during the Improve Tollgate Review.
2. **Compile information** to document progress on meeting those commitments.
3. **Develop graphics or other visual aids** that summarize measures you have put in place that will ensure ongoing monitoring and feedback from the process, as well as what response plans exist for dealing with problems.
4. **Have your team summarize the lessons they've learned** *from the project as a whole.*
   - What they learned about this process, product, or service that represents new knowledge to the organization.

Control Checklist

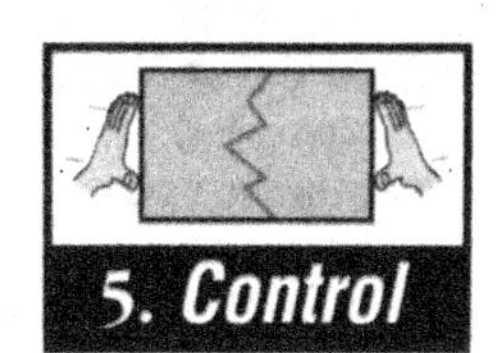

Instructions:

If you can respond "yes" to each statement below, you've completed all key steps in your DMAIC project and are ready to *celebrate* and *maintain* your improvement.

For our project we have ...

1. Compiled results data confirming that our improvement/design has achieved the Goal defined in our DMAIC team charter. YES NO
2. Selected ongoing measures to monitor performance of the process and continued effectiveness of our solution/design. YES NO
3. Determined key charts/graphs for a "process dashboard." YES NO
4. Prepared all essential documention of the revised process, including key procedures and process maps. YES NO
5. Identified an "owner" of the process who will take over responsibility for our solution/design and for managing continuing operations. YES NO
6. Developed (with the process owner) process management charts detailing requirements, measures, and responses to problems in the process. YES NO
7. Updated our storyboard documenting the team's work to date and key learnings. YES NO
8. Forwarded to senior management other issues/opportunities which we were *not* able to address. YES NO
9. Celebrated the hard work and successful efforts of our team! YES NO

Figure 19-8. Control Completion Checklist

CONTROL Tollgate Preparation Worksheet		
Key messages to cover in the Review (several topics are listed; add to the list as appropriate)	**Best way to present this information** (be creative in finding high-impact visuals–handouts, overheads, flipcharts, storyboard, etc.–to use in the presentation)	**Person or persons responsible for this portion of the presentation**
List promises or commitments made in the previous review.	Identify ways to show progress on the issues raised in the previous review.	
Be prepared to describe what measures are in place to maintain the solutions and prevent backsliding.	Consider showing examples of charts, work documentation, etc.	
Summarize lessons learned and gains made in this project.		
Summarize problems/gaps the team was unable to address and/or identify other opportunities discovered through this project.	Revisit Pareto charts or other data displays showing problems that were not part of this project.	

Figure 19-9. Control Tollgate Preparation Worksheet

- What they know about your organization's customers that they didn't know before.
- What they learned about participating in a Six Sigma project, and advice they'd have for others in a similar situation.
- What they learned that they will be able to apply in their daily work.

5. **Identify any opportunities for improvement that were not addressed in the project.**
6. **Complete the rest of the worksheet and prepare an agenda as you have for other tollgates.**

The Control Stage

Chapter 20

Guiding Your Team in the Control Stage

"Are We Really in Control?"

As WITH THE IMPROVE STAGE OF DMAIC, most teams who make it this far show the usual characteristics of a "performing" team: people are working together efficiently, and personalities take a backseat to getting the work done right. What's different now is that this well-oiled team machine needs to come to an end! If team members get along well and have enjoyed the project, it may be a difficult transition to hand off their work to the daily process owners and bring the project to a close.

You can help get your team end gracefully by...

- Asking everyone to reflect on what they will take away from the project, especially methods or tools that will help them in their everyday jobs.
- Identifying ways that members will be involved in maintaining the gains made (and reinforcing what a positive accomplishment this is).
- Identifying who will be the official keeper of the team documentation.
- Discussing ways to share the results of your project within the organization. Nothing serves as a stronger positive reinforcement than when team

members are asked to share their insights into project management, process control, DMAIC, etc., with fellow employees.

Troubleshooting and Problem Prevention for Control

Failure #1: Poor Documentation of Process Improvements

Why this happens: Team members are so familiar with the original problem and the new solution they often assume that those who inherit the improvement are just as familiar. Thus the team writes skimpy directions for the process.

How to prevent it: Assume that those who will be taking on the new process know nothing. Write out complete documentation plus a process map and management chart. Have "newbies" try the new process using the documentation only. When they have problems, revise the documents.

Failure #2: Bad Hand-off from Team to Process Owner

Why this happens: Not including process owner on team; need to create new process owner because of improvements; inadequate preventive planning.

How to prevent it: If the person inheriting the improvement was not on the team from the beginning, they should have been included at the end. Every precaution should be taken to make sure the new process is welcomed by those who must nurture it from now on. The process owner, in turn, needs to create a process management chart to track the new solution. To make sure the improved process does not revert to its bad old ways, the process owner must have a detail-oriented personality somewhere between an IRS auditor and a junkyard dog.

Control Do's and Don'ts

Do

- **Document the improved process.** Keep the documents brief, clear, and visual wherever possible.
- **Balance your measures.** Select key input and process measures to predict and interpret output measures.
- **Create channels of information back from the customer to you.**

Don't

- **File the updated process documents away**. If you do, people will follow their own ideas, variation will increase, problems will occur ... you know the story.
- **Expect the data you collect to confirm your assumptions.** It's easy to accept data that confirms your pet theories. Be prepared—and open-minded—when the data refute what you expected to see.
- **Forget the process maps.** They are worth thousands and thousands of unread words in three-ring binders.

Chapter 21

Six Sigma Process Design/Redesign

Restarting from Scratch

THE MOST COMMON DMAIC PROJECTS are focused on weeding out the few main causes of pain, costs, and defects in a process or product. But these improvement projects won't be sufficient to meet all the demands of change in a business. Some processes are like an old car: you can fix this or that annoying problem and keep the thing on the road, but eventually you realize that you'd be better off junking it and getting another car. In the case of business improvement, of course, you don't get a new car—you create or "design" a new process.

For your DMAIC teams, the need for design or redesign may not be obvious at the outset. You may start off with a root-cause-focused project that's supposed to be finished in three to six months and set out to identify the critical Xs creating the problem. But a couple of issues can prompt a rethinking of the project:

- You find that the gap between real customer requirements and current performance is so wide that "fix it" solutions won't do.
- The number of causes (critical Xs) that combine to drag the process down is so large that (like the old, falling apart car) it's best to replace the whole process with a new one.

In these cases, it's best to start with a clean slate and design a new process from scratch. But don't take this path too quickly or lightly! Design/Redesign efforts can be long, risky, and resource-intensive in comparison to the typical improvement project. They can have much greater overall payoff, but need to be carefully chosen and planned. If a few more inches of duct tape can hold the door on the car, or hold the process together, you may want to do that and learn more before taking on Design.

In general, we encourage eager Black Belts to try to improve what they have in small steps, until there is a good case for starting over from scratch. How do you know when that time has come? There's no easy answer—but a "design from scratch" effort might be worthwhile when...

- The existing or improved process simply is not capable of meeting customer requirements. This can be the conclusion of both statistical calculations of capabilities and the gut feelings of battle-weary managers and Black Belts.
- Customers demand greater flexibility, and the team's improved processes are still too rigid.
- New technologies make obsolete what your company has to offer.
- Changes in government rules and regulations are opportunities or threats, depending on your company's agility.
- New competitors appear with new technologies, ready to eat your lunch.
- Old assumptions become invalid. What happens when a company's success is based on the assumption that energy prices will not double in a year—and they do?

Think Before You Act

If one or more of the situations described above pose a serious threat or major opportunity, your Six Sigma team should certainly consider process design. But you and your Champion or other managers still need to answer a few questions before you embark on the good ship *Redesign*:

- **Are we willing to have a longer lead-time for completion of the new design?** Getting to results takes longer when you are creating a new process compared to fixing an existing one. It is not at all uncommon for process design projects to take a year or more. Can your business and its customers wait that long? Often, teams will defer a design effort until after several waves of DMAIC improvement, or even "quick fixes," have been put in place—that way, they can make immediate improvements that will attract or retain customers and/or provide time and cost savings in terms of increased operational efficiency. An alternative is to manage the design lead time by taking a phased or "multi-generational" approach, where the design effort is implemented incrementally. (See below for more details.)

- **Do we have the necessary people and resources available?** The odds are that process design will require greater investments in customer research, IT, and new people with new assumptions. Are these in the budget? Can a case be made for the investment (*adding* it to the budget)?
- **Are senior managers committed to the effort?** When asked once by the CEO of a large healthcare company how much time he and his managers would have to devote to making reengineering work in his organization, one of the authors answered, "Pat, you'll have to stop going to church for the next three years. It's a full-time job." There was eventually a falling away of the faithful, and the change effort died the death of a thousand mid-management cuts. Redesigning processes requires hard work and hard decisions involving people, their jobs, and, sometimes, their termination. Are the will and commitment at both Black Belt and management levels strong enough for this kind of heavy lifting?
- **Are the team and management willing to accept the risk of design/redesign?** Most innovations fail. That's a fact of business life. The Black Belt and Champion of a design project must be willing and able to manage a type of project that has a high risk of failure. Fortunately, the approaches applied to design in Six Sigma can help manage the risk and raise the probability of success. But there's still a need for people who won't shy away from a challenge and who live by the phrase "No guts, no glory."

None of the above should discourage the Six Sigma team from designing a brand new process, but it should offer plenty of food for thought and discussion before charging off down the path of redesign.

Implementing a Design/Redesign Project

If, after all the discussions are over, the answer is "Let's do it!" then it's time to go to work. You will find much of the ground is similar to the DMAIC process of incremental improvement, but with different emphasis along the way. Recall Figure 2-3 from Chapter 2 that showed the basic DMAIC steps. Here (Figure 21-1) those steps are compared for process improvement vs. design/redesign.

Six Sigma Improvement Processes

	Process Improvement	Process Design/ Redesign
1. Define	✓ Identify the problem ✓ Define requirements ✓ Set goal	✓ Identify specific or broad problems ✓ Define goal/change vision ✓ Clarify scope and customer requirements
2. Measure	✓ Validate problem/ process ✓ Refine problem/goal ✓ Measure key steps/ inputs	✓ Measure performance to requirements ✓ Gather process efficiency data
3. Analyze	✓ Develop causal hypotheses ✓ Identify "vital few" root causes ✓ Validate hypothesis	✓ Identify "best practices" ✓ Assess process design ◆ value/non-value adding ◆ bottlenecks/disconnects ◆ alternate paths ✓ Refine requirements
4. Improve	✓ Develop ideas to remove root causes ✓ Test solutions ✓ Standardize solution/ measure results	✓ Design new process ◆ challenge assumptions ◆ apply creativity ◆ workflow principles ✓ Implement new process, structures, systems
5. Control	✓ Establish standard measures to maintain performance ✓ Correct problems as needed	✓ Establish measures and reviews to maintain performance ✓ Correct problems as needed

Figure 21-1. Comparison of DMAIC steps for Process Improvement and Process Design/Redesign

In general, many of the tools described earlier in this book are also well-suited to redesigning a process, especially process mapping tools, creativity techniques, data collection and control measures, and so on. This chapter highlights additional approaches that relate specifically to rethinking how a process should work.

Before You Begin: Have a Vision

Perhaps the biggest mistake people make before embarking on a process redesign effort is not having a true vision for what they want to accomplish. Most importantly, a vision helps a team look beyond "what is" to "what could be"; it opens possibilities we wouldn't normally consider when surrounded by the way things are today. A vision is also an important touchstone, providing a guide against which new ideas can be evaluated. Everything boils down to a simple question: "Will this help us achieve our vision?"

Visions also play other roles, as shown in Figure 21-2.

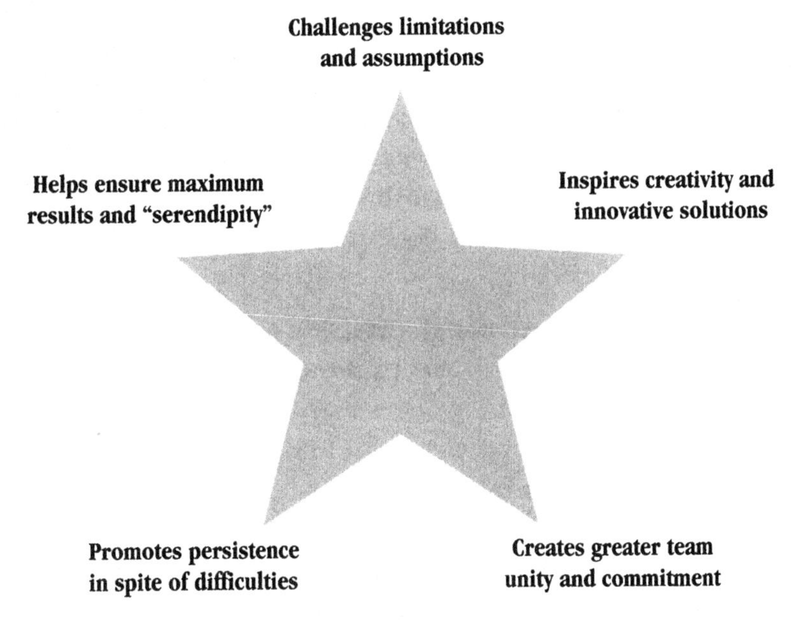

Figure 21-2. Uses for a Vision

Creating a Vision

An inspiring vision that will unify the team isn't something that happens overnight. Generally, it is developed through a series of meetings and discussions among anyone who has a stake in the process being redesigned. Typical elements include:

1. An initial brainstorming session or series of brainstorming sessions (if you need to involve a lot of people), focused on building a description of how the "new world" (process, service, product, capability) will look and feel.
2. Time to mull over and evaluate what was developed through brainstorming. This includes comparing the vision ideas to the company's strategic goals, customer needs, organizational competencies, etc.
3. Having the whole group or a subgroup synthesize the brainstormed ideas into draft visions; then time for everyone to discuss and have input on that draft.
4. Writing the final version that must be agreed to by consensus.

The tricky part in this process is to encourage creativity yet keep everything real; after all, your purpose is to actually redesign a process *now*, not five years in the future! You can help reach this balance by asking participants to brainstorm ideas in the following categories:

See Brainstorming, pp. 54-55; Advanced Creativity techniques, pp. 306-309; Affinity Diagrams, pp. 56-57; and Consensus, pp. 59-62.

- **Customer satisfaction and delight**. What will create 100% loyalty or thrill customers? What *can't* our current process, product, or service do that customers would like it to?
- **Contribution to our core competencies**. How can we become world leaders in this process and its related outputs?
- **Leaps in growth or profitability**. What would it take to make this process the most profitable in the organization?
- **Greater satisfaction for our people**. What will make this a more fun, exciting process to work on?
- **Current and future competitiveness**. What could we do that will leave our competitors eating our dust?

It's helpful to have a vision defined, at least in draft and endorsed by business leaders, before the Design team is formed. However, to thoroughly engage the team in the project, its members should have input to the vision. They need to feel real ownership of the goals they are setting out to achieve.

Take your time (as much as reasonable!) and plan on revisiting the project vision repeatedly throughout the project—both to keep the team inspired and to communicate your goals to others in the organization.

Step 1: Define the Design/Redesign Goal, Scope, and Requirements

While the basic contents of a process design charter are similar to those of an improvement team project charter, the spirit behind it is obviously more far-reaching: to create something new rather than repair something old. The description of the problem the redesign is addressing, as well as the goal to be achieved, will be more vague than those found in process improvements. We don't want to restrict the process redesign team's imagination right out of the starting gate, do we? The business rationale for the redesign, however, should be made clear: create something really capable of delivering high quality or be resigned to being number one or number three in the business.

Under "scope," the redesign team should do its best to identify the precise boundaries of the process being redesigned. This is never easy. It's usually a subjective judgment by informed people. Working from a detailed process map or at least a high-level SIPOC map helps to set the boundaries for change. To set the scope, the team needs to:

1. **Name the process to be redesigned,** not the department to be reorganized. The team might be about to redesign "Service to High-End Customers," not the "Customer Service Department."
2. **Specify the "thing" that is the output of the existing process.** And who is the customer for this thing? At what point does the team cut off its redesign effort?
3. **Identify the starting point of the existing process.** What action begins the process? What key input starts things moving?
4. **Identify the steps that fall between the start and stop.** Is there enough information now to create a SIPOC map?

5. **Examine the scope** defined as a result of these questions. Is it too broad (We only have a year!) or too narrow (We can do this in a week, but the customer won't notice a thing!).

Scoping the project is important because the changed (or brand new) process will probably affect the inputs and outputs of several other processes—a major impact on the organization.

A Multi-Generational Approach

One of the creative ways Six Sigma design teams have made process design and redesign more effective is to segment a large effort into two or more phases or "generations." A major pitfall in the wave of "reengineering" efforts of the 1990s was the huge complexity of the projects. Many elements of a vast process—systems, training, materials, facilities, and so on—would have to merge almost flawlessly for the effort to succeed. Six Sigma design/redesign teams should consider approaches that allow the new design to emerge in phases: for example, start the new process without a big new software component, then add the IT element in a subsequent phase of the effort. It's possible to have several "subteams" handle elements of the design simultaneously, but often doing them sequentially will turn out to be a more effective strategy. It allows each key element of the design (all under the same vision) to be tried and perfected before the next major elements comes on line.

Defining and Revising Customer Requirements and Outputs

A process exists (or should!) because customers require the output of the process. Behind the process to be redesigned are a variety of assumptions that the redesign team must challenge. During the course of their work the team members must step back and ask questions like these about the process output:

- What is the Output of the process today?
- Does the current Output best fulfill the requirements of customers?
- What might we offer customers instead of the current Output?

Regarding Customer Requirements, there are these questions:

- What is there about the current product that makes it useful to customers?
- What Customer Requirements are not being met by today's output?

- What are the needs of the customers of our customers that we can help them with our product?
- How could our product be made more useful to customers?
- How exactly do our customers use our product?

Ask these questions about assumptions concerning your customer requirements:

- What are our assumptions about how the customer uses our product?
- What is our understanding of customer requirements?
- How can we test these assumptions?
- Are there segments of our customer population we are ignoring?

Underlying all these questions is the intention to explore and challenge the assumptions behind accepted products and services. Challenging assumptions is not something most people are good at, so Black Belts will have to remind people at this stage that the team must challenge assumptions early in the Design/Redesign Process.

Step 2: Measure to Establish Baseline Performance

In designing a process, the Six Sigma team is not trying to detect the root causes of problems or variation, so measurement is not as important to developing solutions as it is in a typical DMAIC project. If you're redesigning an existing process—versus creating a brand-new, never-had-it-before process—your team will want to baseline the performance of the existing process so you'll be able to gauge progress later on.

Measuring existing processes against those of other companies (aka benchmarking) is often well worth the effort for design/redesign teams. One of the best ways to challenge complacent assumptions of "We're #1" is to benchmark against other organizations that do well with similar products and services. If your organization has multiple locations, divisions, or business units, your design team may be able to start with *internal* benchmark comparisons, and then look at other companies if additional data is useful.

The design/redesign team should identify measures early on that can be used later to gauge how well the new process is meeting Customer

Requirements. These measures can be used in design of experiments and other pilots or tests of the new process. For example, if the redesign goal is to reduce delivery lead time to one day and cut package damage to one defect per 100,000 packages, these will be key Ys to be measured in testing various delivery and packaging design ideas.

Step 3: Analyze the Critical Elements

The team's objective in process design is to improve and optimize performance on *all* the key variables impacting process efficiency and effectiveness. You're not looking for root causes in the Analyze phase (as you would in an Improve project)—rather, you're seeking to understand the environment and those critical factors that will impact your design. That knowledge gives you the power to create a design that will work *from the start*, rather than after several months or years of agonizing rework and refinement.

The trick in Six Sigma Design/Redesign Projects is to move efficiently through the Analyze phase to actually designing the new process—but not *so* quickly that the design is based on unclear data or faulty assumptions. It's a fuzzy line between too much analysis and not enough. Fortunately, even if you short-cut the analysis somewhat, you can compensate by better testing and enhancing the new process *before* the new process "goes live."

Overview of Analysis Concepts and Tools Used in Design/Redesign

The "Transfer Function"

By now you should be familiar with the core Six Sigma formula expressed algebraically as $\mathbf{Y = f(X)}$ (Y is a function of X); that is, the Output (Y) is determined by the Inputs (Xs). When applied to a real working process or product—where the relationships between the Xs and Ys can actually be quantified and predicted—this is called a "transfer function." (After all, it indicates how changes in Xs "transfer" results to the Ys.) In the Analyze phase of a design project, you will need to start identifying the complete array of Ys and the target performance for each—in essence, putting real numbers behind your project vision and goals.

For example, when a team at Six Sigma Pizza was redesigning processes to add a full range of new "low-fat and healthy" menu options, they identified goals for key Output/Customer Requirements such as order cycle time, price, taste of the food, and accuracy of calorie and fat content claims. They then identified all the "upstream" or X variables they would need to manage to achieve those performance objectives. While they had not yet started to design the process itself, they had a much clearer idea of how different design options would need to be evaluated.

Interface Analysis

No design is implemented in a vacuum. Some of the more influential X variables in your design are the Inputs to your process owned by others: suppliers, other departments, regulators, equipment, computers, etc. For your design to be effective and realistic, it will need to take into account these external variables. Examining the variation and capability of these "feeder" processes may help spark completely new design ideas (for example, new approaches to outsourcing, or insourcing, certain activities). On the other hand, confronted with limitations that you cannot control or change, you may be forced to lower your expectations for what the new design can accomplish. Not fun, but better to recognize reality earlier than later.

Assumption Busting

We've talked about this method before (see p. 309), but this may be the *most* important place for you to try to smash through unwarranted beliefs about "why we have to do things *this* way." Assumptions can be one of the biggest obstacles to developing really innovative new processes. Recognizing *what* your assumptions are in advance will be a big advantage as you begin to consider new approaches to meeting customer needs.

Value-/Non-Value-Added Analysis

When you're redesigning an existing process, value-/non-value-added analysis (covered in Chapters 12 and 13) will give you a clear idea of the types of activities that are essential and those that may be eliminated in the reformulated process.

Cycle Time Analysis

Taking time out of a process is usually one of the keys to reducing variation and cost and increasing flexibility. Redesign teams also can examine the time usage in their existing process to get clues on how to speed up the work. The basic concepts of cycle time analysis were discussed in Chapter 12, pp. 226-228, but they should take on a new rigor when applied in a design/redesign process. What you're looking for are entirely new ways to perform the same work such that wait times, delays, etc., are completely eliminated.

Benchmarking and Best Practices Analysis

For an all-new or "greenfield" design project, you can't analyze an existing set of activities. As a proxy, however, your team can analyze data gleaned from other process examples to see where *those* approaches may be enhanced in your effort.

Using the Tools to Gain Knowledge

Remember that the objective in the Analyze phase of a design project is simply to gain more knowledge that will help you create a highly effective and innovative new process that meets the design vision. The danger of *too much* analysis—especially for a redesign effort—is that you will become *so* familiar with the way the process works now that you have trouble considering new approaches. That's one reason it may be better in some cases to put your analysis resources and energy into more testing and enhancing of your design concepts—which emerge in the Improve phase.

Step 4: Improve: Designing and Implementing the New Process

In the Improve stage the design team must wear many hats and change them often. Not only must they challenge accepted traditions and workflows, they must create new processes that are cost-effective, defect free, and operating at very high levels of performance. All of this in the face of other people who keep saying, "I'm not sure we can do that...."

At the heart of the Improve stage are three steps:

1. Design the new process.
2. Refine the design of the new process.
3. Implement the new process.

In practice, these steps are iterative: your team will come up with an innovative design, test it on a small scale and make refinements, try full-scale implementation, discover areas that need further design ideas, and so on. So don't get discouraged if you end up cycling through these steps several times before getting a process that works in the real world!

4.1: Design the New Process

The hardest part of designing a new process is to forget everything you think you know about the way work is *currently* done. Don't let yourself get trapped in today's methods and assumptions; you need to challenge everything you think is a "given" and look for entirely new ways to deliver services and products to customers. The work of design therefore combines both standard process planning skills and creativity techniques. You should have...

- **Clear goals and objectives**. What exactly is the new process supposed to deliver?
- **Well-defined process scope**. How much should the team bite off? Check with the Project Champion.
- **Practical "how to" skills**. It's one thing to have a creative mind bubble, another to turn it into something that really works.
- **Criteria for success**. Even while getting outside the box, people need one foot in reality. Criteria based on customer requirements and what the organization can really do provide clues for brainstorming sessions.

In addition, push for creativity:

- **Adopt a "bad kid" attitude.** If the "good kids" in the old process agreed on the assumptions it was based on, the "bad kids" must challenge them. Good kids at school memorize all the state capitals. Bad kids ask, "Why do you assume I will be on a quiz show some day and need to know this trivia?" It's not easy to maintain this attitude at most businesses. It's a basic prerequisite for redesigners.
- **Use advanced creativity techniques.** Creativity is hard to define, harder to teach to people. The trick is finding ways to get outside the box long enough to see something new where others don't. Advanced creativity techniques, such as those discussed in Chapter 16, can help.

Options for Designing/Redesigning Processes

Part of your analysis in designing a new process will be how to avoid the problems inherent in your current processes. Here are some design options to consider:

- **Simplification (Figure 21-3a):** The fewer the steps, the fewer the opportunities for defects.

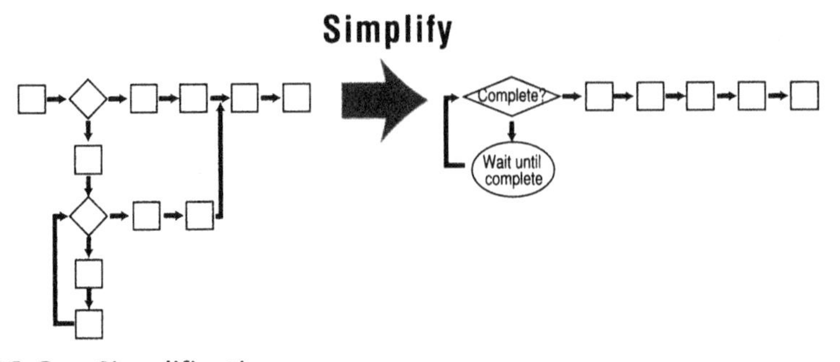

Figure 21-3a. Simplification

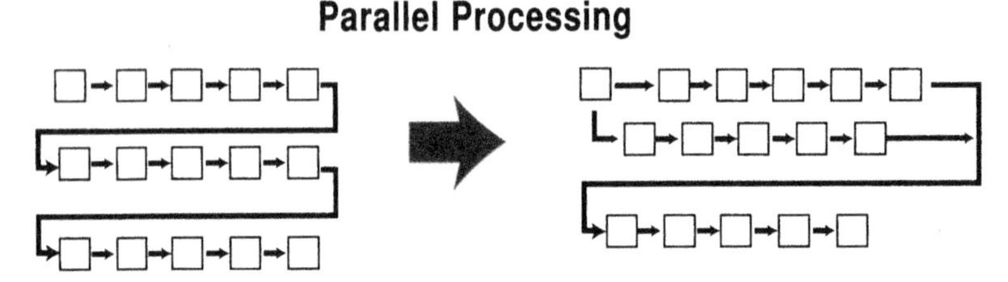

Figure 21-3b. Parallel Processing, where streams of tasks are scheduled to occur simultaneously

- **Straight-line processing:** Having all steps in sequence means fewer coordination issues, but risks delays if one step breaks down.
- **Parallel processing (Figure 21-3b):** Doing a number of process steps at once saves time, but risks problems downstream when one process fails to deliver the correct Output.
- **Alternative paths:** Maintaining a variety of ways to produce different products for different customers usually requires sophisticated tracking systems.
- **Managing bottlenecks:** If there are inevitable "tight" places in a process, flexible staffing may be an option. However, it's best to eliminate bottlenecks and not have to deal with that kind of hiring problem.
- **Front-loaded decision making**: This removes delays downstream by making most decisions upstream. It is really about making decisions to prevent problems from happening, and detailing contingency plans if they happen anyway.
- **Single point/multiple points of contact**: Do you want one staffer to work with individual customers ("Just ask for Bob whenever you call in!") or

have a system large and flexible enough so that every employee can work with any customer ("Just ask for anybody when you call in!")?

There are many variations on these themes, as well as new ones the design team can create for itself. Always the guiding question must be "Which design will work best to optimize customer satisfaction, speed, profitability, etc.—and achieve our design vision?"

4.2: Refine the Design of the New Process

Planning Changes at Six Sigma Pizza

Because order-taking, assembly, and cooking could be significantly more complicated with quite a few new ingredients, the SSP team brought in people from each key step and described its plan for the new process. For most tasks, the guests posed challenging questions that helped the team make critical adjustments to new procedures—and even prompted a major revision of the new order form.

A week after the talk-through, the team actually went to a store and did a deliberate rehearsal of the process, asking each key participant to do the "new" job according to the design and noting issues, concerns, and problems.

Refining the design is about improving the first process redesign, because first ideas are often the worst ideas. There are three broad categories of tools used to refine a process design:

A. Design concept discussions.
B. Process simulations.
C. Small-scale tests (pilot tests).

By combining some of these different methods, your redesign team may avoid the risk of "one big trial" that either works on doesn't work. Here are some examples of each.

A. Design Concept Discussions

- **Design concept "talk-throughs."** This is a decidedly low-tech, but often invaluable way of assessing a design by literally talking out how each step will work (or not work). (See p. 389 for one method of doing a talk-through.)

- **Potential problem analysis.** Subject the new process to the power of negative thinking: regarding every critical step ask…
  - What could go wrong with this step?
  - What would cause it to go wrong?
  - How can we prevent that cause?
  - What do we do if the problem happens anyway?
  - What's the alarm signal to go to our back-up plan?
  - Are there any opportunities we might exploit in the process?

B. Process Simulations

- **Designed experiments.** Using statistical models on the computer, the team can see which process factors work best and worst. Since these are only computer models, they will not draw any fire from opponents.
- **Process simulations.** Advanced computer models can provide simulations of some processes, especially in manufacturing where the inputs are more predictable.

Redesigning processes usually generates more active and passive resistance inside the organization than simple process improvements because they affect more people in more ways. Because of this opposition, piloting the new process and demonstrating its success is paramount to gaining acceptance of the changed process.

C. Small-Scale Tests

There are numerous ways to try out your ideas on a small scale before trying for full-scale implementation. Here are some of the most common methods:

- **Off-line pilot:** Part of an assembly line or office is set aside to test the new process full-scale, but without sending the product to real customers.
- **Set-time pilots:** Pilot versions of the new design are tested at specific times or on specific days, while "normal" process work goes on around it.
- **Selected locations:** Trying out the solution at one or a few sites.
- **Selected process components:** Trying out parts of the process, perhaps to verify results from designed experiments.
- **"Walk-throughs":** Creating a small-scale or full-scale model of all or a portion of a redesigned process and walking through each step to look for glitches and enhancements. (See p. 390 for one walk-through method.)

Conducting a Process Talk-Through

Purpose: To identify potential problems and opportunities in the earliest stages of process planning.

Application: Particularly useful in a process design project, but can also be used in a process improvement effort.

Instructions:

1. **Select participants** (usually team members plus "helpers").
   - Consider all the people involved in the work.
2. **Hold a meeting** with the participants. Set up a recorder and scribe to capture input.
3. **Post the process map** for the designed/redesigned process.
   - Should be at "task" level of detail.
4. **Explain each step.**
   - Requirements of "inputs" to the step.
   - Procedures for the task.
5. **Ask for participant reactions/concerns.** Capture:
   - Ideas for tools, documents, skills needed.
   - Cycle time issues.
   - Potential problems, defects.
   - Measures and performance management options.

Remember: piloting is labor-intensive, with all of the team members involved in measuring, analyzing, and improving the process even as it operates.

Winning Support for the Design/Redesign Plan

As always, the team will have to sell its great new process to buyers who may be less than enthusiastic. The team should try:

- **Strategic personal selling.** Find out who the key decision makers are inside your organization, determine their new idea buying process, and spend time with them pitching your redesign. (Also see stakeholder analysis in Chapters 15 and 16.)

Conducting a Process Walk-Through (Rehearsing a Process)

Purpose: To anticipate process problems in the design stage.

Application: Best for tangible processes and service delivery.

Instructions: *Do this on-site or "on-stage."*

1. **Create or locate real or dummy items** and documents that are representative of what the new process will handle.
2. **Set up a work area that mimics the suggested process design.**
3. **Perform the process as if it were real.** Invite participants to walk the items through the process and do the tasks as best they can in simulated conditions.
4. **Capture reactions and ideas.** At each work "station," have forms or blank sheets of paper available so participants can jot down areas of concern: confusion, disconnects, delays, problems, etc. You might also want to have *process observers* (team members who watch as others test the process).
5. **Hold a final debrief and develop an action plan** for refining and implementing the process.

- **Force field analysis.** This is a simple tool used to identify the forces driving your redesign forward and the forces opposing the change. How can you weaken the opposing forces, rather than simply pushing harder from your side? (See Chapter 16, pp. 317-319.)

Having refined the redesign plan and gathered support for it, the Six Sigma team can now direct its attention to the third step in the Improve stage: implementation.

4.3: Implement the New Process

If all went well, the pilot was the success the team and its Champion hoped for. But, as they say, a single robin doesn't mean winter is over. Pilots are not real life, and redesign teams are not average workers. The shift from pilot to actual rollout in the real world of the company has its own challenges. The team still has to handle the issues of:

- **Measurement:** Is the process operating as predicted? What is the new sigma level? Is the process in control?
- **Training:** People will have to learn the new process and unlearn the old process.
- **Documentation:** "How to" guides, new process maps, videos, lists of FAQs, etc.—all have to be ready for the rollout. (See Chapter 18.)
- **Getting the word out:** If the team hasn't already found a way to publicize the new process, it had better do so before sundown, lest people think the old process and processors are still at work in the salt mine.
- **Ongoing improvement:**
  - **Assessing Moments of Truth.** Here the team examines process points where the paying customer is in direct contact with the company and can draw conclusions about the quality of your service. The best back-room or manufacturing processes won't help if the employee meeting the customer decides to be rude. These Moments of Truth should be refined like gold, until all the impurities have been removed, for this is your chance to shine directly in the customer's presence.
  - **Focus group feedback sessions.** Collect a representative sample of customers and get their feedback on the new process.

The team should expect to spend a lot of time in the Implementation phase, collecting data about the new operation and what exactly it means in terms of the original charter and goal.

Step 5: Control the New Process

Just as in the process improvement form of DMAIC, control in design/redesign effort involves putting measures in place to make sure that the new process is monitored and continuously improved. You may want to review Chapter 18, which covers topics such as:

- **Process hand-off:** Who is the Process Owner now? Some companies actually have the new PO sign a document acknowledging this event.
- **Troubleshooting:** Problems will emerge even as the new process shakes out. "Who ya gonna call?" Probably the Process Owner, not Ghostbusters!

- **Process management chart:** What key measures and plans will the Process Owner want to track and display? (See Chapter 19, pp. 362-363.)
- **Performance management:** Does the new process require new incentives? Does it alter bonus payouts? Is the new process written into performance evaluations?

And so, with a great deal of hard work and good luck, the process redesign team can reflect on its activities, celebrate their success, and dunk the project Champion in the corporate fishpond.

Other Elements of the Design/Redesign Path

The brief overview given here may give a false impression: after all, most of this book was devoted to the basic process improvement form of DMAIC and here is a single chapter devoted to process design/redesign. As noted early in this chapter, however, designing a new process is usually much more time-consuming and challenging than improving an existing one. The purpose of this chapter was to give you a sense of how the basic DMAIC steps can be adapted to help you come up with and implement innovative process design ideas. If you choose the design/redesign path, you should still review the instructions in the appropriate "improvement" DMAIC chapters, since many of those tools will be useful in a redesign effort as well. In particular, pay close attention to tollgate reviews in a design project: the stakes are much higher, so passing through the tollgate should be much more difficult!

Index

L

M

N

O

About the Authors

Peter S. Pande is Founder and President of Pivotal Resources, Inc., an international consulting firm providing Six Sigma implementation training and managing development services to industries from financial services to high technology. Pete has worked in the organization improvement field for over 15 years, supporting change initiatives in a variety of companies, including GE, Citicorp, Chevron, and Read-Rite. He and his wife and two children live in the San Francisco Bay area.

Robert P. Neuman, Ph.D., is a senior consultant and noted speaker in the area of business improvement methods and Six Sigma. Bob's background in Six Sigma and quality systems includes two years with a major California health care system and consulting work with such Pivotal Resources clients as NBC, GE Capital, Cendant, and Auspex Systems. Bob and his wife live in Davis, California.

Roland R. Cavanagh, P.E., is a professional engineer who has an extensive background in improving manufacturing and service business processes. His areas of expertise include Process Measurement Applied Statistics, business reorganization, and Six Sigma methods. He has worked with such organizations as America West Airlines, Commonwealth Edison, GE, and Tencor Investments. Roland and his family make their home outside Chico, California.

CPSIA information can be obtained
at www.ICGtesting.com
Printed in the USA
LVOW04*0212290118
564404LV00009B/41/P